Recent Advances in Antimicrobial Nanodrugs

Recent Advances in Antimicrobial Nanodrugs

Editor

Fu-Gen Wu

MDPI • Basel • Beijing • Wuhan • Barcelona • Belgrade • Manchester • Tokyo • Cluj • Tianjin

Editor
Fu-Gen Wu
School of Biological Science
and Medical Engineering
Southeast University
Nanjing
China

Editorial Office
MDPI
St. Alban-Anlage 66
4052 Basel, Switzerland

This is a reprint of articles from the Special Issue published online in the open access journal *Pharmaceuticals* (ISSN 1424-8247) (available at: www.mdpi.com/journal/pharmaceuticals/special_issues/Antimicrobial_Nanodrugs).

For citation purposes, cite each article independently as indicated on the article page online and as indicated below:

LastName, A.A.; LastName, B.B.; LastName, C.C. Article Title. *Journal Name* **Year**, *Volume Number*, Page Range.

ISBN 978-3-0365-7229-1 (Hbk)
ISBN 978-3-0365-7228-4 (PDF)

© 2023 by the authors. Articles in this book are Open Access and distributed under the Creative Commons Attribution (CC BY) license, which allows users to download, copy and build upon published articles, as long as the author and publisher are properly credited, which ensures maximum dissemination and a wider impact of our publications.
The book as a whole is distributed by MDPI under the terms and conditions of the Creative Commons license CC BY-NC-ND.

Contents

About the Editor vii

Preface to "Recent Advances in Antimicrobial Nanodrugs" ix

Fanqiang Bu, Mengnan Liu, Zixu Xie, Xinyu Chen, Guofeng Li and Xing Wang
Targeted Anti-Biofilm Therapy: Dissecting Targets in the Biofilm Life Cycle
Reprinted from: *Pharmaceuticals* **2022**, *15*, 1253, doi:10.3390/ph15101253 1

Fengming Lin, Zihao Wang and Fu-Gen Wu
Carbon Dots for Killing Microorganisms: An Update since 2019
Reprinted from: *Pharmaceuticals* **2022**, *15*, 1236, doi:10.3390/ph15101236 35

Xiaoyan Wu, Khurram Abbas, Yuxiang Yang, Zijian Li, Antonio Claudio Tedesco and Hong Bi
Photodynamic Anti-Bacteria by Carbon Dots and Their Nano-Composites
Reprinted from: *Pharmaceuticals* **2022**, *15*, 487, doi:10.3390/ph15040487 55

Shuwen Guo, Yuling He, Yuanyuan Zhu, Yanli Tang and Bingran Yu
Combatting Antibiotic Resistance Using Supramolecular Assemblies
Reprinted from: *Pharmaceuticals* **2022**, *15*, 804, doi:10.3390/ph15070804 77

Rihab Lagha, Fethi Ben Abdallah, Amine Mezni and Othman M. Alzahrani
Effect of Plasmonic Gold Nanoprisms on Biofilm Formation and Heat Shock Proteins Expression in Human Pathogenic Bacteria
Reprinted from: *Pharmaceuticals* **2021**, *14*, 1335, doi:10.3390/ph14121335 99

Maher N. Alandiyjany, Ahmed S. Abdelaziz, Ahmed Abdelfattah-Hassan, Wael A. H. Hegazy, Arwa A. Hassan and Sara T. Elazab et al.
Novel In Vivo Assessment of Antimicrobial Efficacy of Ciprofloxacin Loaded Mesoporous Silica Nanoparticles against *Salmonella typhimurium* Infection
Reprinted from: *Pharmaceuticals* **2022**, *15*, 357, doi:10.3390/ph15030357 109

Nashwah G. M. Attallah, Engy Elekhnawy, Walaa A. Negm, Ismail A. Hussein, Fatma Alzahraa Mokhtar and Omnia Momtaz Al-Fakhrany
In Vivo and In Vitro Antimicrobial Activity of Biogenic Silver Nanoparticles against *Staphylococcus aureus* Clinical Isolates
Reprinted from: *Pharmaceuticals* **2022**, *15*, 194, doi:10.3390/ph15020194 127

Long Chen, Xin Fu, Mei Lin and Xingmao Jiang
Azeotropic Distillation-Induced Self-Assembly of Mesostructured Spherical Nanoparticles as Drug Cargos for Controlled Release of Curcumin
Reprinted from: *Pharmaceuticals* **2022**, *15*, 275, doi:10.3390/ph15030275 149

Peiyuan Niu, Jialing Dai, Zeyu Wang, Yueying Wang, Duxiang Feng and Yuanyuan Li et al.
Sensitization of Antibiotic-Resistant Gram-Negative Bacteria to Photodynamic Therapy via Perfluorocarbon Nanoemulsion
Reprinted from: *Pharmaceuticals* **2022**, *15*, 156, doi:10.3390/ph15020156 161

Ștefana Bâlici, Dan Rusu, Emőke Páll, Miuța Filip, Flore Chirilă and Gheorghe Zsolt Nicula et al.
In Vitro Antibacterial Susceptibility of Different Pathogens to Thirty Nano-Polyoxometalates
Reprinted from: *Pharmaceuticals* **2021**, *15*, 33, doi:10.3390/ph15010033 173

About the Editor

Fu-Gen Wu

Fu-Gen Wu is a full professor of Biomedical Engineering at Southeast University (Nanjing, China). He obtained his BS and PhD degrees from Tsinghua University in 2006 and 2011, respectively. After a postdoctoral period at the University of Michigan-Ann Arbor, he joined Southeast University in 2013 and was promoted to the role of professor. His main research interests are cell imaging, biosensing, biomaterials, and nanomedicine.

Preface to "Recent Advances in Antimicrobial Nanodrugs"

Microbial infections have always posed a threat to public health, and the situation is becoming worse as some strains of these microorganisms develop resistance to drugs such as conventional antibiotics. With the rapid development of nanotechnology and nanomaterials, an increasing number of nanodrugs capable of fighting against various microorganisms (e.g., bacteria, fungi, and viruses) have been designed and prepared. The diversified size, shape, and chemical characteristics enable nanomaterials to facilitate molecular interactions with the microorganisms, and the high surface-to-volume ratio allows the incorporation of abundant functional moieties to these nanomaterials, thus promoting multivalent interactions with the microorganism. To date, the most commonly used antimicrobial nanomaterials include conventional metal (Ag, Cu, Zn, and Ti)-containing nanoagents, two-dimensional nanoagents (e.g., graphene materials, layered double hydroxides, transition-metal dichalcogenides, graphitic carbon nitride, MXenes, black phosphorus, and their derivatives), polymeric nanomaterials, nanomicelles and nanovesicles, carbon dots and silicon nanoparticles, aggregation-induced emission (AIE) nanodots, nanocomposite materials, etc. In addition, the nanomaterials with specific light-, heat-, electricity-, magnetic field-, and ultrasound-responsive properties, as well as excellent antimicrobial activity, have also attracted interest from a growing number of researchers. Common examples include nanoagents with a photodynamic therapy capacity and sonodynamic therapy activity. Further, nanozymes with antimicrobial activity have also drawn tremendous research interest from many researchers worldwide. In addition, chemodynamic therapy that utilizes Fenton or Fenton-like reactions to kill microbial cells represents an emerging research direction. Finally, the nanomaterials decorated with conventional antibiotics have also shown potential for achieving an enhanced antimicrobial capacity compared with free antibiotics.

Fu-Gen Wu
Editor

Review

Targeted Anti-Biofilm Therapy: Dissecting Targets in the Biofilm Life Cycle

Fanqiang Bu [†], Mengnan Liu [†], Zixu Xie, Xinyu Chen, Guofeng Li and Xing Wang *

State Key Laboratory of Organic-Inorganic Composites, Beijing Laboratory of Biomedical Materials, Beijing Advanced Innovation Center for Soft Matter Science and Engineering, College of Life Science and Technology, Beijing University of Chemical Technology, Beijing 100029, China
* Correspondence: wangxing@mail.buct.edu.cn
† These authors contributed equally to this work.

Abstract: Biofilm is a crucial virulence factor for microorganisms that causes chronic infection. After biofilm formation, the bacteria present improve drug tolerance and multifactorial defense mechanisms, which impose significant challenges for the use of antimicrobials. This indicates the urgent need for new targeted technologies and emerging therapeutic strategies. In this review, we focus on the current biofilm-targeting strategies and those under development, including targeting persistent cells, quorum quenching, and phage therapy. We emphasize biofilm-targeting technologies that are supported by blocking the biofilm life cycle, providing a theoretical basis for design of targeting technology that disrupts the biofilm and promotes practical application of antibacterial materials.

Keywords: biofilm; microenvironment; biofilm-targeting material; antibacterial

1. Introduction

Biofilms are still considered as a major cause of chronic infections (such as chronic periapical periodontitis, chronic lung infection, infective endocarditis, etc.) [1–5]. Additionally, biofilms cause extensive damage to the marine environment and agriculture [6–10]. There has thus been considerable interest in the biofilm formation mechanism [11–17]. Accumulating evidence suggests that these bacterial resistance phenomena result from the ability of bacteria to enter into a dormant or persistent state in the biofilm [18–23]. The biofilm forms a complex microenvironment and spatial organization structure, such as extreme internal environments and extracellular polymeric substances (EPS), which limit entry of most drugs into the biofilm [24–30]. As such, understanding biofilm formation processes and chronic infections that can benefit from changing treatment, and, thus, tailoring personalized treatment to clinical patients, is paramount in improving the anti-biofilm therapeutic efficacy. Nonetheless, clinical treatment protocols for biofilm infections have not been updated accordingly.

The anti-biofilm strategy was still in an early stage of physical clearance and high-dose continuous administration in early clinical studies [31,32]. At present, most biofilm removal methods or treatment methods approved by the US Food and Drug Administration (FDA) focus on retained medical devices [33]. Research has shown that killing bacteria does not necessarily eradicate biofilms. Therefore, the challenge of residual biofilm, which may trigger chronic infections, must be addressed. A comprehensive understanding of the mechanisms and inherent properties of the biofilm life cycle was required to address this grand challenge (Figure 1). The following four stages can accurately represent the process from biofilm formation to re-spreading:

1. Initial adhesion stage: The reversible adhesions are dominated by Lewis acid–base, van der Waals forces, electrostatic interactions, and hydrophilic–hydrophobic interactions [34,35]. Irreversible adhesion is triggered by the bacteria's own adhesins and

adhesion proteins [36]. Reversible and irreversible adhesion of bacteria to the surface is the main feature of this stage.

2. Early biofilm formation stage: After bacteria adhere to the surface, bacteria activate their own metabolic pathways, which induces the bacteria to secrete metabolites (proteins, polysaccharides, eDNA etc.) to form EPS. At the same time, this also promotes bacteria-to-bacteria adhesion and activates quorum sensing (QS) [37]. Proteins, polysaccharides, eDNA, and QS of bacteria are the main features of this stage.

3. Biofilm maturation stage: A complex spatial structure and a microenvironment with chemical gradients (acidity, hypoxia, high reduction, etc.) are gradually formed with the increase in EPS synthesized by bacteria. At the same time, some bacteria will enter a dormant and persistent state [38]. Therefore, the characteristics of this stage are mainly complex chemical gradient microenvironment, persistent cells, and dormant cells.

4. Biofilm dispersion stage: Bacteria will secrete relevant secretions (enzymes, D-amino acids, surfactants, and other substances) to destroy EPS in response to nutrient deficiencies and accumulation of toxic substances, returning to a planktonic state [39]. This stage is characterized by associated secretions of bacteria and residual biofilm after dispersal.

Depending on the growth environment, biofilm formation changes, resulting in different biofilm spatial structures and bacterial gene expression differences [40–43]. Although the "characteristics" of biofilm have been revealed for many years, the clinical treatment of biofilm infections has not been updated due to the high complexity of biofilms [44,45]. These days, with the rapid growth experienced in materials science, surface-coating and eluting substrates materials are gradually being used clinically to remove biofilms (e.g., antibiotic-loaded bone cement to prevent orthopedic infections) [46,47]. Similarly, studies of biomimicry, surface textures, and chemicals in plants and animals are also promising approaches to preventing microbial adhesion and biofilm formation [48,49]. Using the amino acids and enzymes produced by bacteria to accelerate disintegration of biofilms is also one of the frontiers of anti-biofilm research [50,51]. These materials and technologies are very promising to solve the problem of biofilm infection. Although these studies have some statistical significance, to determine whether these technologies have the potential of being transformed into clinical technologies, researchers have to consider using in vivo or human cell models for further verification. The main reason for this is that most biofilm models are constructed from a single strain in the laboratory, but actual clinical situations may consist of multiple strains or lurking beneath probiotics [52,53]. Relevant studies have pointed out that some strains cannot form biofilm alone, but a variety of strains will help each other to build shelters together [54], such as *Actinomyces naeslundii T14V* and *Streptococcus oralis* (*S. oralis*) 34, which promote symbiosis in saliva to form biofilm [55]. Compared with a biofilm of a single bacterial strain, the harm of a multi-species biofilm to the host will increase exponentially. Interestingly, bacteria will also have hostile relations that try to destroy the enemy's shelter [56], such as *Pseudoalteromonas tunicate*, in the process of biofilm formation, which could inhibit and destroy the biofilm formed by other bacteria [57]. Therefore, the mutual hostility or mutual support in vivo between reference bacteria can provide effective theoretical support for design and development of biofilm-targeting materials.

Several excellent reviews discuss the protective mechanisms of biofilms against bacteria in response to antibiotics, antibacterial agents, and host immunity [17,58–60]. This review focuses on development and design of specific targeted biofilm therapeutic strategies and materials, as well as the challenges faced. A comprehensive description of how the properties of membranes at different stages can be exploited to design targeted materials, current insights into the targeting of EPS matrices, inhibition of chemical gradients and diffusion pathways, etc., as well as drug resistance and tolerance reversal strategies for dormant cells and persistent bacteria in biofilms, are provided. Furthermore, this paper reviews strategies that are expected to improve the efficacy of current clinical treatment modalities or provide new biofilm-targeting technologies, including the targeting adhesin strategy, quorum quenching, phage-targeting strategy, and targeting dormant cells strategy. Herein, we focus on biofilm-specific targeting materials that can be applied clinically. However, not all biofilm-targeting technologies are limited to clinical but also include agriculture, forestry, marine, and other directions. Therefore, we provide a comprehensive list of recent and prospective biofilm targets in Table 1. Finally, we believe that treatment of mature biofilm infections is more similar to treatment of cancer because their microenvironments are extremely similar (such as low pH, oxygen deficiency, overexpression of GSH, abnormal osmotic pressure, etc.) [17,37,38]. However, it may be more difficult than cancer therapy because the life cycle of biofilm is faster and more uncontrollable. It is important to note that more complex tissue structures, such as EPS, QS, and eDNA, are present in biofilms to inhibit therapeutic effects. More importantly, biofilm infections often exist in complex flora, making it more challenging to specifically target pathogen biofilms and achieve clearance [61,62]. Therefore, in the face of various challenges, biofilm-targeting technology has irreplaceable significance.

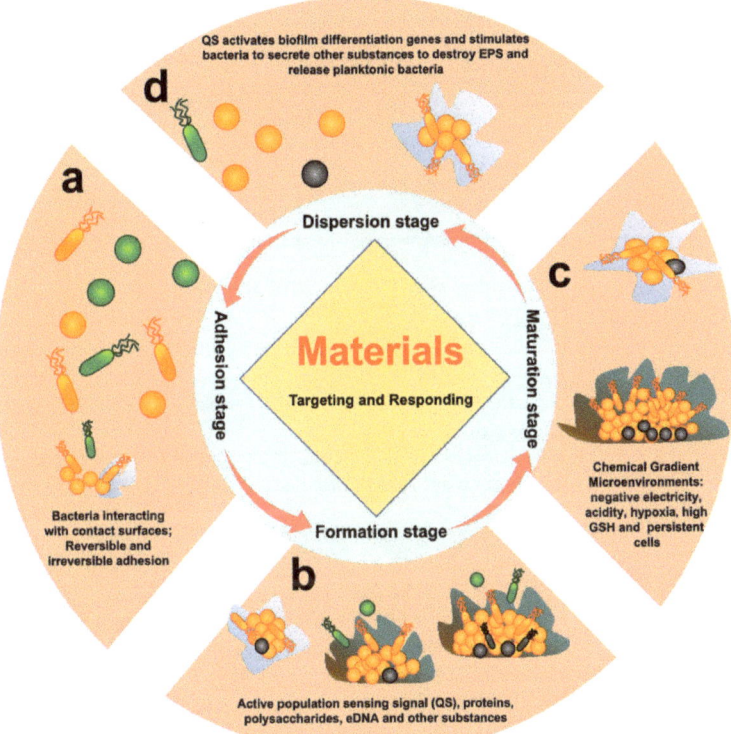

Figure 1. The biofilm life cycle. Different stages of bacterial biofilm formation. (a) Initial adhesion, in which bacteria adhere to surface of tissue through their own adhesins of bacteria; (b) early biofilm development stage, whereby the bacteria begin to divide and produce EPS by quorum sensing, eDNA, polysaccharide, and protein; (c) biofilm maturation stage, in which the biofilm will form a stable 3D structure through EPS, and the internal microenvironment exhibits a certain chemical gradient, such as acidity, hydrogen peroxide (H_2O_2), hypoxia, and overexpressed glutathione (GSH); (d) biofilm dispersion stage, whereby bacteria are oppressed by the extreme microenvironment, and their own secreted enzymes and D-type amino acids lyse the biofilm and return to the state of planktonic bacteria.

Table 1. Characteristics, targets, and targeting advantages of biofilm at different stages.

Biofilm Types	Characteristic	Target	Pros	Cons
Initial adhesion stage	Reversible and Irreversible adhesion.	Adhesin and Adhesion protein.	Prevention preferable to treatment. Will not cause drug resistance. Access not required after implantation.	Stability of surface coatings. Not necessarily kill bacteria. Potential substrate utilization by host.
Early formation stage	Active intercellular Communication and progressive formation of EPS.	QS; Polysaccharide Intracellular Adhesin (PIA); eDNA; Polysaccharides and Proteins.	Molecular medicine. Controlled locally. It will affect metabolism and will not produce drug resistance.	Potential degradation by nucleases, proteins, or enzyme. Highly localized. Composition variability.
Maturation stage	Mature EPS and Gradient chemical microenvironment and Changes in bacterial metabolism.	Hypoxic; Low pH; Negative; Overexpression GSH; H_2O_2; Persistent and dormant bacteria.	Disrupt pathogenic microenvironment. Readily functionalized. Active on dormant cells.	Difficult to simulate in vivo models. Incomplete eradication. Interaction with host.
Dispersion stage	Accumulation of biofilm residues and associated secretions.	Enzymes, D-amino acids; surfactants and others.	Readily combined with antimicrobials. Avoid cell dormancy. High universality.	Low spatiotemporal controllability. Residues to be resolved. Release of pathogens may result in recolonization and acute infection.

2. Strategies of Targeting Initial Adhesion Stage

2.1. Inhibit Biofilm Formation

Although, most of the time, points studied on biofilms were in the middle and late stages of the biofilm life cycle, we believe that precise targeting of adhesion properties during the first stage of biofilm formation is an effective strategy [63]. First, its advantage is avoiding drug resistance, tolerance, impermeability, etc., caused by the middle and late stages of biofilms. Second, early anti-adhesion strategies can not only effectively inhibit formation of biofilms but also achieve preventive effects.

2.1.1. Targeted Adhesin Strategy

Bacterial adhesin plays a key role in bacterial colonization and subsequent infection. Adhesin or adhesion protein could be used as a bacteria–host cell or bacteria–bacteria "bridge". Multiple adhesions are activated and expressed (such as proteins, lipids, and glycopolymers) [64]. Moreover, vitronectin and fibrinogen were also used similarly to adhesin [65]. Linke et al. found that *Yersinia enterocolitica* uses tiny sticky hairs to attach to the target. *Yersinia* adhesin A (*YadA*) protein of bacteria could penetrate two layers of cell membrane without any cell energy [66]. Recent research found that targeting-adhesin strategies were not thought to increase bacterial resistance nor interfere with the bacterial life cycle [67,68]. Therefore, the adhesion stage was a strategic step for bacteria, and antiadhesion therapy could effectively hinder the infection process. Notably, using materials to reduce bacterial adhesion could also promote the host immune system [69,70]. According to these principles, various therapies have been designed. Research by Heras et al. showed that blocking the super adhesion protein (*UpaB*) of bacteria could effectively inhibit bacterial colonization in the host [71]. Zhan et al. synthesized an indole derivative of selenium-containing (SYG-180-2-2, $C_{21}H_{16}N_2OSe$), which could inhibit biofilm by downregulating *icaA* and *icaD* and upregulating *icaR* and *coY*, thereby affecting PIA (*ica*: intercellular adhesin) [72]. Ravi et al. demonstrated that 2-hydroxy-4-methoxyben- zaldehyde (HMB, natural product of *Hemidesmus indicus*) could target the initial cell adhesion of *Staphylococcus epidermidis* (*S. epidermidis*) (Figure 2a) [73]. *Sortase A* (*SrtA*) is able to catalyze the initial adhesion between *Streptococcus mutans* (*S. mutans*) surface protein Pac and lectin. Ma et al. found that myricetin can effectively inhibit *SrtA* (Figure 2b) [74]. Liu et al. confirmed that nucleotide second messengers (such as cyclic adenosine monophosphate (cAMP) and cyclic diguanylate (c-di-GMP)) play an important role in regulating biofilm maintenance. It has been reported that pathogenic bacteria have evolved strategies to manipulate host cAMP concentrations [75]. This discovery provided an important direction for new drug design. Ashraf et al. found that the extract of *Eruca sativa Miller* (*E. sativa*) could effectively target adhesion proteins. A molecular docking analysis of *E. sativa* phytochemicals showed interaction with active site of adhesion proteins *Sortase A*, *EspA*, *OprD*, and *type IV b pilin* of *Staphylococcus aureus* (*S. aureus*), *Escherichia coli* (*E. coli*), *Pseudomonas aeruginosa* (*P. aeruginosa*), and *Salmonella enterica serovar Typhi* (*S. enterica ser. typhi*), respectively [76]. Krachiler et al. designed a functionalized multivalent adhesion molecules (MAM7) adhesive polymer bead that could effectively reduce infection of *P. aeruginosa* in the burn model and promote healing (Figure 2c) [77]. Cardoso et al. used gluconamide moieties to specifically target lipopolysaccharide (LPS) molecules in the outer membrane of *E. coli*, which efficiently prevented non-specific protein adhesion [78].

2.1.2. Interference Adhesion Strategy

Developing adhesion targeting compounds has been a long and in-depth development process. Thus, some scholars have proposed a strategy to interfere with adhesin, which uses compounds as analogues of bacterial adhesin receptors to make bacteria "mistakenly" adhere to the host to achieve anti-adhesion effects [79,80]. A treatment scheme using α-mannoside that interferes with FimH1 for treatment of catheter-related urinary tract infection (CAUTI) was applied [81,82]. In addition, Hartmann et al. used mannose-modified diamond to effectively enrich *E. coli* in sewage, achieving the removal effect [83]. In addition, after the discovery of *PapG* protein, a therapeutic scheme to inhibit *PagG* adhesion with galacto-oligosaccharide was finally formed in the clinical environment [84]. Most pathogens were opportunistic pathogens, and interaction of specific receptors and outer membrane molecules between bacteria and tissue cells was a prerequisite for infection [85]. Thus, interfering with bacterial adhesion is a therapeutic strategy that deserves further investigation. Zhang et al. prepared a nanoparticle coated with the outer membrane of *Helicobacter pylori* (*H. pylori*) (Figure 3a). NPs could compete with bacteria for binding sites on cells and inhibit bacteria from adhering to gastric epithelial cells and stomach tissues [86]. L. Davies et al. identified a 20 kDa peptide binding domain in the 1.5 MDa RTX adhesin of marine bacteria (*Vibrio cholerae* and *Aeromonas veronii*). Researchers used peptide library analysis to obtain a tripeptide that could effectively inhibit pathogen adhesion to the host [87]. Choi et al. found that D-arabinose could inhibit biofilm formation of oral bacteria (*S. oralis*, *Fusobacterium nucleatum* (*F. nucleatum*), and *Porphyromonas gingivalis* (*P. gingivalis*)) and the activity of autoinducer 2 (a QS molecule) [88]. Xu et al. designed G(IIKK)$_3$I-NH$_2$ (G3) based on α- A helical peptide, which inhibited bacterial adhesion and interfered with biofilm formation (Figure 3b) [89].

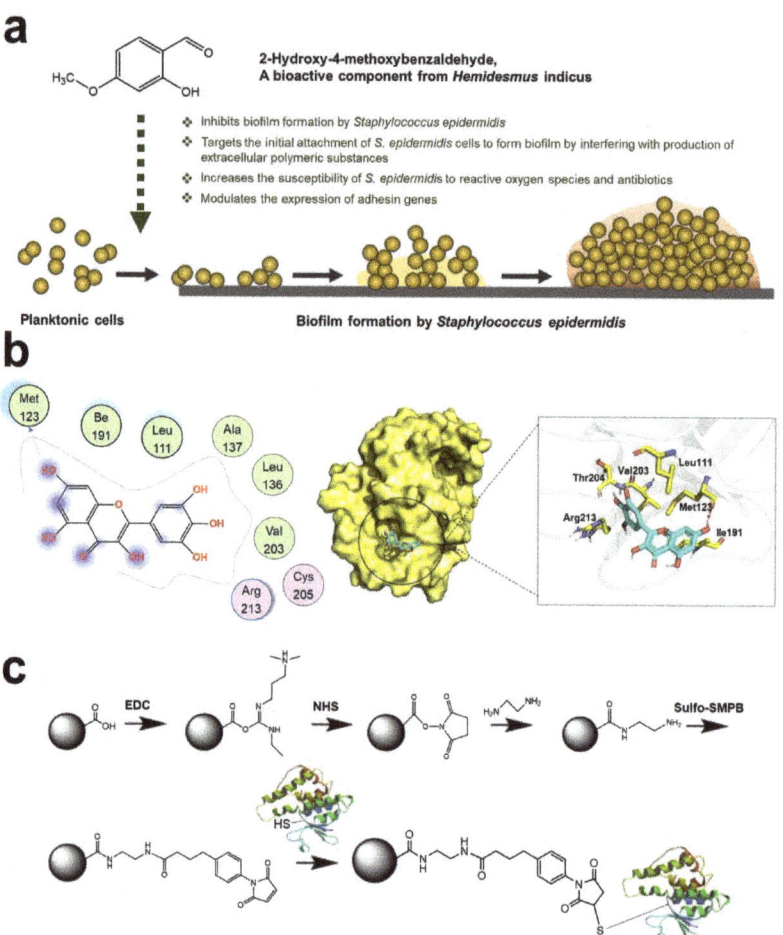

Figure 2. Schematic diagram of targeting adhesin strategy. (**a**) Mechanism of HMB targeting *S.epidermidis*. Reprinted with permission from Ref. [73]. Copyright 2020, Arumugam Veera Ravi. (**b**) Action sites of myricetin and *SrtA*. Reprinted from Ref. [74]. (**c**) Interaction mechanism between inhibitor MAM7 and Glutathione-S-Transferase (GST) fusion protein. Reprinted from Ref. [77].

2.1.3. Surface Anti-Adhesion Strategies

In clinical practice, biofilm infections caused by implants and medical devices often occur. A foreign body implantation is one of the main causes of biofilm infection [44]. To eliminate this biofilm-related infection, only uninfected medical equipment can be used, along with high-dose antibiotic treatment. Shortly afterwards, Khoo and Ji et al. proposed that endowing anti-adhesion performance to medical devices can better inhibit formation of biofilm and greatly reduce the use of antibiotics [46,90]. Based on this theory, a large number of laboratory designs have been proposed. Based on the optimization strategy of film surface morphology and hydrophobicity, Wang et al. designed four membranes with very high antiseptic properties (Figure 4a) [91]. Inspired by hydration ability of zwitterionic brushes, Hong et al. grafted 2-methacryloyloxyethyl phosphate choline (MPC) onto medical devices, which can effectively inhibit formation of biofilm [92]. Wang et al. proposed a stereochemical antibacterial strategy to achieve an anti-adhesion effect through the selective differentiation of L/D molecules by bacteria [93]. Antognazza et al. patterned silk film substrates that could effectively reduce adhesion of bacteria [94]. Leu et al.

modified the polypropylene (PP) surface by reactive ion etching (RIE) technology and reduced the adhesion of *E. coli* on the PP surface, which decreased by 99.6% via pro-hydrophobic interactions [95]. In vitro surface anti-adhesion technology alone does not meet practical clinical needs; it is also vital to address how to apply these techniques in vivo. Didar et al. transferred the topography present with hierarchical polystyrene surfaces onto polydimethylsiloxane (PDMS), which prevents biofilm and thrombosis in vivo (Figure 4b) [96]. Sun et al. integrated highly antibacterial copper nanoparticles (CuNPs) into hydrophilic polydopamine (PDA) coating and finally fixed it on a reverse osmosis (RO) thin-film composite membrane, which could reduce bacterial adhesion and significantly inhibited the formation of biofilm [97]. Ji et al. constructed a multifunctional modified surface multifunctional coating (mPep). Application of mPep in medical catheters in vivo proved to be effective in reducing bacterial adhesion and antibacterial (Figure 4c) [98].

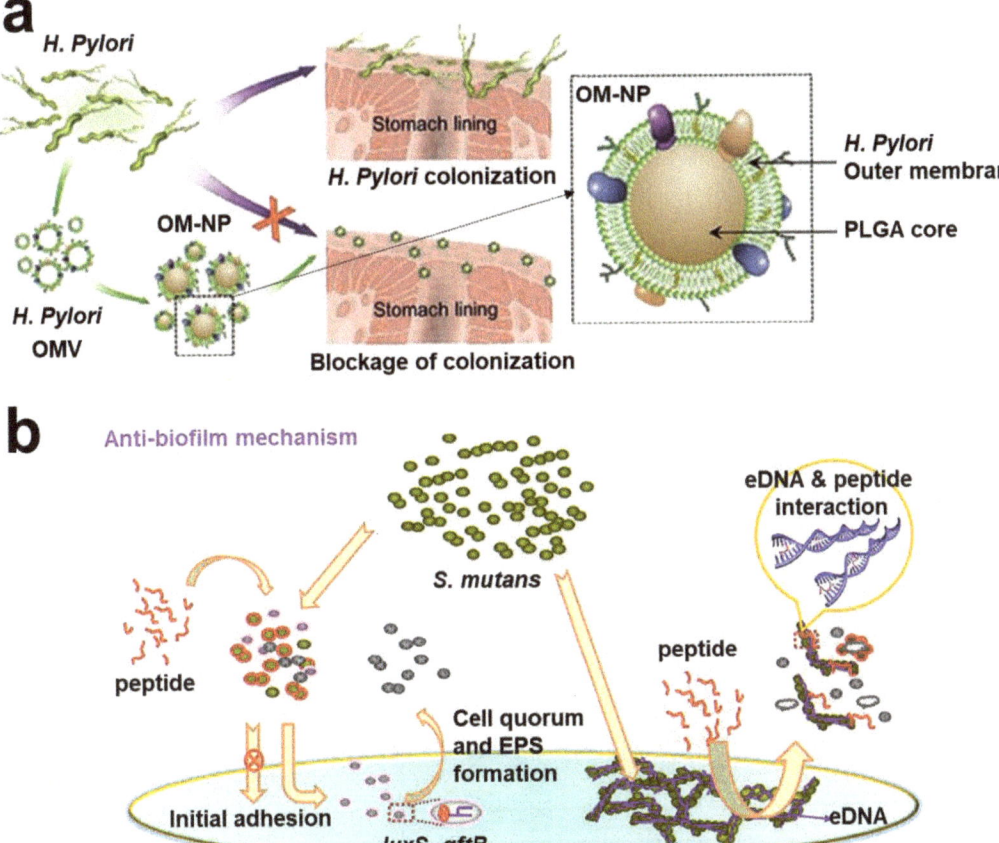

Figure 3. Schematic diagram of interfering with bacterial adhesion strategies. (a) OM-NPs coated with outer membrane of *H. pylori*, which could compete with bacteria for binding sites on cells. Reprinted with permission from Ref. [86]. Copyright 2012, Prof. Dr. Anke Kruege. (b) G3 inhibited bacterial adhesion and interfered with biofilm formation. Reprinted with permission from Ref. [89]. Copyright 2020, Hai Xu.

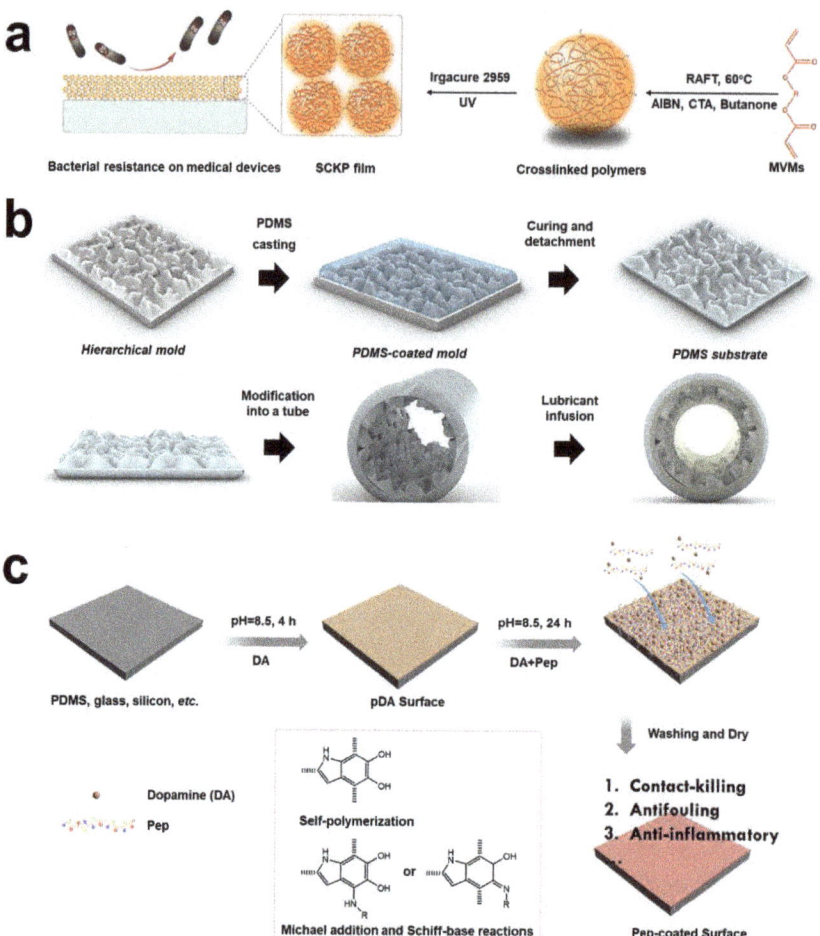

Figure 4. Schematic diagram of surface anti-adhesion technology. (**a**) The anti-adhesion polymers were synthesized by RAFT homopolymerization of MVMs. Reprinted with permission from Ref. [91]. Copyright 2019, Prof. Wenxin Wang. (**b**) Transferring the topography present of hierarchical polystyrene surfaces onto PDMS, forming an anti-adhesion, preventing thrombosis, and flexible biocompatible elastomer. Reprinted with permission from Ref. [96]. Copyright 2022, Tohid F. Didar. (**c**) The catechol, cationic, and anionic units to construct a multifunctional modified surface multifunctional coating (mPep) in medical catheters. Reprinted with permission from Ref. [98]. Copyright 2020, Jian Ji.

On balance, anti-adhesion technology has a "prevention preferable to treatment" advantage in anti-biofilm infection, and it is also one of the essential conditions for food packaging materials and biological storage materials. The potential advantages of targeted adhesion technology as a vaccine or drug remains underexplored in anti-biofilm therapy. Most compounds of targeting adhesins were easily ignored thus far because they did not exhibit specific minimum inhibitory concentration (MIC) and minimum bactericidal concentration (MBC). Identifying their potential to trigger biological function and effect, in comparison with the main studies, such as those involving peptides and antibiotics, would be of value. Therefore, whether in vivo or in vitro, anti-adhesion technology is very promising to achieve clinical transformation.

2.2. Targeting Biofilm Formation Strategy

As planktonic bacteria adhere to tissues or abiotic surfaces, bacteria spontaneously enter the second stage: biofilm formation. At this stage, the bacterial phenotype gradually changes, which causes the bacteria to have mutual adhesion and aggregate, forming small colonies. Bacteria will trigger QS for "communication" during formation of small colonies. Acyl-homoserine lactones (AHL) and autoinducing peptides (AIPs) are signaling molecules that mediate Gram-negative and Gram-positive bacteria, respectively [99]. At the same time, there was a "general language" autoinducer-2 (AI-2) that can mediate both Gram-negative and Gram-positive bacteria [100]. After bacteria receive QS signal molecules, bacteria gradually change their metabolism and participate in biofilm formation, including expression of PIA, bacterial autolysis and death, release of eDNA, and secretion of polysaccharides and proteins [101–103]. Finally, EPS is formed under the joint action of various mechanisms [30]. Therefore, the biological behaviors of the above bacteria can be used as potential targets to provide a theoretical basis for design of targeting materials.

2.2.1. Quorum Quenching

Formation of biofilm is a complex and relatively slow dynamic process. QS is the communication language of bacteria, which can effectively tell bacteria what to do now. At present, it is known that QS molecules could directly regulate bacterial behavior in biofilm. Many studies have reported that some compounds had the ability to quench QS, thereby destroying the biofilm, termed quorum quenching (QQ) [104]. These molecules are called quorum sensing inhibitors (QSIs) [105]. QSIs have been found to destroy the QS process mainly through the following ways thus far: 1. inhibit QS molecular synthesis; 2. simulate QS molecules; 3. degrade QS molecules; 4. chemically modify QS molecules. QSI will not affect DNA and cell division of bacteria, so bacteria rarely develop related drug resistance [106]. Many QSI compounds have been found and synthesized now; therefore, this paper only reviews QSI compounds with targeting functions.

QSI molecules with targeting function mainly have two mechanisms of action: the first is to target QS synthetase to inactivate or degrade QS signal molecules [107,108]. The second is the receptor that targeted QS signaling molecules so that the receptor cannot receive QS molecules or compete with QS molecules [109,110]. The quorum-quenching enzyme (QQE), such as acylase and lactonase, can degrade the QS signal and destroy QS in the extracellular environment. Tzanov et al. found that QQE acyltransferase could reduce the AHL signal (Figure 5a) [111]. The aceleacin A acylase (*Au* AAC) and N-acyl homoserine lactone acyltransferase (*Au* AHLA) have the same effect [112]. These enzymes have a QS targeting function. In addition, accessory gene regulator (*agr*) is the most classic QS system of *S. aureus* [113]. Xu et al. verified that hyperbranched poly-L-lysine (HBPL) inhibited QS mediated by the *agr* system and inhibited expression of QS-related genes (Figure 5b) [114]. Luteolin, as a QSI, also inhibited downregulation of *agrA* gene, but whether it has a targeted effect needs further study [115]. Bendary et al. further proved that zinc oxide nanoparticles (ZnO NPs), Hamamelis tannin (HAM), and protease K could be used as QSIs to downregulate the *agrA* gene, thereby inhibiting formation of biofilm [116]. Pseudomonas quinolone signal (PQS) is bound by cytosolic LysR-type receptor *PqsR* (also known as *MvfR*) [117]. Therefore, *PqsR* antagonists were found [118]. Recent studies have found that quercetin can specifically target the *lasIR* and *rhlIR* systems of *P. aeruginosa* and *LuxS* and *agr* systems of *Listeria monocytogenes* (*L. monocytogenes*), thereby inhibiting the QS system (Figure 5c) [119,120]. Ho et al. found a new lipophilic QSI for destroying biofilm (Figure 5d) [121].

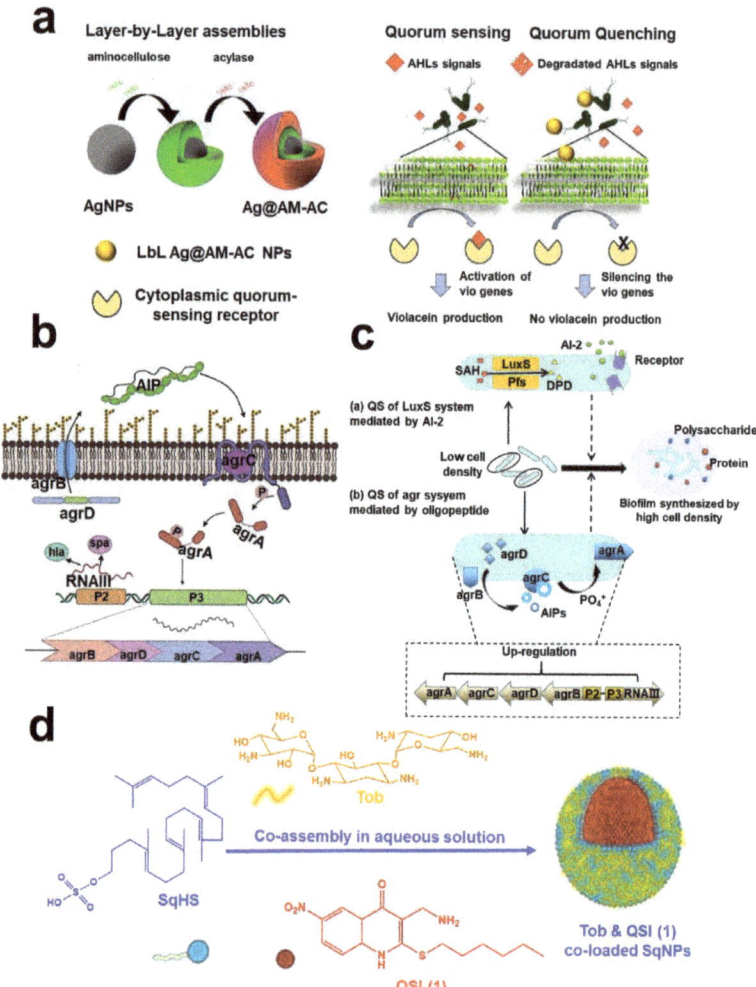

Figure 5. Quorum sensing targeting technology. (**a**) A schematic diagram of QQE acylase and amino-bearing biopolymer AM was covered layer by layer on the AgNPs template. Reprinted from Ref. [111]. (**b**) Schematic diagram of the *agr* QS system and expression of QS-related genes in Methicillin-resistant *Staphylococcus aureus* (*MRSA*). Reprinted with permission from Ref. [114]. Copyright 2022, Feng Xu. (**c**) Two quorum sensing systems (a) *LuxS* system and (b) *agr* system in *L. monocytogenes* could be used as targets of quercetin. Reprinted with permission from Ref. [119]. Copyright 2020, Yong Hong Meng. (**d**) The self-assembling nanoparticles of a squalenyl hydrogen sulfate (SqNPs) composed of a new lipophilic QSI (1), tobramycin, and SqHS. Reprinted from Ref. [121].

2.2.2. Targeted Polysaccharide Strategy

As one of the important components of the protective barrier and biofilm surface, polysaccharides can enhance intercellular adhesion and aggregation of bacteria, promote bacterial immune escape, stabilize, and maintain the biofilm microenvironment, and provide nutrients for bacteria [122–125]. Targeted design of related materials and strategies to target polysaccharides in biofilm are effective methods to remove biofilm. The initial targeting strategy is to inhibit enzymes that produce polysaccharides in bacteria, such as *glucosyltransferases* (*Gtfs*) in Gram-positive *S. mutans* and aggregative exopolysaccharides *Psl* and *Pel* in Gram-negative *P. aeruginosa* (Figure 6a) [126–128]. At present, development of *Gtfs* inhibitors for *S. mutans* is very extensive (Figure 6b) [129,130]. It is worth noting that *Gtfs* inhibitors are also used in developing vaccines [131]. Similarly, the combination of *Gtfs* inhibitors and drugs could demonstrate practical anti-cariogenic efficacy. *Disperse B* (*DspB*), glycoside hydrolase, and monoclonal antibodies are also common mainstream strategies for targeting EPS. Drug delivery systems (DDS) could protect enzymes from the external environment, and enzymes provide DDS targeting specificity [132]. *DspB* can efficiently and specifically hydrolyze poly-beta (1,6)-N-acetyl-glucosamine (PNAG) [133]. Immobilized *DspB-MagR* showed a high inhibitory effect on biofilm [134]. Using enzymes to degrade polysaccharides that disintegrate biofilm was gradually accepted; the related technology was rapidly expanded. Fan et al. devised a method based on α-amylase to develop a microneedle patch for removing biofilms caused by bacterial infections in wounds (Figure 6c) [135]. Therapeutic strategies of *P. aeruginosa* biofilm infections based on enzyme targeted acidic heteropolysaccharide (Alginate) have been reported [136,137]. Lee et al. cloned an alginate lyase *Aly08* from marine bacterium *Vibrio sp.* SY01 [138]. Daboor et al. also purified alginate lyase *Alyp1400* from marine *P. aeruginosa* [139]. The above extracted lyase could form an efficient combination treatment with antibiotics. Zhang et al. further encapsulated alginate lyase and other drugs to form a silver nanocomposite, and successfully eradicated *P. aeruginosa* infection in the lungs of mice [140]. In addition to alginate, *P. aeruginosa* biofilm also contains polysaccharides *Pel* and *Psl*. Drozd et al. fixed *Pel* hydrolase *PelA* on bacterial cellulose, solving the problem of chronic wound infection [141].

2.2.3. Targeted eDNA Strategy

In 1956, Catlin et al. first observed eDNA as one of the structural components of biofilm, which not only proved that eDNA can be separated from the biofilm matrix but also proved that addition of bovine deoxyribonuclease I (DNase I) can significantly reduce the viscosity of biofilm, eventually leading to diffusion [142]. Subsequent studies have proven that anionic eDNA can chelate cations from the immune system and drugs in the biofilm, providing a "protective umbrella" for bacteria [143,144]. When bacteria are hungry, eDNA acts as a nutrient. In addition, eDNA can also increase the hydrophobicity of the cell membrane, making it easier for bacteria to adhere to the cell surface [145,146]. Thereby, eDNAase synergistic therapy is applied and born [147–149]. Based on the above theory, the targeting materials and strategies of eDNA have been put forward successively, and good results have been achieved in removing biofilm and interrupting biofilm formation.

To date, targeted eDNA technology is no longer limited to DNase. Bing et al. designed an eDNAase-simulated artificial enzyme based on graphene-oxide-based naturalistic acid–cerium (IV) composite (GO-NTA-Ce) (Figure 7a) [150]. Qu et al. also designed cerium (IV) complexes (eDNAase mimics) for targeting and hydrolyzing eDNA in biofilm [151]. As the structure and mechanism of eDNA were gradually analyzed, other targeting materials and strategies have emerged. Natural products had always been the first choice for drug research and development. Some natural products with anti-biofilm effects were screened, and it was found that emodin could effectively target eDNA in biofilm [152]. Ramesh et al. reported an amphiphile (C1) with eDNA and membrane targeting, assembling nanoparticles based on human serum albumin for targeting and destroying the biofilm of *S. aureus* (Figure 7b) [153]. Chang et al. screened a fluorescence probe (CDr15), which realized eDNA visualization in *P. aeruginosa* biofilm [154].

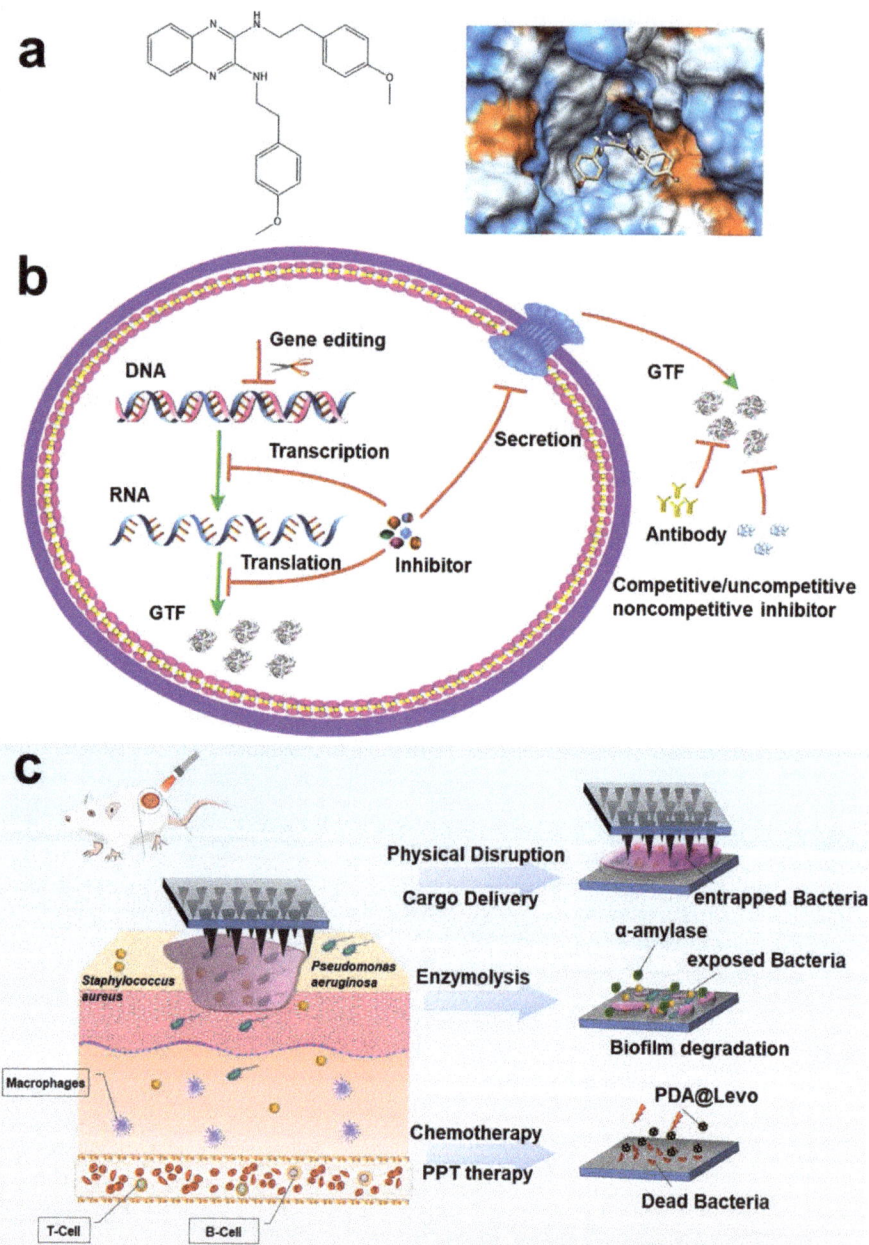

Figure 6. Polysaccharide targeting technology. (a) 2-(4-methoxyphenyl)-N-(3-{[2-(4-methoxyphenyl) ethyl] imino}-1,4-dihydro-2-quinoxalinylidene) ethanamine targeting glucosyltransferase and docking analysis. Reprinted with permission from Ref. [126]. Copyright 2015, Yuqing Li. (b) Schematic diagram of inhibition process of Gtfs inhibitors for S. mutans. Reprinted from Ref. [135]. (c) Schematic diagram of α-amylase-PDA@Levo microneedle patch treating wound biofilm infection in mice. Reprinted with permission from Ref. [135]. Copyright 2022, Daidi Fan.

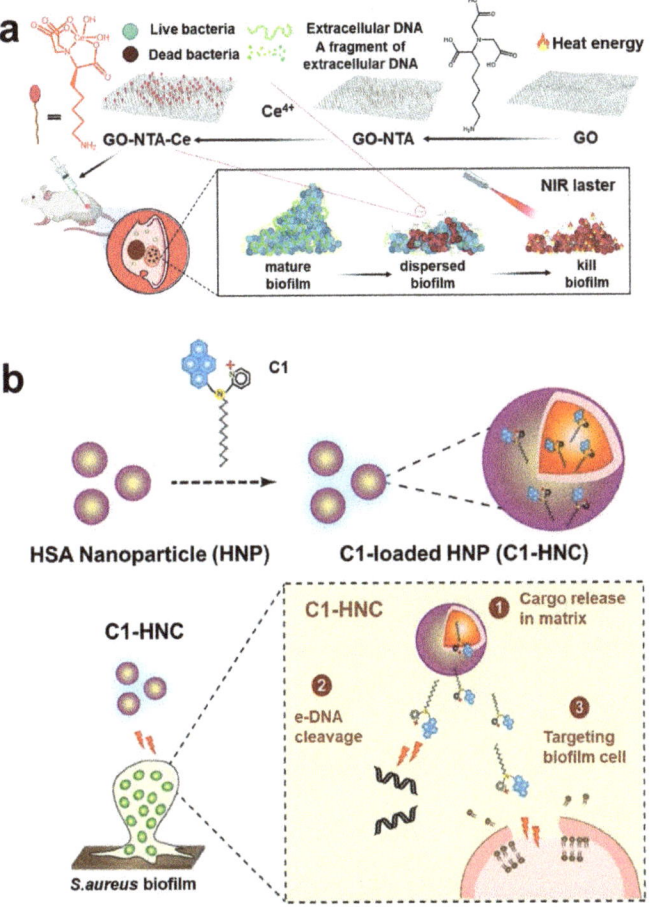

Figure 7. The eDNA targeting technology. (a) GO-NTA-Ce was used to target and destroy biofilm. Reprinted with permission from Ref. [150]. Copyright 2022, Haiwei Ji. (b) Amphiphilic compound C1 with eDNA and membrane-targeting function, assembled with HNP into nanoparticles for targeting and destroying *S. aureus* biofilm. Reprinted with permission from Ref. [153]. Copyright 2016, Prof. Aiyagari Ramesh.

2.2.4. Targeted Protein Strategy

Protein plays an important role in promoting formation of biofilm and maintaining structural stability of biofilm [155,156]. More and more evidence shows that biofilm-associated protein can promote development of bacterial biofilm [157–160]. Interestingly, extracellular proteins do not work alone but jointly with eDNA, polysaccharides, and other components. Some studies have shown that biofilm will spread rapidly after the absence of extracellular proteins in EPS [161,162]. Thus, targeting the protein in biofilm is emerging as a hot research topic. Lin et al. designed a framework nucleic acid delivery that could deliver antisense oligonucleotides to target *S. mutans*, destroying the biofilm (Figure 8a) [163]. The characteristics of carbohydrate–protein interactions were well known. Zhang et al. proposed an inspired nanoplatform composed of spiropyran and galactose. It has dual functions of selectively imaging and eliminating the biofilm in situ [164]. Based on the efficient hydrolysis mechanism of protease to protein, a series of enzyme-functionalized materials were derived. Weldrick et al. introduced a gel carrier nanotechnology based on protease functionality, which, loaded with antibiotics, showed an efficient removal effect on biofilm (Figure 8b) [165]. Devlin introduced that mesoporous silica nanoparticles (MSNs) functionalized by servants could efficiently hydrolyze proteins in *MRSA* biofilm [166]. Curcumin can also target cellular walls and proteins of *Vibrio parahaemolyticus* (*V. parahaemolyticus*) [167].

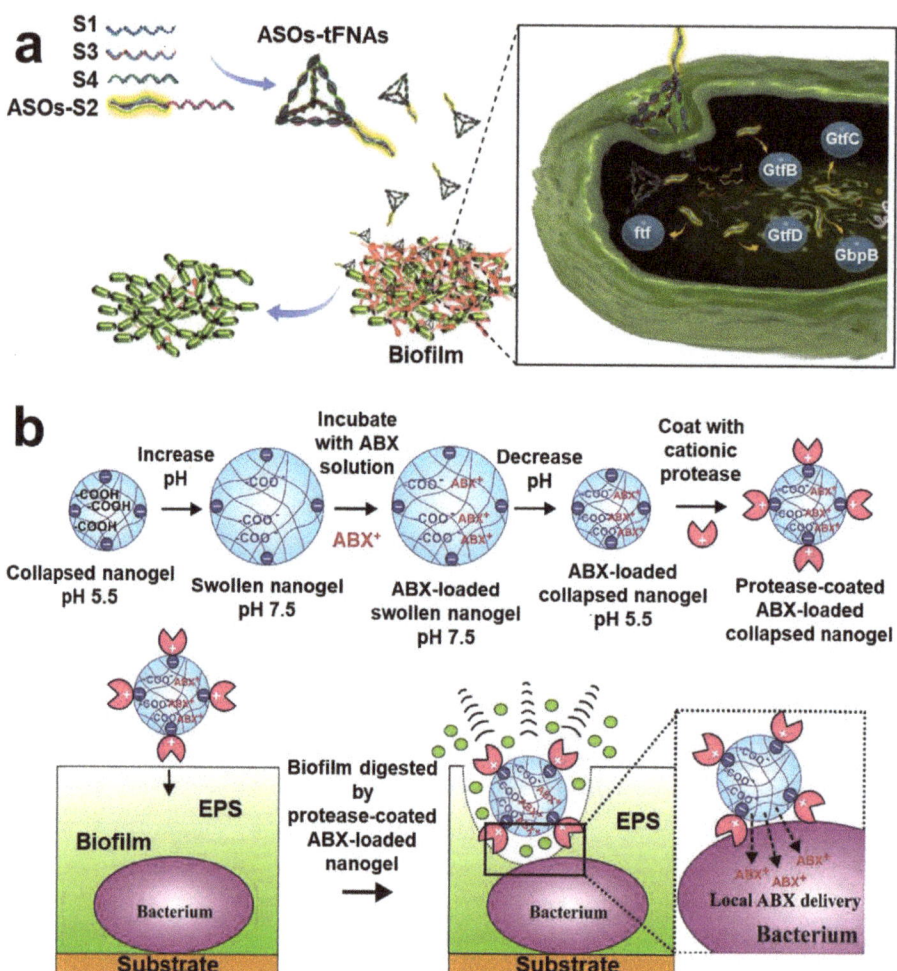

Figure 8. Protein targeting technology. (a) Scheme of a framework nuclear acid delivery that could target *S. mutans* and destroy biofilm bifunction. Reprinted from Ref. [163]. (b) Preparation process and targeting mechanism of gel carrier nanotechnology of protease functionalized. Reprinted with permission from Ref. [165]. Copyright 2019, Vesselin N. Paunov.

Numerous compounds with targeting functions have been synthesized and identified to date. Among them, some compounds were effective in reducing substances in biofilm. However, the future research directions of anti-biofilm molecules with targeting function should include several aspects. First, we should consider the species-specific effects of targeted molecules. They may target a substance in the biofilm of pathogenic bacteria, but they may also have the opposite effect on probiotics. In addition, the effect of targeted molecules on normal cell function should be considered. Second, it is worth noting that the targeted molecules are basically targeting a single substance, but it is worth considering whether the targeted molecules will have a cross-reaction effect on the multi-component aspects of biofilm. Third, most studies on targeted molecules are completed in in vitro biofilm models, which means that they are not necessarily applicable to biofilm produced in vivo. Therefore, future research should focus on use of in vivo models to confirm the anti-biofilm activity of targeted molecules.

2.3. Targeting Strategy for Biofilm Maturation Stage

In the mature stage of biofilm, bacteria will secrete much EPS and then form a dense mushroom-shaped or pile-shaped mature biofilm with 3D structure [59]. Its internal structure is stable and hydrophobic, which can effectively resist external mechanical forces and drug invasion. Due to the dense encapsulation of EPS, the continuous fermentation, and accumulation of bacterial metabolites in biofilm, a unique chemical gradient microenvironment is formed, such as hypoxia, low pH, negative charge, overexpressed GSH, etc. [168]. These extreme microenvironments will cause some bacteria to enter a dormant and persistent state, thereby reducing the sensitivity of bacteria to antimicrobial agents and antibiotics [169,170]. This is also one of the main reasons why mature biofilm infection is difficult to clear.

2.3.1. Targeted Persistent and Dormant Cells Strategy

After the biofilm is formed, the internal chemical gradient environment of biofilm is hostile to bacteria, so bacteria differentiate into different bacterial subpopulations to protect themselves. In 1942, persistent bacteria were first discovered. They will not develop resistance to drugs, but, because of their slow metabolism, or even dormancy, they can avoid being persecuted by drugs [171]. Similar phenomena have been found in clinical treatment [172]. Therefore, in view of these results, it was proposed that this was equivalent to slow and chronic infection [173]. Targeted dormancy, that is, persistent bacteria, is conducive to removal of biofilm and was more conducive to solving the problems of chronic infection and repeated infection.

Typical representatives of dormant bacteria are *Mycobacterium tuberculosis* (Mtb). It is reported that targeting persistent bacilli could effectively improve the treatment success rate and shorten the time after granuloma formation [174]. Dialylquinoline TMC207 could target adenosine triphosphate (ATP) synthase, thereby damaging the lipopeptide of the bacterial membrane to achieve the effect of scavenging persistent Mtb [175]. Based on structure–activity relationships of TMC207 analogs, many derivative compounds have been gradually reported for targeting persistent bacteria (Figure 9a) [176]. Some researchers also found that halogenated phenazine (HP) derivatives can also effectively target persistent bacteria (*MRSA; vancomycin-resistant Enterococcus* (VER); *Mtb*) [177]. The stringent response is an adaptive mechanism controlled by response enzyme (Rel_{Mtb}), which will promote Mtb to enter a persistent state. Using lead compound to target Rel_{Mtb} could directly kill Mtb of a dormant state [178]. Some diterpene analytics can also target Rel_{Msm} and *RelZ* to inhibit formation of persistent cells and biofilm [179]. Narayanan et al. reported that a compound (FNDR-20081) could target maturities *marR* (*Rv0678*, a regulator of *MmpL5*) [180]. Some studies hold that waking up persistent cells is more conducive to killing them than killing them directly [181]. Kim et al. found that adenosine (ADO) can activate ATP and guanosine triphosphate (GTP) synthesis and promote cell respiration, thereby enhancing killing of persistent cells by antibiotics [182]. Rotello et al. proposed a strategy of using biodegradable nanoemulsions to load eugenol and triclosan for synergistic removal of biofilm and persistent cells [183]. In addition, *Acyl* peptide antibiotic ADEP4 is an effective activator of *ClpP* protease, which can adjust persistent MRSA [184]. Yue et al. found that felodipine enhanced the clearance efficiency of aminoglycosides on persistent cells (Figure 9b) [185].

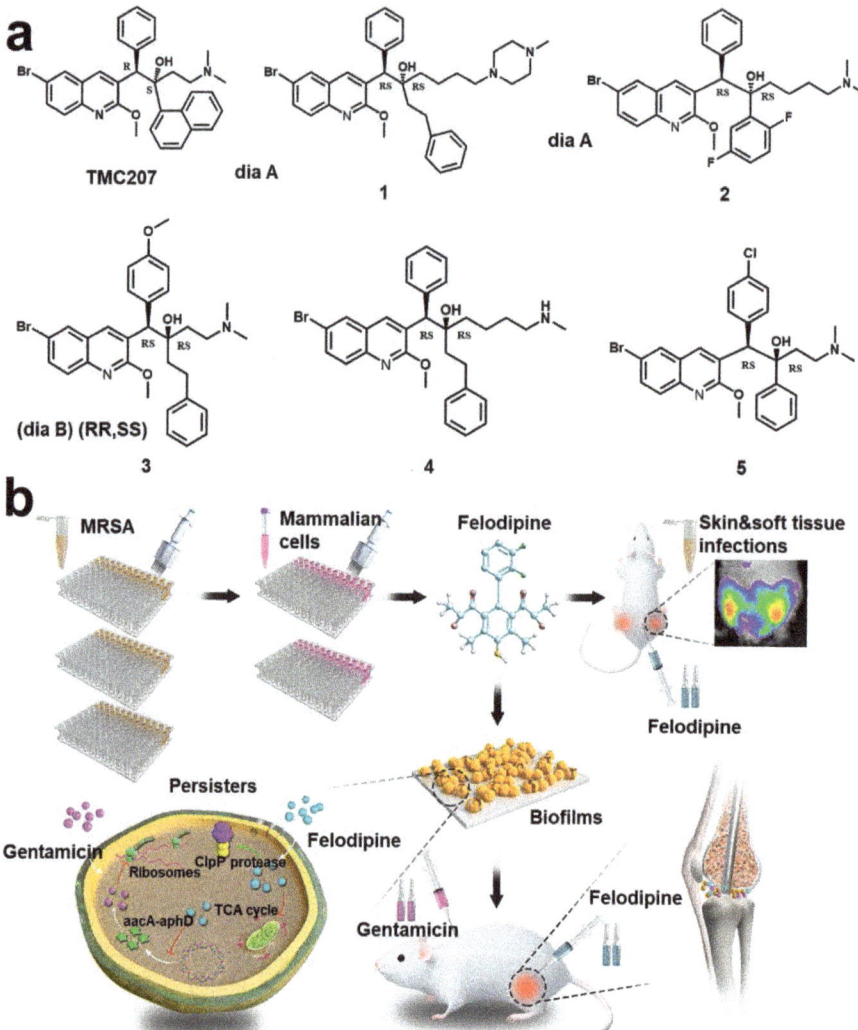

Figure 9. Persistent and dormant cells targeting technology. (a) Compound structure (TMC207) with the function of targeting persistent bacteria and related derivative structure (Compound **1–5**). Reprinted with permission from Ref. [176]. Copyright 2012, Anil Koul. (b) New uses of old drugs using felodipine to regulate bacterial metabolism and improve the clear efficiency of aminoglycosides on persistent cells. Reprinted from Ref. [185].

2.3.2. The Intelligent Release of Microenvironment Response Strategy

Chemical gradient is one of the classic characteristics of biofilm maturity. Thus far, antibacterial materials that use chemical gradient to achieve intelligent release are constantly emerging. Since this review mainly discusses materials and strategies with targeting function, we will briefly introduce this.

Hypoxic

The hypoxic environment will limit metabolism of bacteria, thereby increasing drug resistance [186]. At the same time, it will also enhance the invasion function and virulence factors of bacteria [187]. Therefore, alleviating the hypoxic environment is an effective method to reverse drug resistance of biofilm. Carrying oxygen (O_2) can not only effectively overcome a hypoxic microenvironment but also enhance photodynamic therapy (PDT) [188–190]. It has also been reported that use of catalysts or enzymes to catalyze the endogenous overexpression of H_2O_2 to produce O_2 can also effectively solve the hypoxic microenvironment of biofilm (Figure 10a) [191–195].

Low pH

Lactic acid and acetic acid, which are metabolized by bacteria, will continue to accumulate in the biofilm. At the same time, inflammatory cells continuously release lactic acid, leading to a slight acid phenomenon in the microenvironment of mature biofilm [196,197]. PH-responsive drug delivery systems are widely used in oncology therapy. They are stable in neutral environments but degrade or destroy to release drugs in an acidic environment. Current known degradable bonds that are sensitive to acidity include Schiff bases, esters, ketals, acetals, anhydrides, etc. [198,199]. In addition, using functional groups at a low pH to realize charge reverse and dimensional change is also one of the mainstream strategies in anti-biofilm therapy (Figure 10b) [200–202].

Negative

The negative microenvironment of mature biofilms is primarily caused by eDNA. The negative microenvironment can effectively neutralize invasion of cationic drugs or antibiotic peptides. Using the negative characteristics to design materials and strategies can enhance penetration and retention of materials into biofilm through electrostatic interaction [203,204]. Strategies that exploit negative features are often combined with other targeting strategies to remove biofilms (Figure 11a) [205–207].

Overexpression GSH

In biofilm, GSH acts as an antidote against oxidative stress damage to bacteria from reactive oxygen species. In addition, GSH is a major sulfur source for bacteria, and the sulfur metabolic pathway is one of the main causes of bacterial drug resistance [208]. Therefore, using materials to consume GSH in biofilm may make bacteria unable to maintain redox equilibrium, which is favorable for biofilm removal [209]. Some studies have proposed that using endogenous signal molecule nitric oxide (NO) not only consumes GSH but also disintegrates biofilm and promotes immunity (Figure 11b) [210].

Hydrogen Peroxide

It is understood that endogenous H_2O_2 is over-expressed in the microenvironment of biofilms. As discussed above, H_2O_2 is commonly used as a catalytic substrate to produce O_2 and alleviate a hypoxic microenvironment. H_2O_2 is converted into toxic hydroxyl radicals and superoxide radicals under catalysis of peroxidase (POD) or catalyst [211,212]. This kind of treatment is called chemokinetic therapy (CDT) [213]. This method does not cause bacteria to become resistant.

2.3.3. Other Targeting Strategies

For mature biofilms, in addition to the targeted strategies reviewed above, there are different technical targeting strategies that can still be effective in eradicating biofilms. Rapid developments in biotechnology, nanotechnology, and chemical engineering provide unparalleled flexibility for anti-biofilm technology. Functionalized nanoparticles offer the advantages of controllable structure, morphology, charge, size, target, and optional antibacterial methods. These nanostructures can be used to accurately target and clear the biofilms while avoiding bacterial resistance. We focus on the overall concept and review some nano research in vivo models with clinical potential.

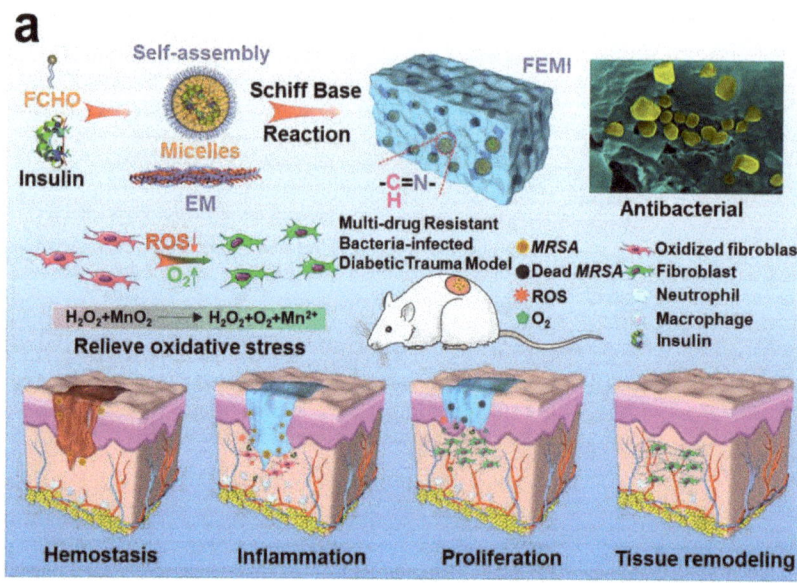
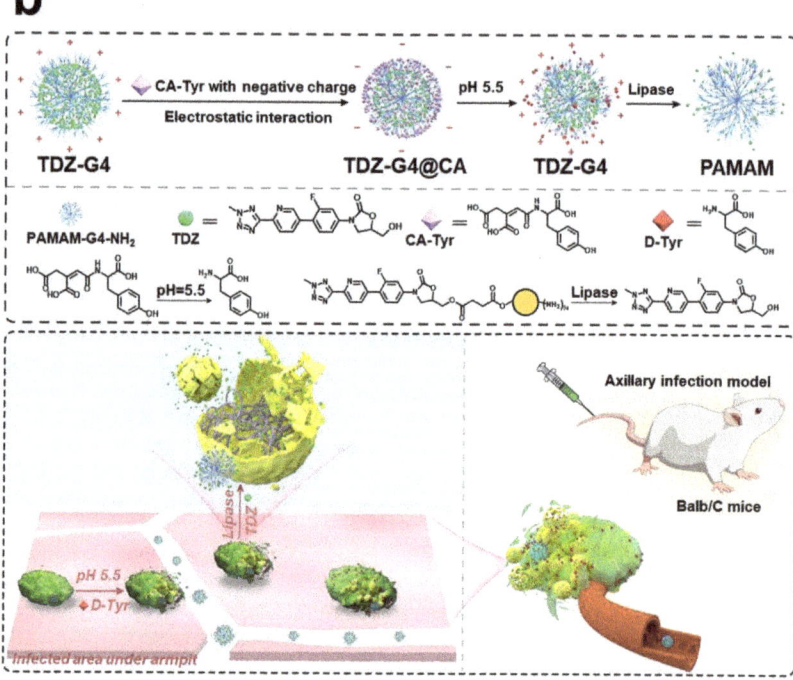

Figure 10. Using hypoxia and low pH to realize intelligent response technology. (**a**) Using Mn^{2+} endogenous overexpression of H_2O_2 to produce O_2 to solve the hypoxia. Reprinted with permission from Ref. [194]. Copyright 2020, Qiuyu Zhang. (**b**) Schematic diagram of functional group protonation in low pH environment to realize charge reversal and intelligent release strategy. Reprinted with permission from Ref. [200]. Copyright 2022, Wei Hong.

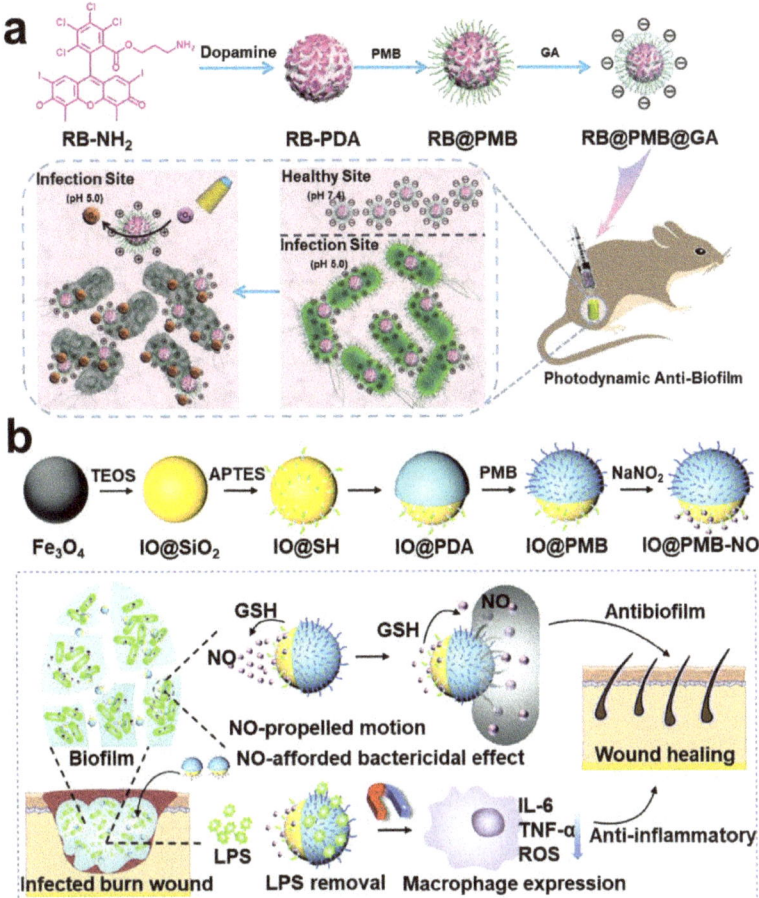

Figure 11. Using negative overexpression GSH and H_2O_2 to realize intelligent response technology. (a) Nanoparticles with charge reversal; the retention capacity of nanoparticles is improved through electrostatic interaction, thereby improving the antibacterial effect. Reprinted with permission from Ref. [206]. Copyright 2021, Fu-Jian Xu. (b) A therapeutic regimen that utilizes NO to deplete GSH and trigger immunotherapy. Reprinted with permission from Ref. [210]. Copyright 2022, Xiaohong Li.

Magnetic Targeting Technology

Iron-based nanoparticles have stable paramagnetic properties. Among them, Fe_3O_4, which is tether-free and harmless to the human body, has been widely applied in construction of *magnetic micro-robots* [214]. Meanwhile, Fe_3O_4 may promote the Fenton reaction, which has certain antibacterial properties. Zhang et al. designed a magnetic micro swarm based on porous Fe_3O_4 masterclass, which showed efficient removal of biofilm (Figure 12a) [215]. Shi et al. loaded glucose-oxidase and L-arginine on $Fe_3O_4@SiO_2$ to deliver nanoparticles to the infected site in mice by magnetic targeting technique. Nanoparticles achieve a cascade reaction to produce NO to eliminate the biofilm infection of drug-resistant bacteria [216]. Escarpa et al. reported a dual-propelled (both catalytic and magnetic) lanbiotic-based Janus micromotor, which can efficiently and selectively capture/inactivate Gram-positive bacteria and biofilms [217].

Phage-Targeting

Bacteriophage–bacteria interaction has been a hot topic and research frontier. Specific targeted function of bacteriophages has been used in most therapeutic areas, such as intestinal infection, intracellular bacterial infection, and liver disease [218–220]. As a result, *phage-targeting techniques* have also appeared in treatment of biofilm infection. Yang et al. designed a strategy of combining phage-guided targeting with AIEgens photodynamic inactivation (PDI) [221]. Sharma et al. found a bacteriophage targeting drug resistance *Enterococcus faecalis* (*E. faecalis*) biofilm; it is worth noting that this phage can be administered orally [222]. Hazan et al. also screened a phage targeting *E. faecalis* biofilm [223]. Wang et al. reported a bacteriophage-photodynamic antibacterial chemotherapy for precise antibacterial and biofilm ablation (Figure 12b) [224]. Hatful et al. reported for the first time the therapeutic effect of bacteriophages on multi-drug-resistant *Mycobacterium chelonae* and described the observed clinical efficacy. The results suggest that bacteriophages are a promising treatment. However, the safety of phage therapy needs to be investigated further [225].

Probiotic Targeting

Since the introduction of probiotic targeted delivery, it has been widely used in a variety of fields, including improving gut flora, oncology, and immunotherapy. In addition, *probiotic delivery techniques* have been widely used in anti-infection applications. This technique not only disintegrates biofilms of pathogenic bacteria but also effectively stimulates the immune system, resulting in a distinct antibacterial–immune combination treatment regimen. Chapman et al. found that four probiotics (*Lactobacillus acidophilus* NCIMB 30184 (PXN 35); *Limosilactobacillus fermentum* NCIMB 30226 (PXN 44); *Lactiplantibacillus plantarum* NCIMB 30187 (PXN 47); and *Lacticaseibacillus rhamnosus* NCIMB 30188 (PXN 54)) could inhibit biofilm formation of pathogenic bacteria through competing for binding sites on the host bladder epithelium, and adhesion of urinary tract pathogens was inhibited [226]. Lorenzo Drago et al. observed two probiotics (*Streptococcus salivarius* 24smb and *S. oralis* 89A) could inhibit biofilm formation of specific pathogens and even disperse their preformed biofilm [227]. Gabriele Meroniet et al. summarized that lactic acid bacteria could inhibit the role of pathogenic bacteria biofilm through multiple pathways [228]. Successive studies of probiotics against pathogenic bacteria have shown that probiotics have the function of targeting and inhibiting disease-causing bacteria pathogenic bacteria and have great potential as drugs or drug vectors [229–231].

Gene Targeting

Gene targeting techniques alter endogenous genes of bacteria by homologous recombination. The effects of this targeting technology could be lasting. In addition to the advantages of directly disintegrating biofilms, it may also directly shadow the dormant cells or newly dividing cells, leading to unique therapeutic effects [232]. Thorsten M. Seyler et al. reported a derivative of PKZ18 (PKZ18-22) for the first time, which can selectively target Gram-positive bacteria [233]. CRISPR interference (CRISPRi) was also one of the main technologies developed in the field of anti-infection [234]. Kimberly A. Kline et al. developed a dual-carrier nisin-inducible CRISPRi system in *E. faecalis* that can target and effectively silence resistance genes via non-template and template chains [235]. In addition, numerous *gene targeting techniques* have been applied in the research and development of antimicrobial drugs [236–238].

Metabolic Targeting

The metabolic pathway of drug resistance has consistently been one of the hotspots of antimicrobial research. It has the potential to reduce bacterial resistance or restore bacterial sensitivity to antimicrobials in a number of ways. Shatalin et al. designed a cystathionine based on bacterial hydrogen sulfide (H_2S) to increase antibiotic resistance γ-*Lyase* (CSE) inhibitor. The inhibitor takes CSE as its target, which inhibits production of H_2S and reduces antibiotic resistance [208]. Other studies have also shown that targeting drug resistance genes can be used to develop new antibacterial drugs [64,239].

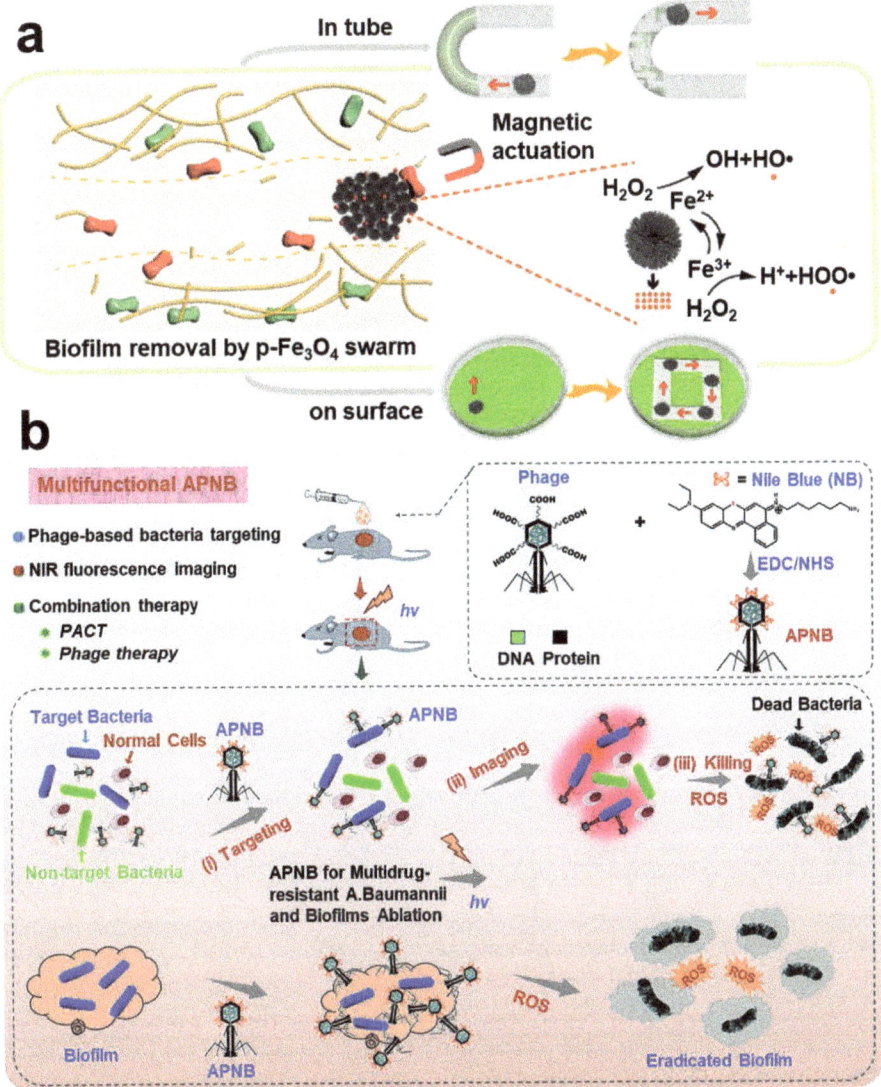

Figure 12. Other targeting strategies in anti-biofilm infection. (**a**) Magnetic targeting nanoparticles prepared by Fe_3O_4 and the Fenton reaction antibacterial mechanism. Reprinted with permission from Ref. [215]. Copyright 2021, Li Zhang. (**b**) Therapeutic scheme for eradicating biofilm using phage-targeting technology and photodynamic therapy. Reprinted from Ref. [224].

Mature biofilms are the model used in most laboratory studies, so there are a myriad of fascinating targeting techniques available at this stage. These studies provide a valuable theoretical basis for clinical transformation, and even some targeting techniques can target specific cell subsets in multi-strain biofilm. The wide development of biofilm-targeting technology should consider the following points: first, biological effects of materials between host and bacteria; second, the whereabouts and potential hazards for host of the materials after antibacterial processes in vivo. The most important point is whether the targeted material has a negative effect on the normal flora during or after treatment.

2.4. Targeted Strategy for Biofilm Dispersion Stage

The biofilm dispersal phase is a unique phase that represented the transition from bacterial biofilms to planktonic bacteria and represented the final step in the biofilm life cycle. Dispersed planktonic bacteria lose their "shelter" and "umbrella", so they are easier to kill. As a result, some studies have considered active dispersal of biofilms as a promising method to control biofilm removal [240]. However, several studies have considered active dispersal of biofilms as a promising method to control biofilm removal. The biofilm should be prevented from entering the dispersion stage [241–243]. Therefore, in this chapter, we discuss the application of targeting technology from two parts: voluntary dispersion biofilm and limited biofilm dispersion.

2.4.1. Active Dispersion Biofilm

After the biofilm has grown to a certain size, the bacteria will actively disrupt the biofilm, thus achieving diffusion. Currently, most studies have proposed various strategies for dispersing biofilm based on the mechanism of bacterial self-degradation of EPS. As the strategy of enzymatic hydrolysis of EPS has been fully discussed above, the discussion will not be repeated in this section. D-type amino acid is one of the main compounds secreted by bacteria in the biofilm dispersion stage [244]. Therefore, therapeutic strategies have been proposed to combine D-type amino acids with drugs. Part of the D-type amino acids was initially used to label peptidoglycan of bacteria, thus achieving effective targeting. However, some D-type amino acids can efficiently cleave EPS in biofilms. Interestingly, this cleavage effect was only directed at the bacterial biofilm and is harmless to normal cells. Cláudia et al. constructed a nanoparticle functionalized with D-amino acids, which can break down the biofilm, thereby improving the bactericidal effect of moxifloxacin in the biofilm [245]. Simple antibiotic-D-amino acid combination therapy could also effectively eradicate biofilm infection of drug-resistant bacteria [246]. Wang et al. constructed a chiral-glutamate-functionalized gold nano bipyramid (Au NBP). The results showed that D-Glu-Au NBPs could more accurately target bacterial cell walls and eliminate biofilms [247]. Li et al. designed a kind of micelle, and the D-Tyrosine loaded on the micelle was released in an acidic environment to decompose the biofilm matrix [248]. Most studies have demonstrated the great clinical value of D-type amino acid, a dispersal factor of bacteria (Figure 13a) [249,250]. Furthermore, Olivier et al. first studied the effect of human hormone atrial natriuretic peptide (hANP) on formation and dispersion of *P. aeruginosa* biofilm [251].

2.4.2. Control Biofilm Dispersion

Some researchers believe that the control of biofilm dispersion is significant compared to active dispersive biofilm techniques. The main reason for this is that the control of biofilm dispersion can be manually controlled both spatially and temporally. Moreover, it can effectively address the problem of secondary infection of biofilm residues. Manju et al. showed that RV_{1717} was a kind of β-D-galactosidase in the cell wall. It has been demonstrated experimentally that RV_{1717} expression is downregulated, which prevents *Mtb* from dispersing from the biofilm in vitro [252]. Kobayashi et al. found that adding Ca^{2+} to the culture medium could counteract the biofilm dispersion mechanism in the study (Figure 13b) [253]. Although the research on regulating of biofilm dispersion is relatively limited, it provides a fresh theoretical basis for development of new drugs.

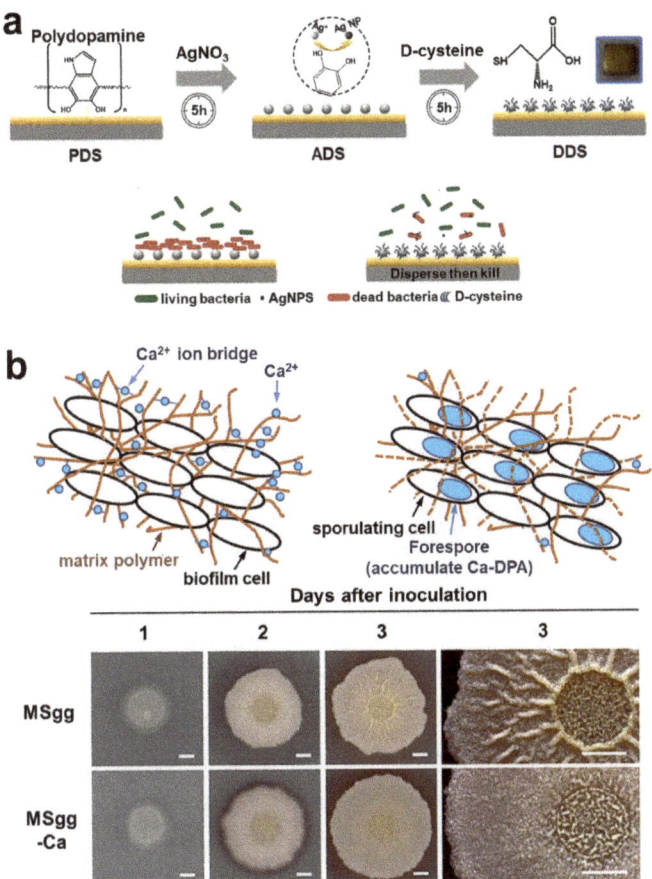

Figure 13. Targeting technology and regulation technology of biofilm dispersion stage. (**a**) The therapeutic strategy of using D-type amino acids to achieve antibacterial and dispersive functions. Reprinted with permission from Ref. [249]. Copyright 2020, Dawei Zhang. (**b**) The strategy of using calcium ions to regulate the biofilm and prevent it from entering the dispersion stage. Reprinted with permission from Ref. [253]. Copyright 2021, Kazuo Kobayashi.

3. Conclusions and Perspectives for Future Research

Tremendous development in bacterial targeting technology has occurred in recent years, including *metabolic targeting, gene targeting, membrane targeting, protein targeting,* and *extracellular matrix targeting*. Compared to conventional antibacterial materials, the targeting technique is more accurate and efficient and, therefore, has better antibacterial properties and ablation efficiency for biofilm. The intensive study of biofilm has greatly accelerated the pace of biofilm-targeting technologies. Targeting techniques have significantly improved biocompatibility by fine-tuning the life-cycle properties of biofilms and related components, combined with diagnostic imaging techniques to unlock high-dimensional multimodal studies. Based on these advantages, biofilm-targeting technology has been applied to ultra-sensitive diagnosis and personalized treatment. This paper reviews the known biofilm-targeting technologies, with a particular focus on targeting dormant cells and the regulation strategy for biofilm life cycle. While significant progress has been made at this stage, as described in this paper, there are still many challenges in clinical targeting technology:

1. The key barrier is the in vivo biofilm model, especially for a multi-species biofilm model. In this regard, substantive research on the targeted techniques should be conducted before entering the clinic; it is extremely important to implement techniques that can accurately target the objective in multi-species biofilm.
2. Further implementation of differential targeting of pathogenic bacteria and probiotics is highly beneficial and is expected to facilitate rapid development of immunotherapies.
3. To clarify the metabolic pathway of targeted techniques under host pathological conditions, it is necessary to develop targeted techniques with long-term visualization or monitoring.
4. Currently, targeting techniques target different phases of biofilms. Could there be a technique to observe the biofilm phase in patients to make treatment plans more effective?
5. The biological effects of targeting technology among materials, cells, and bacteria are very worthy of study.
6. Currently, small molecules of targeted inhibitors have the potential to replace antibiotics for treatment, but antibiotics have a chiral structure. Research on the combination of targeted inhibitors and stereochemistry may be a new generation of antibiotic research and development route.
7. Targeting technology is needed to meet clinical needs. Cost-effective, simplified, and economical amplification preparation strategies need to be widely studied.

In the rapidly evolving antibacterial field, we assume that continuous improvement in biofilm-targeting technology will make it possible to target in an accurate way and introduce single-bacterial targeting technology that is not available at present. This is not only conducive to accurate clinical diagnosis and treatment but also helps to stimulate discovery of new technologies.

Funding: The authors thank the National Natural Science Foundation (52273118, 22275013).

Institutional Review Board Statement: Not applicable.

Informed Consent Statement: Not applicable.

Data Availability Statement: Data sharing not applicable.

Conflicts of Interest: The authors declare no conflict of interest.

References

1. Kolpen, M.; Kragh, K.N.; Enciso, J.B.; Faurholt-Jepsen, D.; Lindegaard, B.; Egelund, G.B.; Jensen, A.V.; Ravn, P.; Mathiesen, I.H.M.; Gheorge, A.G.; et al. Bacterial Biofilms Predominate in Both Acute and Chronic Human Lung Infections. *Thorax* **2022**, *77*, 1015–1022. [CrossRef]
2. Tang, Y.; Huang, Q.X.; Zheng, D.W.; Chen, Y.; Ma, L.; Huang, C.; Zhang, X.Z. Engineered *Bdellovibrio bacteriovorus*: A Countermeasure for Biofilm-Induced Periodontitis. *Mater. Today* **2022**, *53*, 71–83. [CrossRef]
3. Silva, M.D.; Sillankorva, S. Otitis Media Pathogens—A Life Entrapped in Biofilm Communities. *Crit. Rev. Microbiol.* **2019**, *45*, 595–612. [CrossRef]
4. Lerche, C.J.; Schwartz, F.; Theut, M.; Fosbøl, E.L.; Iversen, K.; Bundgaard, H.; Høiby, N.; Moser, C. Anti-Biofilm Approach in Infective Endocarditis Exposes New Treatment Strategies for Improved Outcome. *Front. Cell Dev. Biol.* **2021**, *9*, 643335. [CrossRef]
5. Raheem Lateef Al-awsi, G.; Al-Hadeithi, Z.S.M.; Abdalkareem Jasim, S.; Alkhudhairy, M.K.; Ghasemian, A. Virulence Traits and Plasmid-Mediated Quinolone Resistance among *Aggregatibacter actinomycetemcomitans* from Iraq: Low Rate of Highly Virulent JP2 Genotype. *Microb. Pathog.* **2022**, *164*, 105438. [CrossRef]
6. Xiao, Y.; Jiang, S.C.; Wang, X.; Muhammad, T.; Song, P.; Zhou, B.; Zhou, Y.; Li, Y. Mitigation of Biofouling in Agricultural Water Distribution Systems with Nanobubbles. *Environ. Int.* **2020**, *141*, 105787. [CrossRef]
7. Winters, H.; Eu, H.G.; Li, S.; Alpatova, A.; Alshahri, A.H.; Nasar, N.; Ghaffour, N. Biofouling of Seawater Reverse Osmosis Membranes Caused by Dispersal of Planktonic Particulate Bacterial Aggregates (Protobiofilms) from Rotary Energy Recovery Devices. *Desalination* **2022**, *529*, 115647. [CrossRef]
8. Pichardo-Romero, D.; Garcia-Arce, Z.P.; Zavala-Ramírez, A.; Castro-Muñoz, R. Current Advances in Biofouling Mitigation in Membranes for Water Treatment: An Overview. *Processes* **2020**, *8*, 137. [CrossRef]

9. Ricart, M.; Guasch, H.; Barceló, D.; Brix, R.; Conceição, M.H.; Geiszinger, A.; de Alda, M.J.L.; López-Doval, J.C.; Muñoz, I.; Postigo, C.; et al. Primary and Complex Stressors in Polluted Mediterranean Rivers: Pesticide Effects on Biological Communities. *J. Hydrol.* **2010**, *383*, 52–61. [CrossRef]
10. Zhu, Q.; Gooneratne, R.; Hussain, M.A. Listeria Monocytogenes in Fresh Produce: Outbreaks, Prevalence and Contamination Levels. *Foods* **2017**, *6*, 21. [CrossRef]
11. Habash, M.; Reid, G. Microbial Biofilms: Their Development and Significance for Medical Device-Related Infections. *J. Clin. Pharmacol.* **1999**, *39*, 887–898. [CrossRef] [PubMed]
12. Schramm, A.; De Beer, D.; Gieseke, A.; Amann, R. Microenvironments and Distribution of Nitrifying Bacteria in a Membrane-Bound Biofilm. *Environ. Microbiol.* **2000**, *2*, 680–686. [CrossRef]
13. Kolenbrander, P.E. Oral Microbial Communities: Biofilms, Interactions, and Genetic Systems. *Annu. Rev. Virol.* **2000**, *54*, 413–437. [CrossRef]
14. Periasamy, S.; Joo, H.S.; Duong, A.C.; Bach, T.H.L.; Tan, V.Y.; Chatterjee, S.S.; Cheung, G.Y.C.; Otto, M. How *Staphylococcus aureus* Biofilms Develop Their Characteristic Structure. *Proc. Natl. Acad. Sci. USA* **2012**, *109*, 1281–1286. [CrossRef]
15. Sutherland, I.W. The Biofilm Matrix-An Immobilized but Dynamic Microbial Environment. *Trends Microbiol.* **2001**, *9*, 222–227. [CrossRef]
16. Decho, A.W. Microbial Biofilms in Intertidal Systems: An Overview. *Cont. Shelf Res.* **2000**, *20*, 1257–1273. [CrossRef]
17. Koo, H.; Allan, R.N.; Howlin, R.P.; Stoodley, P.; Hall-Stoodley, L. Targeting Microbial Biofilms: Current and Prospective Therapeutic Strategies. *Nat. Rev. Microbiol.* **2017**, *15*, 740–755. [CrossRef]
18. Lewis, K. Multidrug Tolerance of Biofilms and Persister Cells. *Curr. Top. Microbiol. Immunol.* **2008**, *322*, 107–131. [CrossRef]
19. Allison, K.R.; Brynildsen, M.P.; Collins, J.J. Metabolite-Enabled Eradication of Bacterial Persisters by Aminoglycosides. *Nature* **2011**, *473*, 216–220. [CrossRef]
20. Keren, I.; Shah, D.; Spoering, A.; Kaldalu, N.; Lewis, K. Specialized Persister Cells and the Mechanism of Multidrug Tolerance in *Escherichia coli*. *J. Bacteriol.* **2004**, *186*, 8172–8180. [CrossRef]
21. Fry, D.E. Antimicrobial Peptides. *Surg. Infect.* **2018**, *19*, 804–811. [CrossRef] [PubMed]
22. Lewis, K. Persister Cells, Dormancy and Infectious Disease. *Nat. Rev. Microbiol.* **2007**, *5*, 48–56. [CrossRef] [PubMed]
23. Høiby, N.; Bjarnsholt, T.; Givskov, M.; Molin, S.; Ciofu, O. Antibiotic Resistance of Bacterial Biofilms. *Int. J. Antimicrob. Agents* **2010**, *35*, 322–332. [CrossRef]
24. Hunter, R.C.; Beveridge, T.J. Application of a PH-Sensitive Fluoroprobe (C-SNARF-4) for PH Microenvironment Analysis in *Pseudomonas aeruginosa* Biofilms. *Appl. Environ. Microbiol.* **2005**, *71*, 2501–2510. [CrossRef] [PubMed]
25. Koo, H.; Falsetta, M.L.; Klein, M.I. The Exopolysaccharide Matrix: A Virulence Determinant of Cariogenic Biofilm. *J. Dent. Res.* **2013**, *92*, 1065–1073. [CrossRef] [PubMed]
26. Bjarnsholt, T.; Alhede, M.; Alhede, M.; Eickhardt-Sørensen, S.R.; Moser, C.; Kühl, M.; Jensen, P.Ø.; Høiby, N. The In Vivo Biofilm. *Trends Microbiol.* **2013**, *21*, 466–474. [CrossRef] [PubMed]
27. Hammer, B.K.; Bassler, B.L. Quorum Sensing Controls Biofilm Formation in *Vibrio cholerae*. *Mol. Microbiol.* **2003**, *50*, 101–104. [CrossRef]
28. Vu, B.; Chen, M.; Crawford, R.J.; Ivanova, E.P. Bacterial Extracellular Polysaccharides Involved in Biofilm Formation. *Molecules* **2009**, *14*, 2535–2554. [CrossRef]
29. Hall-Stoodley, L.; Stoodley, P. Evolving Concepts in Biofilm Infections. *Cell. Microbiol.* **2009**, *11*, 1034–1043. [CrossRef]
30. Flemming, H.C.; Wingender, J.; Szewzyk, U.; Steinberg, P.; Rice, S.A.; Kjelleberg, S. Biofilms: An Emergent Form of Bacterial Life. *Nat. Rev. Microbiol.* **2016**, *14*, 563–575. [CrossRef]
31. Bjarnsholt, T.; Ciofu, O.; Molin, S.; Givskov, M.; Høiby, N. Applying Insights from Biofilm Biology to Drug Development-Can a New Approach Be Developed? *Nat. Rev. Drug Discov.* **2013**, *12*, 791–808. [CrossRef] [PubMed]
32. O'Gara, J.P.; Humphreys, H. *Staphylococcus epidermidis* Biofilms: Importance and Implications. *J. Med. Microbiol.* **2001**, *50*, 582–587. [CrossRef] [PubMed]
33. Zachary, K.Z.; Mattew, L.B. Antimicrobial and Antifouling Strategies for Polymeric Medical Devices. *ACS Macro. Lett.* **2018**, *7*, 16–25. [CrossRef]
34. Dunne, W.M. Bacterial adhesion: Seen any good biofilms lately? *Clin. Microbiol. Rev.* **2002**, *15*, 155–166. [CrossRef]
35. Fu, J.; Zhang, Y.; Zhang, W.; Shu, G.; Lin, J.; Xu, F.; Tang, H.; Peng, G.; Zhao, L.; Chen, S.; et al. Strategies for Interfering With Bacterial Early Stage Biofilms. *Clin. Front. Microbiol.* **2021**, *12*, 675843. [CrossRef]
36. Musa, H.M.; Aisha, L.I.; Xiao, F.; Yachong, G.; Yiyan, Y.; Xu, J.; Junzhi, Q.; Xiong, G.; Tianpei, H. Beyond Risk: Bacterial Biofilms and Their Regulating Approaches. *Front. Microbiol.* **2020**, *11*, 928. [CrossRef]
37. Armbruster, C.; Parsek, M. New insight into the early stages of biofilm formation. *Proc. Natl. Acad. Sci. USA* **2018**, *115*, 4317–4319. [CrossRef] [PubMed]
38. Conlon, B.P.; Nakayasu, E.S.; Fleck, L.E.; LaFleur, M.D.; Isabella, V.M.; Coleman, S.N.; Smith, R.D.; Adkins, J.N.; Lewis, K. Activated ClpP kills persisters and eradicates a chronic biofilm infection. *Nat. Prod. Rep.* **2013**, *503*, 365–370. [CrossRef]
39. Rumbaugh, K.P.; Sauer, K. Biofilm dispersion. *Nat. Rev. Microbiol.* **2020**, *18*, 571–586. [CrossRef]
40. Kindler, O.; Pulkkinen, O.; Cherstvy, A.G.; Metzler, R. Burst Statistics in an Early Biofilm Quorum Sensing Model: The Role of Spatial Colony-Growth Heterogeneity. *Sci. Rep.* **2019**, *9*, 12077. [CrossRef]

41. Sharma, A.; Wood, K.B. Spatial Segregation and Cooperation in Radially Expanding Microbial Colonies under Antibiotic Stress. *ISME J.* **2021**, *15*, 3019–3033. [CrossRef] [PubMed]
42. Ruhal, R.; Kataria, R. Biofilm Patterns in Gram-Positive and Gram-Negative Bacteria. *Microbiol. Res.* **2021**, *251*, 126829. [CrossRef]
43. Nadell, C.D.; Drescher, K.; Foster, K.R. Spatial Structure, Cooperation and Competition in Biofilms. *Nat. Rev. Microbiol.* **2016**, *14*, 589–600. [CrossRef]
44. Wu, H.; Moser, C.; Wang, H.Z.; Høiby, N.; Song, Z.J. Strategies for Combating Bacterial Biofilm Infections. *Int. J. Oral Sci.* **2015**, *7*, 1–7. [CrossRef] [PubMed]
45. Doub, J.B. Bacteriophage Therapy for Clinical Biofilm Infections: Parameters That Influence Treatment Protocols and Current Treatment Approaches. *Antibiotics* **2020**, *9*, 799. [CrossRef] [PubMed]
46. Huang, D.N.; Wang, J.; Ren, K.F.; Ji, J. Functionalized Biomaterials to Combat Biofilms. *Biomater. Sci.* **2020**, *8*, 4052–4066. [CrossRef]
47. Contreras-García, A.; Bucioa, E.; Brackmanc, G.; Coenyec, T.; Concheirob, A.; Alvarez-Lorenzob, C. Biofilm Inhibition and Drug-Eluting Properties of Novel DMAEMA-Modified Polyethylene and Silicone Rubber Surfaces. *Biofouling* **2011**, *27*, 123–135. [CrossRef]
48. Yang, K.; Shi, J.; Wang, L.; Chen, Y.; Liang, C.; Yang, L.; Wang, L.N. Bacterial Anti-Adhesion Surface Design: Surface Patterning, Roughness and Wettability: A Review. *J. Mater. Sci. Technol.* **2022**, *99*, 82–100. [CrossRef]
49. Busscher, H.J.; van der Mei, H.C.; Subbiahdoss, G.; Jutte, P.C.; van den Dungen, J.J.A.M.; Zaat, S.A.J.; Schultz, M.J.; Grainger, D.W. Biomaterial-Associated Infection: Locating the Finish Line in the Race for the Surface. *Sci. Transl. Med.* **2012**, *4*, 153rv10. [CrossRef]
50. Chen, Z.; Wang, Z.; Ren, J.; Qu, X. Enzyme Mimicry for Combating Bacteria and Biofilms. *Acc. Chem. Res.* **2018**, *51*, 789–799. [CrossRef]
51. Xiao, X.; Zhao, W.; Liang, J.; Sauer, K.; Libera, M. Self-Defensive Antimicrobial Biomaterial Surfaces. *Colloids Surf. B* **2020**, *192*, 110989. [CrossRef] [PubMed]
52. Steenackers, H.P.; Parijs, I.; Foster, K.R.; Vanderleyden, J. Experimental Evolution in Biofilm Populations. *FEMS Microbiol. Rev.* **2016**, *40*, 373–397. [CrossRef] [PubMed]
53. Thaarup, I.C.; Iversen, A.K.S.; Lichtenberg, M.; Bjarnsholt, T.; Jakobsen, T.H. Biofilm Survival Strategies in Chronic Wounds. *Microorganisms* **2022**, *10*, 775. [CrossRef] [PubMed]
54. Yamada, M.; Ikegami, A.; Kuramitsu, H.K. Synergistic Biofilm Formation by *Treponema denticola* and *Porphyromonas gingivalis*. *FEMS Microbiol. Lett.* **2005**, *250*, 271–277. [CrossRef] [PubMed]
55. Rickard, A.H.; Palmer, R.J.; Blehert, D.S.; Campagna, S.R.; Semmelhack, M.F.; Egland, P.G.; Bassler, B.L.; Kolenbrander, P.E. Autoinducer 2: A Concentration-Dependent Signal for Mutualistic Bacterial Biofilm Growth. *Mol. Microbiol.* **2006**, *60*, 1446–1456. [CrossRef]
56. Kumada, M.; Motegi, M.; Nakao, R.; Yonezawa, H.; Yamamura, H.; Tagami, J.; Senpuku, H. Inhibiting Effects of *Enterococcus faecium* Non-Biofilm Strain on *Streptococcus mutans* Biofilm Formation. *J. Microbiol. Immunol. Infect.* **2009**, *42*, 188–196.
57. Rao, D.; Webb, J.S.; Kjelleberg, S. Competitive Interactions in Mixed-Species Biofilms Containing the Marine Bacterium *Pseudoalteromonas tunicata*. *Appl. Environ. Microbiol.* **2005**, *71*, 1729–1736. [CrossRef]
58. Srinivasan, R.; Santhakumari, S.; Poonguzhali, P.; Geetha, M.; Dyavaiah, M.; Xiangmin, L. Bacterial Biofilm Inhibition: A Focused Review on Recent Therapeutic Strategies for Combating the Biofilm Mediated Infections. *Front. Microbiol.* **2021**, *12*, 676458. [CrossRef]
59. Karygianni, L.; Ren, Z.; Koo, H.; Thurnheer, T. Biofilm Matrixome: Extracellular Components in Structured Microbial Communities. *Trends Microbiol.* **2020**, *28*, 668–681. [CrossRef]
60. Uruén, C.; Chopo-Escuin, G.; Tommassen, J.; Mainar-Jaime, R.C.; Arenas, J. Antibiotics Biofilms as Promoters of Bacterial Antibiotic Resistance and Tolerance. *Antibiotics* **2020**, *10*, 3. [CrossRef]
61. Jiang, Q.; Yu, Y.; Xu, R.; Zhang, Z.; Liang, C.; Sun, H.; Deng, F.; Yu, X. The Temporal Shift of Peri-Implant Microbiota during the Biofilm Formation and Maturation in a Canine Model. *Microb. Pathog.* **2021**, *158*, 105100. [CrossRef]
62. Urwin, L.; Okurowska, K.; Crowther, G.; Roy, S.; Garg, P.; Karunakaran, E.; MacNeil, S.; Partridge, L.J.; Green, L.R.; Monk, P.N. Corneal Infection Models: Tools to Investigate the Role of Biofilms in Bacterial Keratitis. *Cells* **2020**, *9*, 2450. [CrossRef] [PubMed]
63. Ahmed, W.; Zhai, Z.; Gao, C. Adaptive Antibacterial Biomaterial Surfaces and Their Applications. *Mater. Today Bio* **2019**, *2*, 100017. [CrossRef] [PubMed]
64. Wu, S.; Zhang, J.; Peng, Q.; Liu, Y.; Lei, L.; Zhang, H. The Role of *Staphylococcus aureus* YycFG in Gene Regulation, Biofilm Organization and Drug Resistance. *Antibiotics* **2021**, *10*, 1555. [CrossRef] [PubMed]
65. Solanki, V.; Tiwari, M.; Tiwari, V. Host-Bacteria Interaction and Adhesin Study for Development of Therapeutics. *Int. J. Biol. Macromol.* **2018**, *112*, 54–64. [CrossRef]
66. Chauhan, N.; Hatlem, D.; Orwick-Rydmark, M.; Schneider, K.; Floetenmeyer, M.; van Rossum, B.; Leo, J.C.; Linke, D. Insights into the Autotransport Process of a Trimeric Autotransporter, Yersinia Adhesin A (YadA). *Mol. Microbiol.* **2019**, *111*, 844–862. [CrossRef] [PubMed]
67. Asadi, A.; Razavi, S.; Talebi, M.; Gholami, M. A Review on Anti-Adhesion Therapies of Bacterial Diseases. *Infection* **2019**, *47*, 13–23. [CrossRef]

68. Monserrat-Martinez, A.; Gambin, Y.; Sierecki, E. Thinking Outside the Bug: Molecular Targets and Strategies to Overcome Antibiotic Resistance. *Int. J. Mol. Sci.* **2019**, *20*, 1255. [CrossRef]
69. Filipović, U.; Dahmane, R.G.; Ghannouchi, S.; Zore, A.; Bohinc, K. Bacterial Adhesion on Orthopedic Implants. *Adv. Colloid Interface Sci.* **2020**, *283*, 102228. [CrossRef]
70. Grosheva, I.; Zheng, D.; Levy, M.; Polansky, O.; Lichtenstein, A.; Golani, O.; Dori-Bachash, M.; Moresi, C.; Shapiro, H.; del Mare-Roumani, S.; et al. High-Throughput Screen Identifies Host and Microbiota Regulators of Intestinal Barrier Function. *Gastroenterology* **2020**, *159*, 1807–1823. [CrossRef]
71. Paxman, J.J.; Lo, A.W.; Sullivan, M.J.; Panjikar, S.; Kuiper, M.; Whitten, A.E.; Wang, G.; Luan, C.H.; Moriel, D.G.; Tan, L.; et al. Unique Structural Features of a Bacterial Autotransporter Adhesin Suggest Mechanisms for Interaction with Host Macromolecules. *Nat. Commun.* **2019**, *10*, 1967. [CrossRef] [PubMed]
72. Rao, L.; Sheng, Y.; Zhang, J.; Xu, Y.; Yu, J.; Wang, B.; Zhao, H.; Wang, X.; Guo, Y.; Wu, X.; et al. Small-Molecule Compound SYG-180-2-2 to Effectively Prevent the Biofilm Formation of Methicillin-Resistant *Staphylococcus aureus*. *Front. Microbiol.* **2022**, *12*, 770657. [CrossRef] [PubMed]
73. Kannappan, A.; Durgadevi, R.; Srinivasan, R.; Lagoa, R.J.L.; Packiavathy, I.A.S.V.; Pandian, S.K.; Veera Ravi, A. 2-Hydroxy-4-Methoxybenzaldehyde from *Hemidesmus indicus* Is Antagonistic to *Staphylococcus epidermidis* Biofilm Formation. *Biofouling* **2020**, *36*, 549–563. [CrossRef] [PubMed]
74. Hu, P.; Lv, B.; Yang, K.; Lu, Z.; Ma, J. Discovery of Myricetin as an Inhibitor against *Streptococcus mutans* and an Anti-Adhesion Approach to Biofilm Formation. *Int. J. Med. Microbiol.* **2021**, *311*, 151512. [CrossRef]
75. Liu, C.; Sun, D.; Liu, J.; Chen, Y.; Zhou, X.; Ru, Y.; Zhu, J.; Liu, W. CAMP and C-Di-GMP Synergistically Support Biofilm Maintenance through the Direct Interaction of Their Effectors. *Nat. Commun.* **2022**, *13*, 1493. [CrossRef]
76. Awadelkareem, A.M.; Al-Shammari, E.; Elkhalifa, A.O.; Adnan, M.; Siddiqui, A.J.; Mahmood, D.; Azad, Z.R.A.A.; Patel, M.; Mehmood, K.; Danciu, C.; et al. Anti-Adhesion and Antibiofilm Activity of Eruca Sativa Miller Extract Targeting Cell Adhesion Proteins of Food-Borne Bacteria as a Potential Mechanism: Combined In Vitro-In Silico Approach. *Plants* **2022**, *11*, 610. [CrossRef]
77. Huebinger, R.M.; Stones, D.H.; de Souza Santos, M.; Carlson, D.L.; Song, J.; Vaz, D.P.; Keen, E.; Wolf, S.E.; Orth, K.; Krachler, A.M. Targeting Bacterial Adherence Inhibits Multidrug-Resistant *Pseudomonas aeruginosa* Infection Following Burn Injury. *Sci. Rep.* **2016**, *6*, 39341. [CrossRef]
78. Capeletti, L.B.; de Oliveira, J.F.A.; Loiola, L.M.D.; Galdino, F.E.; da Silva Santos, D.E.; Soares, T.A.; de Oliveira Freitas, R.; Cardoso, M.B. Gram-Negative Bacteria Targeting Mediated by Carbohydrate–Carbohydrate Interactions Induced by Surface-Modified Nanoparticles. *Adv. Funct. Mater.* **2019**, *29*, 1904216. [CrossRef]
79. Sarshar, M.; Behzadi, P.; Ambrosi, C.; Zagaglia, C.; Palamara, A.T.; Scribano, D. FimH and Anti-Adhesive Therapeutics: A Disarming Strategy against Uropathogens. *Antibiotics* **2020**, *9*, 397. [CrossRef]
80. Cusumano, Z.T.; Klein, R.D.; Hultgren, S.J. Innovative Solutions to Sticky Situations: Antiadhesive Strategies for Treating Bacterial Infections. *Microbiol. Spectr.* **2016**, *4*, 4.2.07. [CrossRef]
81. Schembri, M.A.; Hasman, H.; Klemm, P. Expression and Purification of the Mannose Recognition Domain of the FimH Adhesin. *FEMS Microbiol. Lett.* **2000**, *188*, 147–151. [CrossRef] [PubMed]
82. Foroogh, N.; Rezvan, M.; Ahmad, K.; Mahmood, S. Structural and Functional Characterization of the FimH Adhesin of Uropathogenic *Escherichia coli* and Its Novel Applications. *Microb. Pathog.* **2021**, *161*, 105288. [CrossRef] [PubMed]
83. Hartmann, M.; Betz, P.; Sun, Y.; Gorb, S.N.; Lindhorst, T.K.; Krueger, A. Saccharide-Modified Nanodiamond Conjugates for the Efficient Detection and Removal of Pathogenic Bacteria. *Chemistry* **2012**, *18*, 6485–6492. [CrossRef] [PubMed]
84. Bernardi, A.; Jiménez-Barbero, J.; Casnati, A.; de Castro, C.; Darbre, T.; Fieschi, F.; Finne, J.; Funken, H.; Jaeger, K.E.; Lahmann, M.; et al. Multivalent Glycoconjugates as Anti-Pathogenic Agents. *Chem. Soc. Rev.* **2013**, *42*, 4709–4727. [CrossRef] [PubMed]
85. Speziale, P.; Arciola, C.R.; Pietrocola, G. Fibronectin and Its Role in Human Infective Diseases. *Cells* **2019**, *8*, 1516. [CrossRef]
86. Zhang, Y.; Chen, Y.; Lo, C.; Zhuang, J.; Angsantikul, P.; Zhang, Q.; Wei, X.; Zhou, Z.; Obonyo, M.; Fang, R.H.; et al. Inhibition of Pathogen Adhesion by Bacterial Outer Membrane-Coated Nanoparticles. *Angew. Chem. Int. Ed. Engl.* **2019**, *58*, 11404–11408. [CrossRef]
87. Guo, S.; Zahiri, H.; Stevens, C.; Spaanderman, D.C.; Milroy, L.G.; Ottmann, C.; Brunsveld, L.; Voets, I.K.; Davies, P.L. Molecular Basis for Inhibition of Adhesin-Mediated Bacterial-Host Interactions through a Peptide-Binding Domain. *Cell Rep.* **2021**, *37*, 110002. [CrossRef]
88. An, S.J.; Namkung, J.U.; Ha, K.W.; Jun, H.K.; Kim, H.Y.; Choi, B.K. Inhibitory Effect of D-Arabinose on Oral Bacteria Biofilm Formation on Titanium Discs. *Anaerobe* **2022**, *75*, 102533. [CrossRef]
89. Zhang, J.; Chen, C.; Chen, J.; Zhou, S.; Zhao, Y.; Xu, M.; Xu, H. Dual Mode of Anti-Biofilm Action of G3 against *Streptococcus mutans*. *ACS Appl. Mater. Interfaces* **2020**, *12*, 27866–27875. [CrossRef]
90. Khoo, X.; Grinstaff, M.W. Novel Infection-Resistant Surface Coatings: A Bioengineering Approach. *MRS Bull.* **2011**, *36*, 357–366. [CrossRef]
91. Xu, Q.; A, S.; Venet, M.; Gao, Y.; Zhou, D.; Wang, W.; Zeng, M.; Rotella, C.; Li, X.; Wang, X.; et al. Bacteria-Resistant Single Chain Cyclized/Knotted Polymer Coatings. *Angew. Chem. Int. Ed. Engl.* **2019**, *58*, 10616–10620. [CrossRef] [PubMed]
92. Choi, W.; Jin, J.; Park, S.; Kim, J.Y.; Lee, M.J.; Sun, H.; Kwon, J.S.; Lee, H.; Choi, S.H.; Hong, J. Quantitative Interpretation of Hydration Dynamics Enabled the Fabrication of a Zwitterionic Antifouling Surface. *ACS Appl. Mater. Interfaces* **2020**, *12*, 7951–7965. [CrossRef] [PubMed]

93. Zhang, P.; Li, J.; Yang, M.; Huang, L.; Bu, F.; Xie, Z.; Li, G.; Wang, X. Inserting Menthoxytriazine into Poly (Ethylene Terephthalate) for Inhibiting Microbial Adhesion. *ACS Biomater. Sci. Eng.* **2022**, *8*, 570–578. [CrossRef] [PubMed]
94. Tullii, G.; Donini, S.; Bossio, C.; Lodola, F.; Pasini, M.; Parisini, E.; Galeotti, F.; Antognazza, M.R. Micro-And Nanopatterned Silk Substrates for Antifouling Applications. *ACS Appl. Mater. Interfaces* **2020**, *12*, 5437–5446. [CrossRef]
95. Kayes, M.I.; Galante, A.J.; Stella, N.A.; Haghanifar, S.; Shanks, R.M.Q.; Leu, P.W. Stable Lotus Leaf-Inspired Hierarchical, Fluorinated Polypropylene Surfaces for Reduced Bacterial Adhesion. *React. Funct. Polym.* **2018**, *128*, 40–46. [CrossRef]
96. Khan, S.; Jarad, N.A.; Ladouceur, L.; Rachwalski, K.; Bot, V.; Shakeri, A.; Maclachlan, R.; Sakib, S.; Weitz, J.I.; Brown, E.D.; et al. Transparent and Highly Flexible Hierarchically Structured Polydimethylsiloxane Surfaces Suppress Bacterial Attachment and Thrombosis Under Static and Dynamic Conditions. *Small* **2022**, *18*, 2108112. [CrossRef]
97. Liu, C.; He, Q.; Song, D.; Jackson, J.; Faria, A.F.; Jiang, X.; Li, X.; Ma, J.; Sun, Z. Electroless Deposition of Copper Nanoparticles Integrates Polydopamine Coating on Reverse Osmosis Membranes for Efficient Biofouling Mitigation. *Water. Res.* **2022**, *217*, 118375. [CrossRef]
98. Gao, Q.; Li, X.; Yu, W.; Jia, F.; Yao, T.; Jin, Q.; Ji, J. Fabrication of Mixed-Charge Polypeptide Coating for Enhanced Hemocompatibility and Anti-Infective Effect. *ACS Appl. Mater. Interfaces* **2020**, *12*, 2999–3010. [CrossRef]
99. Waters, C.M.; Bassler, B.L. Quorum Sensing: Cell-to-Cell Communication in Bacteria. *Annu. Rev. Cell Dev. Biol.* **2005**, *21*, 319–346. [CrossRef]
100. Spangler, J.R.; Dean, S.N.; Leary, D.H.; Walper, S.A. Response of *Lactobacillus plantarum* WCFS1 to the Gram-Negative Pathogen-Associated Quorum Sensing Molecule N-3-Oxododecanoyl Homoserine Lactone. *Front. Microbiol.* **2019**, *10*, 715. [CrossRef]
101. Hense, B.A.; Schuster, M. Core Principles of Bacterial Autoinducer Systems. *Microbiol. Mol. Biol. Rev.* **2015**, *79*, 153–169. [CrossRef] [PubMed]
102. Mukherjee, S.; Bassler, B.L. Bacterial Quorum Sensing in Complex and Dynamically Changing Environments. *Nat. Rev. Microbiol.* **2019**, *17*, 371–382. [CrossRef] [PubMed]
103. Papenfort, K.; Bassler, B.L. Quorum Sensing Signal-Response Systems in Gram-Negative Bacteria. *Nat. Rev. Microbiol.* **2016**, *14*, 576–588. [CrossRef] [PubMed]
104. Kalia, V.C. Quorum Sensing Inhibitors: An Overview. *Biotechnol. Adv.* **2013**, *31*, 224–245. [CrossRef]
105. Rasmussen, T.B.; Givskov, M. Quorum-Sensing Inhibitors as Anti-Pathogenic Drugs. *Int. J. Med. Microbiol.* **2006**, *296*, 149–161. [CrossRef]
106. Saipriya, K.; Swathi, C.H.; Ratnakar, K.S.; Sritharan, V. Quorum-Sensing System in *Acinetobacter baumannii*: A Potential Target for New Drug Development. *J. Appl. Microbiol.* **2020**, *128*, 15–27. [CrossRef]
107. Christensen, Q.H.; Grove, T.L.; Booker, S.J.; Greenberg, E.P. A High-Throughput Screen for Quorum-Sensing Inhibitors That Target Acyl-Homoserine Lactone Synthases. *Proc. Natl. Acad. Sci. USA* **2013**, *110*, 13815–13820. [CrossRef]
108. Piewngam, P.; Chiou, J.; Chatterjee, P.; Otto, M. Alternative Approaches to Treat Bacterial Infections: Targeting Quorum-Sensing. *Expert Rev. Anti-Infect. Ther.* **2020**, *18*, 499–510. [CrossRef]
109. Tan, S.Y.Y.; Chua, S.L.; Chen, Y.; Rice, S.A.; Kjelleberg, S.; Nielsen, T.E.; Yang, L.; Givskov, M. Identification of Five Structurally Unrelated Quorum-Sensing Inhibitors of *Pseudomonas aeruginosa* from a Natural-Derivative Database. *Antimicrob. Agents Chemother.* **2013**, *57*, 5629–5641. [CrossRef]
110. Chen, G.; Swem, L.R.; Swem, D.L.; Stauff, D.L.; O'Loughlin, C.T.; Jeffrey, P.D.; Bassler, B.L.; Hughson, F.M. A Strategy for Antagonizing Quorum Sensing. *Mol. Cell* **2011**, *42*, 199–209. [CrossRef]
111. Ivanova, A.; Ivanova, K.; Tied, A.; Heinze, T.; Tzanov, T. Layer-By-Layer Coating of Aminocellulose and Quorum Quenching Acylase on Silver Nanoparticles Synergistically Eradicate Bacteria and Their Biofilms. *Adv. Funct. Mater.* **2020**, *30*, 2001284. [CrossRef]
112. De Celis, M.; Serrano-Aguirre, L.; Belda, I.; Liébana-García, R.; Arroyo, M.; Marquina, D.; de la Mata, I.; Santos, A. Acylase Enzymes Disrupting Quorum Sensing Alter the Transcriptome and Phenotype of *Pseudomonas aeruginosa*, and the Composition of Bacterial Biofilms from Wastewater Treatment Plants. *Sci Total. Environ.* **2021**, *799*, 149401. [CrossRef]
113. Kim, M.K. *Staphylococcus aureus* Toxins: From Their Pathogenic Roles to Anti-Virulence Therapy Using Natural Products. *Biotechnol. Bioproc. E* **2019**, *24*, 424–435. [CrossRef]
114. Lu, H.; Tu, C.; Zhou, T.; Zhang, W.; Zhan, Y.; Ding, J.; Wu, X.; Yang, Z.; Cao, W.; Deng, L.; et al. A ROS-Scavenging Hydrogel Loaded with Bacterial Quorum Sensing Inhibitor Hyperbranched Poly-L-Lysine Promotes the Wound Scar-Free Healing of Infected Skin in Vivo. *Chem. Eng. J.* **2022**, *436*, 135130. [CrossRef]
115. Yuan, Q.; Feng, W.; Wang, Y.; Wang, Q.; Mou, N.; Xiong, L.; Wang, X.; Xia, P.; Sun, F. Luteolin Attenuates the Pathogenesis of *Staphylococcus aureus* by Interfering with the Agr System. *Microb. Pathog.* **2022**, *165*, 105496. [CrossRef]
116. Abd El-Hamid, M.I.; El-Naenaeey, E.S.Y.; Kandeel, T.M.; Hegazy, W.A.H.; Mosbah, R.A.; Nassar, M.S.; Bakhrebah, M.A.; Abdulaal, W.H.; Alhakamy, N.A.; Bendary, M.M. Promising Antibiofilm Agents: Recent Breakthrough against Biofilm Producing Methicillin-Resistant *Staphylococcus aureus*. *Antibiotics* **2020**, *9*, 667. [CrossRef]
117. Welsh, M.A.; Blackwell, H.E. Chemical Probes of Quorum Sensing: From Compound Development to Biological Discoveries. *FEMS Microbiol. Rev.* **2016**, *40*, 774–794. [CrossRef]
118. Lu, C.; Maurer, C.K.; Kirsch, B.; Steinbach, A.; Hartmann, R.W. Overcoming the Unexpected Functional Inversion of a PqsR Antagonist in *Pseudomonas aeruginosa*: An in Vivo Potent Antivirulence Agent Targeting Pqs Quorum Sensing. *Angew. Chem. Int. Ed. Engl.* **2014**, *53*, 1109–1112. [CrossRef]

119. Wei, L.N.; Shi, C.Z.; Luo, C.X.; Hu, C.Y.; Meng, Y.H. Phloretin Inhibits Biofilm Formation by Affecting Quorum Sensing under Different Temperature. *LWT* **2020**, *131*, 109668. [CrossRef]
120. Ouyang, J.; Feng, W.; Lai, X.; Chen, Y.; Zhang, X.; Rong, L.; Sun, F.; Chen, Y. Quercetin Inhibits *Pseudomonas aeruginosa* Biofilm Formation via the Vfr-Mediated LasIR System. *Microb. Pathog.* **2020**, *149*, 104291. [CrossRef]
121. Ho, D.K.; Murgia, X.; de Rossi, C.; Christmann, R.; Hüfner de Mello Martins, A.G.; Koch, M.; Andreas, A.; Herrmann, J.; Müller, R.; Empting, M.; et al. Squalenyl Hydrogen Sulfate Nanoparticles for Simultaneous Delivery of Tobramycin and an Alkylquinolone Quorum Sensing Inhibitor Enable the Eradication of P. aeruginosa Biofilm Infections. *Angew. Chem. Int. Ed. Engl.* **2020**, *59*, 10292–10296. [CrossRef] [PubMed]
122. Tomlin, H.; Piccinini, A.M. A Complex Interplay between the Extracellular Matrix and the Innate Immune Response to Microbial Pathogens. *Immunology* **2018**, *155*, 186–201. [CrossRef] [PubMed]
123. Mühlenbruch, M.; Grossart, H.P.; Eigemann, F.; Voss, M. Mini-Review: Phytoplankton-Derived Polysaccharides in the Marine Environment and Their Interactions with Heterotrophic Bacteria. *Environ. Microbiol.* **2018**, *20*, 2671–2685. [CrossRef] [PubMed]
124. McKee, L.S.; la Rosa, S.L.; Westereng, B.; Eijsink, V.G.; Pope, P.B.; Larsbrink, J. Polysaccharide Degradation by the Bacteroidetes: Mechanisms and Nomenclature. *Environ. Microbiol. Rep.* **2021**, *13*, 559–581. [CrossRef] [PubMed]
125. Whitfield, C.; Wear, S.S.; Sande, C. Assembly of Bacterial Capsular Polysaccharides and Exopolysaccharides. *Annu. Rev. Microbiol.* **2020**, *74*, 521–543. [CrossRef]
126. Ren, Z.; Cui, T.; Zeng, J.; Chen, L.; Zhang, W.; Xu, X.; Cheng, L.; Li, M.; Li, J.; Zhou, X.; et al. Molecule Targeting Glucosyltransferase Inhibits *Streptococcus mutans* Biofilm Formation and Virulence. *Antimicrob. Agents. Chemother.* **2015**, *60*, 126–135. [CrossRef]
127. Falsetta, M.L.; Klein, M.I.; Lemos, J.A.; Silva, B.B.; Agidi, S.; Scott-Anne, K.K.; Koo, H. Novel Antibiofilm Chemotherapy Targets Exopolysaccharide Synthesis and Stress Tolerance in *Streptococcus mutans* to Modulate Virulence Expression In Vivo. *Antimicrob. Agents. Chemother.* **2012**, *56*, 6201–6211. [CrossRef]
128. Laverty, G.; Gorman, S.P.; Gilmore, B.F. Biomolecular Mechanisms of Pseudomonas aeruginosa and *Escherichia coli* Biofilm Formation. *Pathogens* **2014**, *3*, 596–632. [CrossRef]
129. De Araujo Lopes, A.C.U.; Lobo, C.I.V.; Ribeiro, S.M.; da Silva Colin, J.; Constantino, V.C.N.; Canonici, M.M.; Barbugli, P.A.; Klein, M.I. Distinct Agents Induce *Streptococcus mutans* Cells with Altered Biofilm Formation Capacity. *Microbiol. Spectr.* **2022**, *10*, e00650-22. [CrossRef]
130. Zhang, Q.; Ma, Q.; Wang, Y.; Wu, H.; Zou, J. Molecular Mechanisms of Inhibiting Glucosyltransferases for Biofilm Formation in *Streptococcus mutans*. *Int. J. Oral Sci.* **2021**, *13*, 30. [CrossRef]
131. Jiang, H.; Hu, Y.; Yang, M.; Liu, H.; Jiang, G. Enhanced Immune Response to a Dual-Promoter Anti-Caries DNA Vaccine Orally Delivered by Attenuated *Salmonella typhimurium*. *Immunobiology* **2017**, *222*, 730–737. [CrossRef] [PubMed]
132. Thorn, C.R.; Howell, P.L.; Wozniak, D.J.; Prestidge, C.A.; Thomas, N. Enhancing the Therapeutic Use of Biofilm-Dispersing Enzymes with Smart Drug Delivery Systems. *Advanced Adv. Drug Deliv. Rev.* **2021**, *179*, 113916. [CrossRef]
133. Donelli, G.; Francolini, I.; Romoli, D.; Guaglianone, E.; Piozzi, A.; Ragunath, C.; Kaplan, J.B. Synergistic Activity of Dispersin B and Cefamandole Nafate in Inhibition of Staphylococcal Biofilm Growth on Polyurethanes. *Antimicrob. Agents. Chemother.* **2007**, *51*, 2733–2740. [CrossRef]
134. Liu, Z.; Zhao, Z.; Zeng, K.; Xia, Y.; Xu, W.; Wang, R.; Guo, J.; Xie, H. Functional Immobilization of a Biofilm-Releasing Glycoside Hydrolase Dispersin B on Magnetic Nanoparticles. *Appl. Biochem. Biotechnol.* **2022**, *194*, 737–747. [CrossRef] [PubMed]
135. Yu, X.; Zhao, J.; Fan, D. A Dissolving Microneedle Patch for Antibiotic/Enzymolysis/Photothermal Triple Therapy against Bacteria and Their Biofilms. *Chem. Eng. J.* **2022**, *437*, 135475. [CrossRef]
136. Maunders, E.; Welch, M. Matrix Exopolysaccharides; the Sticky Side of Biofilm Formation. *FEMS Microbiol. Lett.* **2017**, *364*, fnx120. [CrossRef]
137. Patel, K.K.; Tripathi, M.; Pandey, N.; Agrawal, A.K.; Gade, S.; Anjum, M.M.; Tilak, R.; Singh, S. Alginate Lyase Immobilized Chitosan Nanoparticles of Ciprofloxacin for the Improved Antimicrobial Activity against the Biofilm Associated Mucoid P. aeruginosa Infection in Cystic Fibrosis. *Int. J. Pharm.* **2019**, *563*, 30–42. [CrossRef]
138. Li, S.; Wang, Y.; Li, X.; Lee, B.S.; Jung, S.; Lee, M.S. Enhancing the Thermo-Stability and Anti-Biofilm Activity of Alginate Lyase by Immobilization on Low Molecular Weight Chitosan Nanoparticles. *Int. J. Mol. Sci.* **2019**, *20*, 4565. [CrossRef]
139. Daboor, S.M.; Rohde, J.R.; Cheng, Z. Disruption of the Extracellular Polymeric Network of *Pseudomonas aeruginosa* Biofilms by Alginate Lyase Enhances Pathogen Eradication by Antibiotics. *J. Cyst. Fibros.* **2021**, *20*, 264–270. [CrossRef]
140. Wan, B.; Zhu, Y.; Tao, J.; Zhu, F.; Chen, J.; Li, L.; Zhao, J.; Wang, L.; Sun, S.; Yang, Y.; et al. Alginate Lyase Guided Silver Nanocomposites for Eradicating *Pseudomonas aeruginosa* from Lungs. *ACS Appl. Mater. Interfaces* **2020**, *12*, 9050–9061. [CrossRef]
141. Szymańska, M.; Karakulska, J.; Sobolewski, P.; Kowalska, U.; Grygorcewicz, B.; Böttcher, D.; Bornscheuer, U.T.; Drozd, R. Glycoside Hydrolase (PelAh) Immobilization Prevents *Pseudomonas aeruginosa* Biofilm Formation on Cellulose-Based Wound Dressing. *Carbohydr. Polym.* **2020**, *246*, 116625. [CrossRef] [PubMed]
142. Catlin, B.W. Extracellular deoxyribonucleic acid of bacteria and a deoxyribonuclease inhibitor. *Science* **1956**, *124*, 441–442. [CrossRef] [PubMed]
143. Sarkar, S. Release Mechanisms and Molecular Interactions of *Pseudomonas aeruginosa* Extracellular DNA. *Appl. Microbiol. Biotechnol.* **2020**, *104*, 6549–6564. [CrossRef] [PubMed]

144. Lewenza, S.; Johnson, L.; Charron-Mazenod, L.; Hong, M.; Mulcahy-O'Grady, H. Extracellular DNA Controls Expression of *Pseudomonas aeruginosa* Genes Involved in Nutrient Utilization, Metal Homeostasis, Acid PH Tolerance and Virulence. *J. Med. Microbiol.* **2020**, *69*, 895–905. [CrossRef] [PubMed]
145. Yang, Y.; Li, M.; Zheng, X.; Ma, H.; Nerenberg, R.; Chai, H. Extracellular DNA Plays a Key Role in the Structural Stability of Sulfide-Based Denitrifying Biofilms. *Sci. Total Environ.* **2022**, *838*, 155822. [CrossRef]
146. Peng, N.; Cai, P.; Mortimer, M.; Wu, Y.; Gao, C.; Huang, Q. The Exopolysaccharide-EDNA Interaction Modulates 3D Architecture of *Bacillus subtilis* Biofilm. *BMC Microbiol.* **2020**, *20*, 115. [CrossRef]
147. Tan, Y.; Ma, S.; Leonhard, M.; Moser, D.; Haselmann, G.M.; Wang, J.; Eder, D.; Schneider-Stickler, B. Enhancing Antibiofilm Activity with Functional Chitosan Nanoparticles Targeting Biofilm Cells and Biofilm Matrix. *Carbohydr. Polym.* **2018**, *200*, 35–42. [CrossRef]
148. Panlilio, H.; Rice, C.v. The Role of Extracellular DNA in the Formation, Architecture, Stability, and Treatment of Bacterial Biofilms. *Biotechnol. Bioeng.* **2021**, *118*, 2129–2141. [CrossRef]
149. Okshevsky, M.; Regina, V.R.; Meyer, R.L. Extracellular DNA as a Target for Biofilm Control. *Curr. Opin. Biotechnol.* **2015**, *33*, 73–80. [CrossRef]
150. Hu, H.; Kang, X.; Shan, Z.; Yang, X.; Bing, W.; Wu, L.; Ge, H.; Ji, H. A DNase-Mimetic Artificial Enzyme for the Eradication of Drug-Resistant Bacterial Biofilm Infections. *Nanoscale* **2022**, *14*, 2676–2685. [CrossRef]
151. Liu, Z.; Wang, F.; Ren, J.; Qu, X. A Series of MOF/Ce-Based Nanozymes with Dual Enzyme-like Activity Disrupting Biofilms and Hindering Recolonization of Bacteria. *Biomaterials* **2019**, *208*, 21–31. [CrossRef] [PubMed]
152. Yan, X.; Gu, S.; Shi, Y.; Cui, X.; Wen, S.; Ge, J. The Effect of Emodin on *Staphylococcus aureus* Strains in Planktonic Form and Biofilm Formation In Vitro. *Arch. Microbiol.* **2017**, *199*, 1267–1275. [CrossRef] [PubMed]
153. Thiyagarajan, D.; Das, G.; Ramesh, A. Extracellular-DNA-Targeting Nanomaterial for Effective Elimination of Biofilm. *ChemNanoMat* **2016**, *2*, 879–887. [CrossRef]
154. Kwon, H.Y.; Kim, J.Y.; Liu, X.; Lee, J.Y.; Yam, J.K.H.; Dahl Hultqvist, L.; Xu, W.; Rybtke, M.; Tolker-Nielsen, T.; Heo, W.; et al. Visualizing Biofilm by Targeting EDNA with Long Wavelength Probe CDr15. *Biomater. Sci.* **2019**, *7*, 3594–3598. [CrossRef]
155. Erskine, E.; MacPhee, C.E.; Stanley-Wall, N.R. Functional Amyloid and Other Protein Fibers in the Biofilm Matrix. *J. Mol. Biol.* **2018**, *430*, 3642–3656. [CrossRef]
156. Kavanaugh, J.S.; Flack, C.E.; Lister, J.; Ricker, E.B.; Ibberson, C.B.; Jenul, C.; Moormeier, D.E.; Delmain, E.A.; Bayles, K.W.; Horswill, A.R. Identification of Extracellular DNA-Binding Proteins in the Biofilm Matrix. *mBio* **2019**, *10*, e01137-19. [CrossRef] [PubMed]
157. Shukla, S.K.; Subba Rao, T. *Staphylococcus aureus* Biofilm Removal by Targeting Biofilm-Associated Extracellular Proteins. *Indian J. Med. Res.* **2017**, *146* (Suppl. 1), S1–S8. [CrossRef]
158. Latasa, C.; Solano, C.; Penadés, J.R.; Lasa, I. Biofilm-Associated Proteins. *Comptes Rendus Biol.* **2006**, *329*, 849–857. [CrossRef] [PubMed]
159. Fong, J.N.C.; Yildiz, F.H. Biofilm Matrix Proteins. *Microbiol. Spectr.* **2015**, *3*, 3.2.28. [CrossRef]
160. Branda, S.S.; Chu, F.; Kearns, D.B.; Losick, R.; Kolter, R. A Major Protein Component of the *Bacillus subtilis* Biofilm Matrix. *Mol. Microbiol.* **2006**, *59*, 1229–1238. [CrossRef]
161. Rocco, C.J.; Davey, M.E.; Bakaletz, L.O.; Goodman, S.D. Natural Antigenic Differences in the Functionally Equivalent Extracellular DNABII Proteins of Bacterial Biofilms Provide a Means for Targeted Biofilm Therapeutics. *Mol. Oral Microbiol.* **2017**, *32*, 118–130. [CrossRef] [PubMed]
162. Li, Y.H.; Tang, N.; Aspiras, M.B.; Lau, P.C.Y.; Lee, J.H.; Ellen, R.P.; Cvitkovitch, D.G. A Quorum-Sensing Signaling System Essential for Genetic Competence in *Streptococcus mutans* Is Involved in Biofilm Formation. *J. Bacteriol.* **2002**, *184*, 2699–2708. [CrossRef]
163. Zhang, Y.; Xie, X.; Ma, W.; Zhan, Y.; Mao, C.; Shao, X.; Lin, Y. Multi-Targeted Antisense Oligonucleotide Delivery by a Framework Nucleic Acid for Inhibiting Biofilm Formation and Virulence. *Nanomicro Lett.* **2020**, *12*, 74. [CrossRef] [PubMed]
164. Ma, Z.; Li, J.; Bai, Y.; Zhang, Y.; Sun, H.; Zhang, X. A Bacterial Infection-Microenvironment Activated Nanoplatform Based on Spiropyran-Conjugated Glycoclusters for Imaging and Eliminating of the Biofilm. *Chem. Eng. J.* **2020**, *399*, 125787. [CrossRef]
165. Weldrick, P.J.; Hardman, M.J.; Paunov, V.N. Enhanced Clearing of Wound-Related Pathogenic Bacterial Biofilms Using Protease-Functionalized Antibiotic Nanocarriers. *ACS Appl. Mater. Interfaces* **2019**, *11*, 43902–43919. [CrossRef] [PubMed]
166. Devlin, H.; Fulaz, S.; Hiebner, D.W.; O'gara, J.P.; Casey, E. Enzyme-Functionalized Mesoporous Silica Nanoparticles to Target *Staphylococcus aureus* and Disperse Biofilms. *Int. J. Nanomed.* **2021**, *16*, 1929–1942. [CrossRef]
167. Chen, B.; Huang, J.; Li, H.; Zeng, Q.H.; Wang, J.J.; Liu, H.; Pan, Y.; Zhao, Y. Eradication of Planktonic *Vibrio parahaemolyticus* and Its Sessile Biofilm by Curcumin-Mediated Photodynamic Inactivation. *Food Control* **2020**, *113*, 107181. [CrossRef]
168. Jo, J.; Price-Whelan, A.; Dietrich, L.E.P. Gradients and Consequences of Heterogeneity in Biofilms. *Nat. Rev. Microbiol.* **2022**, *20*, 593–607. [CrossRef]
169. Jung, S.H.; Ryu, C.M.; Kim, J.S. Bacterial Persistence: Fundamentals and Clinical Importance. *J. Microbiol.* **2019**, *57*, 829–835. [CrossRef]
170. Soares, A.; Alexandre, K.; Etienne, M. Tolerance and Persistence of *Pseudomonas aeruginosa* in Biofilms Exposed to Antibiotics: Molecular Mechanisms, Antibiotic Strategies and Therapeutic Perspectives. *Front. Microbiol.* **2020**, *11*, 2057. [CrossRef]
171. Alexander, C.; Guru, A.; Pradhan, P.; Mallick, S.; Mahanandia, N.C.; Subudhi, B.B.; Beuria, T.K. MazEF-rifampicin interaction suggests a mechanism for rifampicin induced inhibition of persisters. *BMC Mol. Cell Biol.* **2020**, *21*, 73. [CrossRef] [PubMed]

172. Siqueira, J.F.; Rôças, I.N. Clinical Implications and Microbiology of Bacterial Persistence after Treatment Procedures. *J. Endod.* **2008**, *34*, 1291–1301.e3. [CrossRef] [PubMed]
173. Cardona, P.J.; Ruiz-Manzano, J. On the Nature of *Mycobacterium tuberculosis*-Latent Bacilli. *Eur. Respir. J.* **2004**, *24*, 1044–1051. [CrossRef] [PubMed]
174. Prabowo, S.A.; Gröschel, M.I.; Schmidt, E.D.L.; Skrahina, A.; Mihaescu, T.; Hastürk, S.; Mitrofanov, R.; Pimkina, E.; Visontai, I.; de Jong, B.; et al. Targeting Multidrug-Resistant Tuberculosis (MDR-TB) by Therapeutic Vaccines. *Med. Microbiol. Immunol.* **2013**, *202*, 95–104. [CrossRef] [PubMed]
175. Koul, A.; Dendouga, N.; Vergauwen, K.; Molenberghs, B.; Vranckx, L.; Willebrords, R.; Ristic, Z.; Lill, H.; Dorange, I.; Guillemont, J.; et al. Diarylquinolines Target Subunit c of Mycobacterial ATP Synthase. *Nat. Chem. Biol.* **2007**, *3*, 323–324. [CrossRef]
176. Balemans, W.; Vranckx, L.; Lounis, N.; Pop, O.; Guillemont, J.; Vergauwen, K.; Mol, S.; Gilissen, R.; Motte, M.; Lançois, D.; et al. Novel Antibiotics Targeting Respiratory ATP Synthesis in Gram-Positive Pathogenic Bacteria. *Antimicrob. Agents Chemother.* **2012**, *56*, 4131–4139. [CrossRef]
177. Garrison, A.T.; Abouelhassan, Y.; Norwood, V.M.; Kallifidas, D.; Bai, F.; Nguyen, M.T.; Rolfe, M.; Burch, G.M.; Jin, S.; Luesch, H.; et al. Structure-Activity Relationships of a Diverse Class of Halogenated Phenazines That Targets Persistent, Antibiotic-Tolerant Bacterial Biofilms and *Mycobacterium tuberculosis*. *J. Med. Chem.* **2016**, *59*, 3808–3825. [CrossRef]
178. Dutta, N.K.; Klinkenberg, L.G.; Vazquez, M.J.; Segura-Carro, D.; Colmenarejo, G.; Ramon, F.; Rodriguez-Miquel, B.; Mata-Cantero, L.; de Francisco, E.P.; Chuang, Y.M.; et al. Inhibiting the Stringent Response Blocks *Mycobacterium tuberculosis* Entry into Quiescence and Reduces Persistence. *Sci. Adv.* **2019**, *5*, eaav2104. [CrossRef]
179. Tkachenko, A.G.; Kashevarova, N.M.; Sidorov, R.Y.; Nesterova, L.Y.; Akhova, A.v.; Tsyganov, I.v.; Vaganov, V.Y.; Shipilovskikh, S.A.; Rubtsov, A.E.; Malkov, A.v. A Synthetic Diterpene Analogue Inhibits Mycobacterial Persistence and Biofilm Formation by Targeting (p)PpGpp Synthetases. *Cell Chem. Biol.* **2021**, *28*, 1420–1432.e9. [CrossRef]
180. Kaur, P.; Potluri, V.; Ahuja, V.K.; Naveenkumar, C.N.; Krishnamurthy, R.V.; Gangadharaiah, S.T.; Shivarudraiah, P.; Eswaran, S.; Nirmal, C.R.; Mahizhaveni, B.; et al. A Multi-Targeting Pre-Clinical Candidate against Drug-Resistant Tuberculosis. *Tuberculosis* **2021**, *129*, 102104. [CrossRef]
181. Roy, S.; Bahar, A.A.; Gu, H.; Nangia, S.; Sauer, K.; Ren, D. Persister Control by Leveraging Dormancy Associated Reduction of Antibiotic Efflux. *PLoS Pathog.* **2021**, *17*, e1010144. [CrossRef] [PubMed]
182. Kitzenberg, D.A.; Lee, J.S.; Mills, K.B.; Kim, J.-S.; Liu, L.; Vázquez-Torres, A.; Colgan, S.P.; Kao, D.J. Adenosine Awakens Metabolism to Enhance Growth-Independent Killing of Tolerant and Persister Bacteria across Multiple Classes of Antibiotics. *mBio* **2022**, *13*, e00480-22. [CrossRef] [PubMed]
183. Nabawy, A.; Makabenta, J.M.; Schmidt-Malan, S.; Park, J.; Li, C.H.; Huang, R.; Fedeli, S.; Chattopadhyay, A.N.; Patel, R.; Rotello, V.M. Dual Antimicrobial-Loaded Biodegradable Nanoemulsions for Synergistic Treatment of Wound Biofilms. *J. Control. Release* **2022**, *347*, 379–388. [CrossRef] [PubMed]
184. Ju, Y.; An, Q.; Zhang, Y.; Sun, K.; Bai, L.; Luo, Y. Recent Advances in Clp Protease Modulation to Address Virulence, Resistance and Persistence of MRSA Infection. *Drug Discov. Today* **2021**, *26*, 2190–2197. [CrossRef]
185. Zhang, S.; Qu, X.; Jiao, J.; Tang, H.; Wang, M.; Wang, Y.; Yang, H.; Yuan, W.; Yue, B. Felodipine Enhances Aminoglycosides Efficacy against Implant Infections Caused by Methicillin-Resistant *Staphylococcus aureus*, Persisters and Biofilms. *Bioact. Mater.* **2022**, *14*, 272–289. [CrossRef]
186. Walters, M.C.; Roe, F.; Bugnicourt, A.; Franklin, M.J.; Stewart, P.S. Contributions of Antibiotic Penetration, Oxygen Limitation, and Low Metabolic Activity to Tolerance of *Pseudomonas aeruginosa* Biofilms to Ciprofloxacin and Tobramycin. *Antimicrob. Agents Chemother.* **2003**, *47*, 317–323. [CrossRef]
187. Marteyn, B.; West, N.P.; Browning, D.F.; Cole, J.A.; Shaw, J.G.; Palm, F.; Mounier, J.; Prévost, M.C.; Sansonetti, P.; Tang, C.M. Modulation of Shigella Virulence in Response to Available Oxygen in Vivo. *Nature* **2010**, *465*, 355–358. [CrossRef]
188. Bai, Y.; Hu, Y.; Gao, Y.; Wei, X.; Li, J.; Zhang, Y.; Wu, Z.; Zhang, X. Oxygen Self-Supplying Nanotherapeutic for Mitigation of Tissue Hypoxia and Enhanced Photodynamic Therapy of Bacterial Keratitis. *ACS Appl. Mater. Interfaces* **2021**, *13*, 33790–33801. [CrossRef]
189. Zou, L.; Hu, D.; Wang, F.; Jin, Q.; Ji, J. The Relief of Hypoxic Microenvironment Using an O_2 Self-Sufficient Fluorinated Nanoplatform for Enhanced Photodynamic Eradication of Bacterial Biofilms. *Nano Res.* **2022**, *15*, 1636–1644. [CrossRef]
190. Hu, D.; Zou, L.; Yu, W.; Jia, F.; Han, H.; Yao, K.; Jin, Q.; Ji, J. Relief of Biofilm Hypoxia Using an Oxygen Nanocarrier: A New Paradigm for Enhanced Antibiotic Therapy. *Adv. Sci.* **2020**, *7*, 2000398. [CrossRef]
191. Deng, Q.; Sun, P.; Zhang, L.; Liu, Z.; Wang, H.; Ren, J.; Qu, X. Porphyrin MOF Dots–Based, Function-Adaptive Nanoplatform for Enhanced Penetration and Photodynamic Eradication of Bacterial Biofilms. *Adv. Funct. Mater.* **2019**, *29*, 1903018. [CrossRef]
192. Xiu, W.; Gan, S.; Wen, Q.; Qiu, Q.; Dai, S.; Dong, H.; Li, Q.; Yuwen, L.; Weng, L.; Teng, Z.; et al. Biofilm Microenvironment-Responsive Nanotheranostics for Dual-Mode Imaging and Hypoxia-Relief-Enhanced Photodynamic Therapy of Bacterial Infections. *Research* **2020**, *2020*, 9426453. [CrossRef] [PubMed]
193. Xu, Q.; Hua, Y.; Zhang, Y.; Lv, M.; Wang, H.; Pi, Y.; Xie, J.; Wang, C.; Yong, Y. A Biofilm Microenvironment-Activated Single-Atom Iron Nanozyme with NIR-Controllable Nanocatalytic Activities for Synergetic Bacteria-Infected Wound Therapy. *Adv. Healthc. Mater.* **2021**, *10*, 2101374. [CrossRef] [PubMed]
194. Wang, S.; Zheng, H.; Zhou, L.; Cheng, F.; Liu, Z.; Zhang, H.; Wang, L.; Zhang, Q. Nanoenzyme-Reinforced Injectable Hydrogel for Healing Diabetic Wounds Infected with Multidrug Resistant Bacteria. *Nano Lett.* **2020**, *20*, 5149–5158. [CrossRef]

195. Zhang, S.; Chai, Q.; Man, Z.; Tang, C.; Li, Z.; Zhang, J.; Xu, H.; Xu, X.; Chen, C.; Liu, Y.; et al. Bioinspired Nano-Painting on Orthopedic Implants Orchestrates Periprosthetic Anti-Infection and Osseointegration in a Rat Model of Arthroplasty. *Chem. Eng. J.* **2022**, *435*, 134848. [CrossRef]
196. Scalise, A.; Bianchi, A.; Tartaglione, C.; Bolletta, E.; Pierangeli, M.; Torresetti, M.; Marazzi, M.; di Benedetto, G. Microenvironment and microbiology of skin wounds: The role of bacterial biofilms and related factors. *Semin. Vasc. Surg.* **2015**, *28*, 151–159. [CrossRef]
197. Wilton, M.; Charron-Mazenod, L.; Moore, R.; Lewenza, S. Extracellular DNA Acidifies Biofilms and Induces Aminoglycoside Resistance in *Pseudomonas aeruginosa*. *Antimicrob. Agents Chemother.* **2015**, *60*, 544–553. [CrossRef]
198. Chen, H.; Yang, J.; Sun, L.; Zhang, H.; Guo, Y.; Qu, J.; Jiang, W.; Chen, W.; Ji, J.; Yang, Y.W.; et al. Synergistic Chemotherapy and Photodynamic Therapy of Endophthalmitis Mediated by Zeolitic Imidazolate Framework-Based Drug Delivery Systems. *Small* **2019**, *15*, 1903880. [CrossRef]
199. Wang, D.Y.; Yang, G.; van der Mei, H.C.; Ren, Y.; Busscher, H.J.; Shi, L. Liposomes with Water as a PH-Responsive Functionality for Targeting of Acidic Tumor and Infection Sites. *Angew. Chem. Int. Ed. Engl.* **2021**, *60*, 17714–17719. [CrossRef]
200. Jiang, X.; Li, W.; Chen, X.; Wang, C.; Guo, R.; Hong, W. On-Demand Multifunctional Electrostatic Complexation for Synergistic Eradication of MRSA Biofilms. *ACS Appl. Mater. Interfaces* **2022**, *14*, 10200–10211. [CrossRef]
201. Gao, Y.; Wang, J.; Chai, M.; Li, X.; Deng, Y.; Jin, Q.; Ji, J. Size and Charge Adaptive Clustered Nanoparticles Targeting the Biofilm Microenvironment for Chronic Lung Infection Management. *ACS Nano* **2020**, *14*, 5686–5699. [CrossRef] [PubMed]
202. Wu, J.; Li, F.; Hu, X.; Lu, J.; Sun, X.; Gao, J.; Ling, D. Responsive Assembly of Silver Nanoclusters with a Biofilm Locally Amplified Bactericidal Effect to Enhance Treatments against Multi-Drug-Resistant Bacterial Infections. *ACS Cent. Sci.* **2019**, *5*, 1366–1376. [CrossRef] [PubMed]
203. Chen, H.; Cheng, J.; Cai, X.; Han, J.; Chen, X.; You, L.; Xiong, C.; Wang, S. PH-Switchable Antimicrobial Supramolecular Hydrogels for Synergistically Eliminating Biofilm and Promoting Wound Healing. *ACS Appl. Mater. Interfaces* **2022**, *14*, 18120–18132. [CrossRef]
204. Fulaz, S.; Devlin, H.; Vitale, S.; Quinn, L.; O'gara, J.P.; Casey, E. Tailoring Nanoparticle-Biofilm Interactions to Increase the Efficacy of Antimicrobial Agents against *Staphylococcus aureus*. *Int. J. Nanomed.* **2020**, *15*, 4779–4791. [CrossRef] [PubMed]
205. Yin, M.; Qiao, Z.; Yan, D.; Yang, M.; Yang, L.; Wan, X.; Chen, H.; Luo, J.; Xiao, H. Ciprofloxacin Conjugated Gold Nanorods with PH Induced Surface Charge Transformable Activities to Combat Drug Resistant Bacteria and Their Biofilms. *Mater. Sci. Eng. C Mater. Biol. Appl.* **2021**, *128*, 112292. [CrossRef] [PubMed]
206. Wu, S.; Xu, C.; Zhu, Y.; Zheng, L.; Zhang, L.; Hu, Y.; Yu, B.; Wang, Y.; Xu, F.J. Biofilm-Sensitive Photodynamic Nanoparticles for Enhanced Penetration and Antibacterial Efficiency. *Adv. Funct. Mater.* **2021**, *31*, 2103591. [CrossRef]
207. Tian, S.; Su, L.; Liu, Y.; Cao, J.; Yang, G.; Ren, Y.; Huang, F.; Liu, J.; An, Y.; van der Mei, H.C.; et al. Self-Targeting, Zwitterionic Micellar Dispersants Enhance Antibiotic Killing of Infectious Biofilms—An Intravital Imaging Study in Mice. *Sci. Adv.* **2020**, *6*, eabb1112. [CrossRef]
208. Shatalin, K.; Nuthanakanti, A.; Kaushik, A.; Shishov, D.; Peselis, A.; Shamovsky, I.; Pani, B.; Lechpammer, M.; Vasilyev, N.; Shatalina, E.; et al. Inhibitors of Bacterial H$_2$S Biogenesis Targeting Antibiotic Resistance and Tolerance. *Science* **2021**, *372*, 1169–1175. [CrossRef]
209. Zeng, J.; Wang, Y.; Sun, Z.; Chang, H.; Cao, M.; Zhao, J.; Lin, K.; Xie, Y. A Novel Biocompatible PDA/IR820/DAP Coating for Antibiotic/Photodynamic/Photothermal Triple Therapy to Inhibit and Eliminate *Staphylococcus aureus* Biofilm. *Chem. Eng. J.* **2020**, *394*, 125017. [CrossRef]
210. Peng, J.; Xie, S.; Huang, K.; Ran, P.; Wei, J.; Zhang, Z.; Li, X. Nitric Oxide-Propelled Nanomotors for Bacterial Biofilm Elimination and Endotoxin Removal to Treat Infected Burn Wounds. *J. Mater. Chem. B* **2022**, *10*, 4189–4202. [CrossRef]
211. Huang, Y.; Liu, Y.; Shah, S.; Kim, D.; Simon-Soro, A.; Ito, T.; Hajfathalian, M.; Li, Y.; Hsu, J.C.; Nieves, L.M.; et al. Precision Targeting of Bacterial Pathogen via Bi-Functional Nanozyme Activated by Biofilm Microenvironment. *Biomaterials* **2021**, *268*, 120581. [CrossRef] [PubMed]
212. Wang, Y.; Shen, X.; Ma, S.; Guo, Q.; Zhang, W.; Cheng, L.; Ding, L.; Xu, Z.; Jiang, J.; Gao, L. Oral Biofilm Elimination by Combining Iron-Based Nanozymes and Hydrogen Peroxide-Producing Bacteria. *Biomater. Sci.* **2020**, *8*, 2447–2458. [CrossRef]
213. Xu, M.; Hu, Y.; Xiao, Y.; Zhang, Y.; Sun, K.; Wu, T.; Lv, N.; Wang, W.; Ding, W.; Li, F.; et al. Near-Infrared-Controlled Nanoplatform Exploiting Photothermal Promotion of Peroxidase-like and OXD-like Activities for Potent Antibacterial and Anti-Biofilm Therapies. *ACS Appl. Mater. Interfaces* **2020**, *12*, 50260–50274. [CrossRef] [PubMed]
214. Chen, X.Z.; Hoop, M.; Mushtaq, F.; Siringil, E.; Hu, C.; Nelson, B.J.; Pané, S. Recent Developments in Magnetically Driven Micro- and Nanorobots. *Appl. Mater. Today* **2017**, *9*, 37–48. [CrossRef]
215. Dong, Y.; Wang, L.; Yuan, K.; Ji, F.; Gao, J.; Zhang, Z.; Du, X.; Tian, Y.; Wang, Q.; Zhang, L. Magnetic Microswarm Composed of Porous Nanocatalysts for Targeted Elimination of Biofilm Occlusion. *ACS Nano* **2021**, *15*, 5056–5067. [CrossRef]
216. Yang, G.; Wang, D.Y.; Liu, Y.; Huang, F.; Tian, S.; Ren, Y.; Liu, J.; An, Y.; van der Mei, H.C.; Busscher, H.J.; et al. In-Biofilm Generation of Nitric Oxide Using a Magnetically-Targetable Cascade-Reaction Container for Eradication of Infectious Biofilms. *Bioact. Mater.* **2022**, *14*, 321–334. [CrossRef]
217. Yuan, K.; Jurado-Sánchez, B.; Escarpa, A. Dual-Propelled Lanbiotic Based Janus Micromotors for Selective Inactivation of Bacterial Biofilms. *Angew. Chem. Int. Ed. Engl.* **2021**, *60*, 4915–4924. [CrossRef]

218. Javaudin, F.; Latour, C.; Debarbieux, L.; Lamy-Besnier, Q. Intestinal Bacteriophage Therapy: Looking for Optimal Efficacy. *Clin. Microbiol. Rev.* **2021**, *34*, e0013621. [CrossRef]
219. Yan, W.; Banerjee, P.; Xu, M.; Mukhopadhyay, S.; Ip, M.; Carrigy, N.B.; Lechuga-Ballesteros, D.; To, K.K.W.; Leung, S.S.Y. Formulation Strategies for Bacteriophages to Target Intracellular Bacterial Pathogens. *Adv. Drug Deliv. Rev.* **2021**, *176*, 113864. [CrossRef]
220. Wang, Z.; Guo, K.; Liu, Y.; Huang, C.; Wu, M. Dynamic Impact of Virome on Colitis and Colorectal Cancer: Immunity, Inflammation, Prevention and Treatment. *Semin. Cancer Biol.* **2021**, in press. [CrossRef]
221. He, X.; Yang, Y.; Guo, Y.; Lu, S.; Du, Y.; Li, J.J.; Zhang, X.; Leung, N.L.C.; Zhao, Z.; Niu, G.; et al. Phage-Guided Targeting, Discriminative Imaging, and Synergistic Killing of Bacteria by AIE Bioconjugates. *J. Am. Chem. Soc.* **2020**, *142*, 3959–3969. [CrossRef] [PubMed]
222. Bhardwaj, S.B.; Mehta, M.; Sood, S.; Sharma, J. Isolation of a Novel Phage and Targeting Biofilms of Drug-Resistant Oral Enterococci. *J. Glob. Infect. Dis.* **2020**, *12*, 11–15. [CrossRef] [PubMed]
223. Khalifa, L.; Brosh, Y.; Gelman, D.; Coppenhagen-Glazer, S.; Beyth, S.; Poradosu-Cohen, R.; Que, Y.A.; Beyth, N.; Hazan, R. Targeting *Enterococcus faecalis* Biofilms with Phage Therapy. *Appl. Environ. Microbiol.* **2015**, *81*, 2696–2705. [CrossRef] [PubMed]
224. Ran, B.; Yuan, Y.; Xia, W.; Li, M.; Yao, Q.; Wang, Z.; Wang, L.; Li, X.; Xu, Y.; Peng, X. A Photo-Sensitizable Phage for Multidrug-Resistant *Acinetobacter baumannii* therapy and Biofilm Ablation. *Chem. Sci.* **2021**, *12*, 1054–1061. [CrossRef]
225. Little, J.S.; Dedrick, R.M.; Freeman, K.G.; Cristinziano, M.; Smith, B.E.; Benson, C.A.; Jhaveri, T.A.; Baden, L.R.; Solomon, D.A.; Hatfull, G.F. Bacteriophage Treatment of Disseminated Cutaneous *Mycobacterium chelonae* Infection. *Nat. Commun.* **2022**, *13*, 2313. [CrossRef]
226. Chapman, C.M.C.; Gibson, G.R.; Rowland, I. Effects of Single- and Multi-Strain Probiotics on Biofilm Formation and In Vitro Adhesion to Bladder Cells by Urinary Tract Pathogens. *Anaerobe* **2014**, *27*, 71–76. [CrossRef] [PubMed]
227. Bidossi, A.; de Grandi, R.; Toscano, M.; Bottagisio, M.; de Vecchi, E.; Gelardi, M.; Drago, L. Probiotics *Streptococcus salivarius* 24SMB and *Streptococcus oralis* 89a Interfere with Biofilm Formation of Pathogens of the Upper Respiratory Tract. *BMC Infect. Dis.* **2018**, *18*, 653. [CrossRef]
228. Meroni, G.; Panelli, S.; Zuccotti, G.; Bandi, C.; Drago, L.; Pistone, D. Probiotics as Therapeutic Tools against Pathogenic Biofilms: Have We Found the Perfect Weapon? *Microbiol. Res.* **2021**, *12*, 916–937. [CrossRef]
229. Santos, R.A.; Oliva-Teles, A.; Pousão-Ferreira, P.; Jerusik, R.; Saavedra, M.J.; Enes, P.; Serra, C.R. Isolation and Characterization of Fish-Gut *Bacillus* spp. as Source of Natural Antimicrobial Compounds to Fight Aquaculture Bacterial Diseases. *Mar. Biotechnol.* **2021**, *23*, 276–293. [CrossRef]
230. Ibeiro, F.C.; Iglesias, M.C.; Barros, P.P.; Santos, S.S.F.; Jorge, A.O.C.; Leão, M.V.P. *Lactobacillus rhamnosus* Interferes with *Candida albicans* Adherence and Biofilm Formation: A Potential Alternative Treatment of Candidiasis. *Austin J. Pharmacol. Ther.* **2021**, *9*, 1133. [CrossRef]
231. Shokouhfard, M.; Kermanshahi, R.K.; Feizabadi, M.M.; Teimourian, S.; Safari, F. *Lactobacillus* spp. Derived Biosurfactants Effect on Expression of Genes Involved in Proteus Mirabilis Biofilm Formation. *Infect. Genet. Evol.* **2022**, *100*, 105264. [CrossRef]
232. Frohlich, K.M.; Weintraub, S.F.; Bell, J.T.; Todd, G.C.; Väre, V.Y.P.; Schneider, R.; Kloos, Z.A.; Tabe, E.S.; Cantara, W.A.; Stark, C.J.; et al. Discovery of Small-Molecule Antibiotics against a Unique TRNA-Mediated Regulation of Transcription in Gram-Positive Bacteria. *ChemMedChem* **2019**, *14*, 758–769. [CrossRef] [PubMed]
233. Seyler, T.M.; Moore, C.; Kim, H.; Ramachandran, S.; Agris, P.F. A New Promising Anti-Infective Agent Inhibits Biofilm Growth by Targeting Simultaneously a Conserved RNA Function That Controls Multiple Genes. *Antibiotics* **2021**, *10*, 41. [CrossRef] [PubMed]
234. Peters, J.M.; Colavin, A.; Shi, H.; Czarny, T.L.; Larson, M.H.; Wong, S.; Hawkins, J.S.; Lu, C.H.S.; Koo, B.M.; Marta, E.; et al. A Comprehensive, CRISPR-Based Functional Analysis of Essential Genes in Bacteria. *Cell* **2016**, *165*, 1493–1506. [CrossRef] [PubMed]
235. Afonina, I.; Ong, J.; Chua, J.; Lu, T.; Kline, K.A. Multiplex Crispri System Enables the Study of Stage-Specific Biofilm Genetic Requirements in *Enterococcus faecalis*. *mBio* **2020**, *11*, e01101-20. [CrossRef]
236. Asaad, A.M.; Soma, S.A.; Ajlan, E.; Awad, S.M. Epidemiology of Biofilm Producing *Acinetobacter baumannii* Nosocomial Isolates from a Tertiary Care Hospital in Egypt: A Cross-Sectional Study. *Infect. Drug Resist.* **2021**, *14*, 709–717. [CrossRef]
237. Subramanian, D.; Natarajan, J. Integrated Meta-Analysis and Machine Learning Approach Identifies Acyl-CoA Thioesterase with Other Novel Genes Responsible for Biofilm Development in *Staphylococcus aureus*. *Infect. Genet. Evol.* **2021**, *88*, 104702. [CrossRef]
238. Abdulhaq, N.; Nawaz, Z.; Zahoor, M.A.; Siddique, A.B. Association of Biofilm Formation with Multi Drug Resistance in Clinical Isolates of *Pseudomonas aeruginosa*. *EXCLI J.* **2020**, *19*, 201–208. [CrossRef]
239. Wu, S.; Liu, Y.; Lei, L.; Zhang, H. Antisense YycG Modulates the Susceptibility of *Staphylococcus aureus* to Hydrogen Peroxide via the SarA. *BMC Microbiol.* **2021**, *21*, 160. [CrossRef]
240. Wille, J.; Coenye, T. Biofilm Dispersion: The Key to Biofilm Eradication or Opening Pandora's Box? *Biofilm* **2020**, *2*, 100027. [CrossRef]
241. Petrova, O.E.; Sauer, K. Escaping the Biofilm in More than One Way: Desorption, Detachment or Dispersion. *Curr. Opin. Microbiol.* **2016**, *30*, 67–78. [CrossRef] [PubMed]
242. Fux, C.A.; Wilson, S.; Stoodley, P. Detachment Characteristics and Oxacillin Resistance of *Staphylococcus aureus* Biofilm Emboli in an In Vitro Catheter Infection Model. *J. Bacteriol.* **2004**, *186*, 4486–4491. [CrossRef] [PubMed]

243. Zhang, K.; Li, X.; Yu, C.; Wang, Y. Promising Therapeutic Strategies Against Microbial Biofilm Challenges. *Front. Cell. Infect. Microbiol.* **2020**, *10*, 359. [CrossRef] [PubMed]
244. Kolodkin-Gal, I.; Romero, D.; Cao, S.; Clardy, J.; Kolter, R.; Losick, R. D-Amino Acids Trigger Biofilm Disassembly. *Science* **2010**, *328*, 624–627. [CrossRef] [PubMed]
245. Pinto, R.M.; Monteiro, C.; Costa Lima, S.A.; Casal, S.; van Dijck, P.; Martins, M.C.L.; Nunes, C.; Reis, S. N-Acetyl-l-Cysteine-Loaded Nanosystems as a Promising Therapeutic Approach Toward the Eradication of *Pseudomonas aeruginosa* Biofilms. *ACS Appl. Mater. Interfaces* **2021**, *13*, 42329–42343. [CrossRef] [PubMed]
246. Chang, R.Y.K.; Li, M.; Chow, M.Y.T.; Ke, W.R.; Tai, W.; Chan, H.K. A Dual Action of D-Amino Acids on Anti-Biofilm Activity and Moisture-Protection of Inhalable Ciprofloxacin Powders. *Eur. J. Pharm. Biopharm.* **2022**, *173*, 132–140. [CrossRef]
247. Zhang, M.; Zhang, H.; Feng, J.; Zhou, Y.; Wang, B. Synergistic Chemotherapy, Physiotherapy and Photothermal Therapy against Bacterial and Biofilms Infections through Construction of Chiral Glutamic Acid Functionalized Gold Nanobipyramids. *Chem. Eng. J.* **2020**, *393*, 124778. [CrossRef]
248. Chen, M.; Wei, J.; Xie, S.; Tao, X.; Zhang, Z.; Ran, P.; Li, X. Bacterial Biofilm Destruction by Size/Surface Charge-Adaptive Micelles. *Nanoscale* **2019**, *11*, 1410–1422. [CrossRef]
249. Huang, L.; Lou, Y.; Zhang, D.; Ma, L.; Qian, H.; Hu, Y.; Ju, P.; Xu, D.; Li, X. D-Cysteine Functionalised Silver Nanoparticles Surface with a "Disperse-Then-Kill" Antibacterial Synergy. *Chem. Eng. J.* **2020**, *381*, 122662. [CrossRef]
250. Liu, X.; Li, Z.; Fan, Y.; Lekbach, Y.; Song, Y.; Xu, D.; Zhang, Z.; Ding, L.; Wang, F. A Mixture of D-Amino Acids Enhances the Biocidal Efficacy of CMIT/MIT Against Corrosive *Vibrio harveyi* Biofilm. *Front. Microbiol.* **2020**, *11*, 557435. [CrossRef]
251. Louis, M.; Clamens, T.; Tahrioui, A.; Desriac, F.; Rodrigues, S.; Rosay, T.; Harmer, N.; Diaz, S.; Barreau, M.; Racine, P.J.; et al. *Pseudomonas aeruginosa* Biofilm Dispersion by the Human Atrial Natriuretic Peptide. *Adv. Sci.* **2022**, *9*, 2103262. [CrossRef] [PubMed]
252. Bharti, S.; Maurya, R.K.; Venugopal, U.; Singh, R.; Akhtar, M.S.; Krishnan, M.Y. Rv1717 Is a Cell Wall-Associated β-Galactosidase of *Mycobacterium tuberculosis* That Is Involved in Biofilm Dispersion. *Front. Microbiol.* **2021**, *11*, 611122. [CrossRef] [PubMed]
253. Nishikawa, M.; Kobayashi, K. Calcium Prevents Biofilm Dispersion in *Bacillus subtilis*. *J. Bacteriol.* **2021**, *203*, e00114-21. [CrossRef] [PubMed]

pharmaceuticals

Review

Carbon Dots for Killing Microorganisms: An Update since 2019

Fengming Lin, Zihao Wang and Fu-Gen Wu *

State Key Laboratory of Bioelectronics, School of Biological Science and Medical Engineering, Southeast University, 2 Sipailou Road, Nanjing 210096, China
* Correspondence: wufg@seu.edu.cn

Abstract: Frequent bacterial/fungal infections and occurrence of antibiotic resistance pose increasing threats to the public and thus require the development of new antibacterial/antifungal agents and strategies. Carbon dots (CDs) have been well demonstrated to be promising and potent antimicrobial nanomaterials and serve as potential alternatives to conventional antibiotics. In recent years, great efforts have been made by many researchers to develop new carbon dot-based antimicrobial agents to combat microbial infections. Here, as an update to our previous relevant review (C 2019, 5, 33), we summarize the recent achievements in the utilization of CDs for microbial inactivation. We review four kinds of antimicrobial CDs including nitrogen-doped CDs, metal-containing CDs, antibiotic-conjugated CDs, and photoresponsive CDs in terms of their starting materials, synthetic route, surface functionalization, antimicrobial ability, and the related antimicrobial mechanism if available. In addition, we summarize the emerging applications of CD-related antimicrobial materials in medical and industry fields. Finally, we discuss the existing challenges of antimicrobial CDs and the future research directions that are worth exploring. We believe that this review provides a comprehensive overview of the recent advances in antimicrobial CDs and may inspire the development of new CDs with desirable antimicrobial activities.

Keywords: antibacterial; bactericidal; disinfection; carbon nanodots; carbonized polymer dots

1. Introduction

Owing to the long-term use and overuse of antibiotics, pathogens have become resistant to almost all existing traditional antibiotics by mutating or acquiring drug-resistant genes from other organisms. It is urgently necessary to develop novel effective antimicrobial compounds as potent alternatives to the conventional small-molecule antibiotics to address the issue of microbial drug resistance. The great advancement of nanoscience and nanotechnology has offered a new solution for the development of antimicrobial materials. Several types of nanomaterials are known to exhibit antibacterial properties. Particularly, inorganic metal and metal oxide nanoparticles have been intensively investigated for their potential use as antimicrobial agents [1,2]. Although these metal (e.g., Au and Ag) and metal oxide (e.g., Fe_2O_3, CuO, and ZnO) nanoparticles possess antimicrobial activities, the release of metal ions may cause nonspecific biological toxicity, which urgently requires the development of safer antimicrobial nanomaterials [3,4]. Among the large variety of antimicrobial nanomaterials, carbon dots (CDs) have received ever-increasing attention, mainly due to their easy preparation and functionalization, great water dispersity, and satisfactory biocompatibility. One appealing merit for CDs is that their property and function can be easily manipulated during the synthesis or post-modification stage, which is highly useful for antibacterial applications. CDs are zero-dimensional carbonaceous nanoparticles with sizes no more than 20 nm, also termed "carbonized polymer dots", "carbon quantum dots", or "carbon nanodots" [5–11]. CDs can be prepared from a wide variety of natural materials such as biomass and waste, and a huge array of chemical agents [12,13]. There are two well-known CD preparation strategies: bottom-up strategy and top-down strategy. The synthetic approaches for CDs include hydrothermal/solvothermal reaction, pyrolysis, sonication, microwave irradiation, etc. [12]. The broad applications of CDs in sensing [12,14–16], optoelectronics [17], energy [18], catalysis [12,19], and nanomedicine [20–24], have been demonstrated since their discovery in 2004 [25].

Currently, three antimicrobial mechanisms have been reported for CDs, including cell wall/membrane disruption, reactive oxygen species (ROS) generation, and DNA damage [23]. The inhibitory action of CDs on microorganisms depends on the composition, size, shape, and surface chemistry of CDs. It is extremely difficult to explain the antimicrobial mechanisms of CDs without performing careful structural characterizations of the CDs. Specifically, the catalytic activity, the crystallographic structure, the surface state (defect or functionalization), and charge transfer are important factors that contribute to the antimicrobial activity of CDs. However, currently, except for the several studies that mentioned the effect of surface functionalization on the antimicrobial activity of CDs [7,24,26], detailed evaluations of the other factors are still lacking in the current CD-based antimicrobial studies. As a result, more attention should be paid to the investigations of the effect of the other factors on the antimicrobial activity of CDs in the future.

In 2019, we have reviewed the advancements of CDs in sensing and killing microorganisms in terms of their preparation, functionalization, toxicity, and underlying antimicrobial mechanism [16]. Nevertheless, the past three years have witnessed the booming applications of CDs in the antimicrobial field. Hence, an update on this topic is essential to embrace the latest progress in this field. Numerous CDs have been reported for antimicrobial therapy in recent years, and they can be classified into four types including nitrogen-doped CDs, metal-containing CDs, antibiotic-conjugated CDs, and photoresponsive CDs (Scheme 1). We discuss their raw materials, synthetic approaches, modification methods, antimicrobial abilities, and the related antimicrobial mechanisms if available in detail below. Furthermore, we also introduce the advances in the applications of the antimicrobial CDs in industry and medicine. Finally, we discuss the current limitations of antimicrobial CDs and propose some research directions.

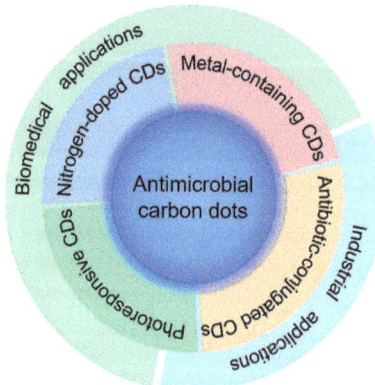

Scheme 1. Scheme illustrating the different types of antimicrobial CDs for biomedical and industrial applications.

2. Antimicrobial CDs

2.1. Nitrogen-Doped CDs

2.1.1. Nitrogen-Doped CDs Derived from Biomass

When we prepared our previous review in 2019, a large number of fluorescent CDs synthesized from different natural sources had been harnessed for microbial imaging, but few CDs derived from biomass had been utilized as antimicrobial agents [16]. Nevertheless, in the last three years, the development of green synthetic methods for fabricating antimicrobial CDs from different natural carbon precursors has attracted considerable interest (Table 1), and such a method is facile, cost-effective, and eco-friendly. CDs prepared from *Lawsonia inermis* (Henna) [27], oyster mushroom (*Pleurotus species*) [28], osmanthus leaves [29], tea leaves [29], *Ananas comosus* waste peels [30], *Impatiens balsamina* L. stems [31], *Aloe vera* leaves [32], medicinal turmeric leaves (*Curcuma longa*) [33], rosemary leaves [34], sugarcane bagasse pulp [35], waste tea extract [36], waste jute caddies [37], *Forsythia* [38], and *Artemisia argyi* leaves [39], have been reported to possess antimicrobial activities. It is interesting to find that all these biomass-derived CDs contain the nitrogen element, which might be from the proteins, animo acids, and nucleic acids in biomass. For instance, Wang et al. synthesized CDs (ACDs) from *Artemisia argyi* leaves through a smoking simulation approach (Figure 1A) [39]. ACDs have spherical morphology with a diameter of 2–5 nm (Figure 1B). ACDs displayed selective antibacterial ability toward Gram-negative bacteria like *Escherichia coli* (*E. coli*), ampicillin-resistant *E. coli* (ARE *E. coli*), kanamycin-resistant *E coli* (KRE *E. coli*), *Pseudomonas aeruginosa* (*P. aeruginosa*), and *Proteusbacillus vulgaris* (*P. vulgaris*), but not Gram-positive bacteria such as *Staphylococcus aureus* (*S. aureus*) and *Bacillus subtilis* (*B. subtilis*) (Figure 1C). ACDs could kill 100% Gram-negative bacteria at 150 µg mL^{-1}. The antibacterial mechanism study demonstrated that ACDs could only disrupt the cell walls of *E. coli* rather than those of *S. aureus*, since according to the high magnification images in (d and h), only the *E. coli* cells showed shrunken and damaged cell structures (Figure 1D). In addition, ACDs could change the secondary structure and thus the activity of cell wall-related enzymes in Gram-negative bacteria. More interestingly, ACDs could strongly prevent the biofilm formation of *E. coli*. The development of ACDs is of great value for the treatment of infections associated closely with Gram-negative bacteria. Saravanan et al. prepared CDs by one-step hydrothermal treatment of medicinal turmeric leaves (*Curcuma longa*) [33]. The as-prepared CDs exhibited antibacterial activities toward both Gram-negative bacteria (*E. coli* and *Klebsiella pneumoniae* (*K. pneumoniae*)) and Gram-positive bacteria (*S. aureus* and *Staphylococcus epidermidis* (*S. epidermidis*)) due to their release of ROS. Collectively, these naturally derived CDs from biomass represent potent candidates as new antimicrobial agents to combat antibiotic-resistance of microorganisms.

Table 1. Nitrogen-doped CDs derived from biomass for killing microorganisms.

Raw Materials	Preparation Method	Size * (nm)	Charge (mV)	QY (%)	Ref.
Lawsonia inermis (Henna)	Hydrothermal treatment	3–7	−39	28.7	[27]
Oyster mushroom	Hydrothermal treatment	8	–	–	[28]
Osmanthus leaves	Hydrothermal treatment	4–9	−20	–	[29]
Tea leaves	Hydrothermal treatment	3–7	−20	–	[29]
Ananas comosus waste peels	Hydrothermal treatment	2.4 ± 0.5	–	10.65	[30]
Impatiens balsamina L. stems	Hydrothermal treatment	2–4.5	22.47	54	[31]
Aloe vera leaves	Hydrothermal treatment	10–20	–	–	[32]
Medicinal turmeric leaves	Hydrothermal treatment	1.5–4.0	−7	–	[33]
Rosemary leaves	Hydrothermal treatment	16.1 ± 4.6	–	–	[34]
Sugarcane bagasse pulp	Hydrothermal treatment	1.7 ± 0.2	–	17.98	[35]
Waste tea extract	Hydrothermal treatment	0.85	–	3.26	[36]
Waste jute caddies	Hydrothermal treatment	6.05	–	14.5	[37]
Forsythia	Microwave treatment	2.6	–	–	[38]
Artemisia argyi leaves	Smoking simulation method	2–5	–	–	[39]

* Size means the diameter distribution (or average diameter) of CDs which was determined from the corresponding TEM result.

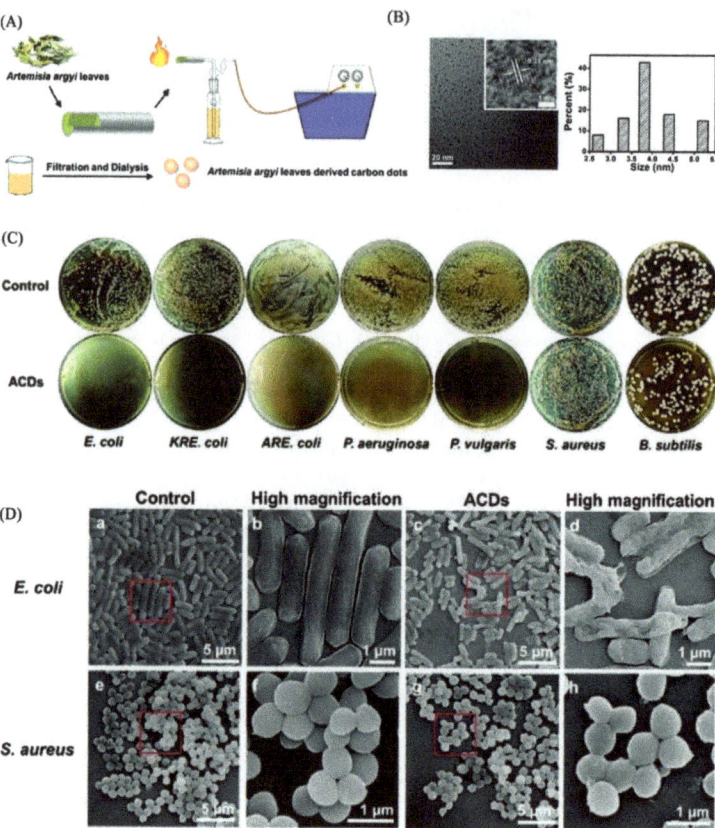

Figure 1. Biomass-derived nitrogen-doped antimicrobial CDs. (**A**) Scheme depicting the synthesis of ACDs. (**B**) **Left**: Transmission electron microscopy (TEM) image and high-resolution TEM image (inset) of ACDs. **Right**: corresponding size distribution. (**C**) Effect of 150 µg mL^{-1} ACDs on the growth of *E. coli*, kanamycin-resistant *E. coli* (*KRE. coli*), ampicillin-resistant *E. coli* (*ARE. coli*), *P. aeruginosa, P. vulgaris, S. aureus*, and *B. subtilis*. (**D**) Scanning electron microscopy (SEM) images of *E. coli* incubated without ACDs (a and b) and with ACDs (c and d), and *S. aureus* incubated without ACDs (e and f) and with ACDs (g and h). ACDs could only disrupt the cell walls of *E. coli* rather than those of *S. aureus*. (**A–D**) Reproduced with permission from [39]. Copyright 2020, The Royal Society of Chemistry.

2.1.2. Nitrogen-Doped CDs Derived from Nitrogen-Containing Compounds

In addition to biomass, nitrogen-doped CDs can also be prepared from nitrogenous compounds such as proteins [40], amino acids [41–44], natural amines [45–47], quaternized compounds [48,49], polyethyleneimine (PEI) [50–53], diethylenetriamine (DETA) [45,54], 2,2′-(ethylenedioxy)-bis(ethylamine) [55], *p*-phenylenediamine [56], and *m*-phenylenediamine [57] for antimicrobial purposes.

Nitrogen-doped CDs have been obtained using protein (protamine sulfate) as the raw material. Zhao et al. reported the simple and fast synthesis of multifunctional blue-emitting protamine sulfate (PS)-based CDs (PS-CDs) by a one-step microwave-mediated approach (Figure 2A) [40]. PS-CDs featured great antibacterial efficacy against *S. aureus* with a minimum inhibitory concentration (MIC) of 25 µg mL^{-1} and methicillin-resistant *S. aureus* (MRSA) with an MIC of 37.5 µg mL^{-1}. Furthermore, PS-CDs displayed high water-dispersity, low cytotoxicity, and excellent blood compatibility. The authors also revealed the antibacterial mechanism of PS-CDs: PS-CDs bound to bacteria by electrostatic interaction, damaged cell membrane, entered the cells, and disrupted the normal survival functions of the bacteria. To conclude, this work employs protein as a raw material to prepare CDs that can be internalized easily by bacteria to realize bactericidal effect.

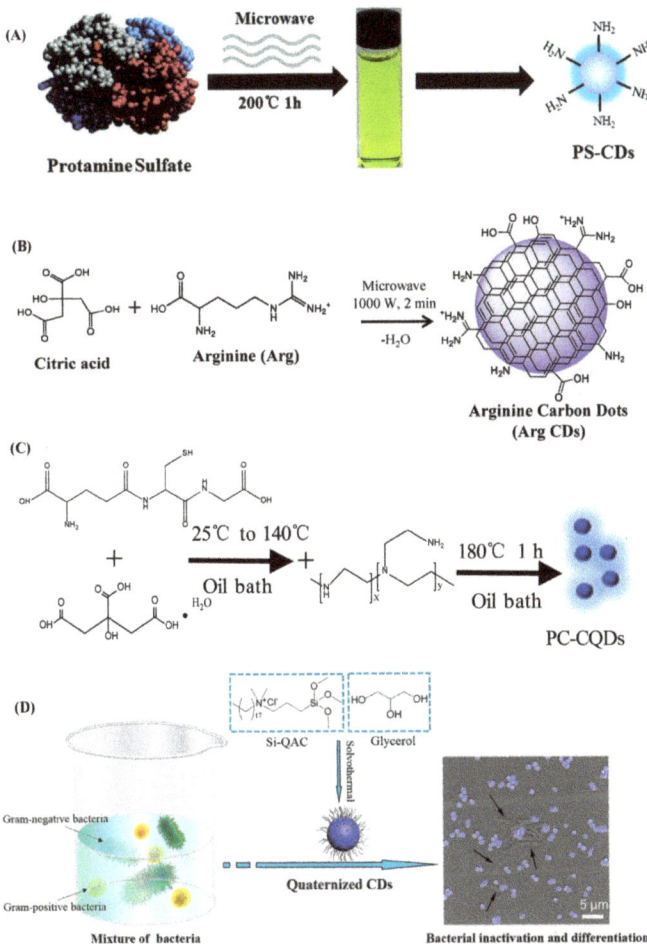

Figure 2. Schematic diagram showing the preparation of nitrogen-doped CDs with antimicrobial activities from (A) protamine sulfate (PS), (B) arginine, (C) L-glutathione, and (D) Si-QAC. (A) Reproduced with permission from [40]. Copyright 2021, The Royal Society of Chemistry. (B) Reproduced with permission from [41]. Copyright 2020, MDPI. (C) Reproduced with permission from [46]. Copyright 2021, Elsevier B.V. (D) Reproduced with permission from [48]. Copyright 2019, Elsevier Ltd.

Several CDs prepared from amino acids such as arginine [41], alanine [42,43], L-tryptophan [58,59], cysteine [43,44], lysine [59], and arginine [59] have been reported to possess antimicrobial activities. Suner et al. synthesized nitrogen-doped arginine CDs (termed Arg CDs) utilizing citric acid as the carbon source and arginine as the amine source by a microwave-mediated approach (Figure 2B) [41]. Arg CDs displayed an MIC of 6.250 mg mL^{-1} against *S. aureus*. To enhance the antimicrobial activity of Arg CDs, two nanocomposites, Arg-Ag CDs and Arg-Cu CDs, were synthesized by generating Ag and Cu nanoparticles (NPs) within Arg CDs. Arg-Ag CDs and Arg-Cu CDs exhibited an MIC of

0.062 and 0.625 mg mL^{-1} against *S. aureus*, respectively. In addition, Arg-Ag CD possessed 0.125 and 0.312 mg mL^{-1} minimum bactericidal concentration (MBC) values against *S. aureus* and *E. coli*, respectively. Pandey et al. prepared CDs from citric acid and β-alanine through a microwave-mediated method [42]. The as-synthesized CDs repressed the growth of diverse Gram-negative bacteria such as *E. coli*, *Salmonella*, *Pseudomonas*, *Agrobacterium*, and *Pectobacterium* species.

Besides proteins and amino acids, other natural amines have been explored as precursors to construct antimicrobial CDs, such as histamine [45], cadaverine [45], putrescine dihydrochloride [45], spermine tetrahydrochloride [45], L-glutathione [46], and spermidine [47]. As an example, Hao et al. prepared positively charged CQDs (PC-CQDs) from citric acid and L-glutathione (Figure 2C) [46]. PC-CQDs exhibited high antibacterial activity against *S. aureus*, *E. coli*, *P. aeruginosa*, and MRSA. PC-CQDs strongly attached to the bacterial cell surfaces due to their small size and the surface groups –NH$_2$ and –NH, entered the cells, and induced the conformational change of DNA and the production of ROS, leading to the rupture of the bacterial cells. It is worth noting that PC-CQDs did not cause detectable drug resistance or hemolysis. No drug resistance was observed in *S. aureus*, *E. coli*, and MRSA incubated with PC-CQDs for over 30 days. Moreover, PC-CQDs were deployed for the antibacterial treatment of mixed *S. aureus*- and *E. coli*-infected wounds in rats with low in vivo toxicity, showing the same therapeutic effect as the traditional antibiotic levofloxacin hydrochloride.

In addition to natural amines as introduced above, other nitrogen-containing agents have also been chosen to prepare nitrogen-doped CDs as antimicrobial agents, including quaternized compounds [48,49], PEI [50–53], DETA [45,54], 2,2′-(ethylenedioxy)-bis(ethylamine) [55], *p*-phenylenediamine [56], *m*-phenylenediamine [57], etc. Our group prepared quaternized CDs through one-step solvothermal treatment of glycerol and dimethyloctadecyl[3-(trimethoxysilyl)propyl]ammonium chloride (Si-QAC) for selective Gram-positive bacterial inactivation (Figure 2D) [48]. The synthesized quaternized CDs could selectively interact with the Gram-positive bacteria due to the distinct surfaces of Gram-positive and Gram-negative bacteria. The quaternized CDs possessed a zeta potential of +33 mV, ensuring their successful electrostatic interaction with negatively charged bacterial cells, while the presence of long alkyl chains in the CDs enabled them to interact with the bacterial cells via hydrophobic interaction. Therefore, the CDs could firmly adhere to (or insert into) the bacterial cell surface which changed the charge balance of the bacterial surface, resulting in the inactivation of Gram-positive bacteria both in vitro and in vivo. At the same time, the CDs featured strong fluorescence emission, which was utilized for fast Gram-type identification. In this way, the CDs may hold the potential for treating Gram-positive bacteria-caused infections. More interestingly, in a later study, we have demonstrated that the CDs showed excellent biofilm penetration capacity due to their small size and could effectively inhibit the biofilm formation and eradicate the formed biofilms [22]. Thus, the CDs represent a highly efficient strategy to combat biofilm-involved infections. In another report, Li et al. fabricated different types of polyamine-modified carbon quantum dots (CQDs) including CQD$_{600}$, CQD$_{1w}$, and CQD$_{2.5w}$, via simple hydrothermal treatment of citric acid and branched PEI (bPEI) with different molecular weights [50]. CQD$_{2.5w}$ possessed higher antibacterial and antibiofilm activities against *S. aureus* and *E. coli* than CQD$_{1w}$ and CQD$_{600}$, since the larger molecular weight of bPEI yielded a larger amount of protonated amines on the surface of the CQDs, giving rise to enhanced electrostatic interaction between CQDs and bacterial cells, and the longer surface corona of the CQDs making their penetration into biofilms easier. Additionally, all the three CQDs had negligible cytotoxicity. However, the reason why longer surface corona of CQDs can give rise to their easier penetration into biofilm was not explained, which requires future investigation. Further, Zhao et al. developed nitrogen-doped CDs (NCQDs) as an antimicrobial nanoagent against *Staphylococcus* to treat infected wounds [54]. NCQDs were made from D(+)-glucose monohydrate and DETA by a heat fusion method. NCQDs displayed antibacterial activity toward *Staphylococcus*, especially against MRSA. Transmission electron microscopy (TEM) analysis showed that

NCQDs could damage the cell structures of *S. aureus* and MRSA, but not *E. coli*. NCQDs exhibited the same therapeutic effect as vancomycin in the treatment of MRSA-infected wounds with negligible toxicity to the main rat organs such as heart, kidney, liver, lung, and spleen.

2.2. Metal-Containing CDs

There are different types of metal-containing CDs that can be used as antimicrobial agents: metal ion-doped CDs [36,60–62], metal nanoparticle-decorated CDs [41,60,63–65], CD/metal oxide nanocomposites [66–70], and CD/metal sulfide nanocomposites [71]. First, CDs doped with metal ions such as Cu^{2+} [36] and Ag^+ [60] have been explored as antimicrobial materials. Qing et al. reported water-dispersible Cu^{2+}-doped CDs (Cu^{2+}–CDs) for antibacterial application [36]. Cu^{2+}–CDs were prepared by one-step hydrothermal carbonization of cupric acetate monohydrate ($Cu(Ac)_2 \bullet H_2O$) and waste tea extract. Cu^{2+}–CDs showed an inhibitory effect on *S. aureus* with an MIC of 0.156 mg/mL. Moreover, Cu^{2+}–CDs possessed low cytotoxicity and appealing biocompatibility. Second, CDs can serve as reducing and stabilizing agents to generate metal nanoparticles on their surface, forming metal nanoparticle-modified CDs for killing microorganisms [41,60,63–65]. For instance, antimicrobial silver nanoparticle-decorated CDs (CD-2) were prepared using a two-step method, in which CDs (PEI-CD) were first obtained by hydrothermal treatment of PEI, and Ag^+ was then reduced to Ag NPs in the presence of formaldehyde and the resultant Ag NPs were bound onto the surface of PEI-CD to produce CD-2 (Figure 3A) [60]. By inducing membrane disruption and intracellular DNA/protein damage, CD-2 displayed high and broad-spectrum antimicrobial activities against Gram-positive bacteria (*S. aureus*), Gram-negative bacteria (*E. coli*, *P. aeruginosa*, and *P. vulgaris*), and fungi (*S. cerevisiae*), with excellent biocompatibility. Third, CDs were integrated with metal oxide to form CD/metal oxide nanocomposites [66–70]. As an example, Gao et al. fabricated $CD/ZnO/ZnAl_2O_4$ nanocomposites which possessed an excellent antibacterial property with antibacterial ratios of 97% and 94% against *S. aureus* and *E. coli*, respectively [68]. Within the antibacterial concentration range, the nanocomposites were nontoxic to human cells. As shown in Figure 3B, the authors proposed a dual-mode antibacterial mechanism for the nanocomposite. On the one hand, the binding of the $CD/ZnO/ZnAl_2O_4$ nanocomposites to the surface of the bacteria blocked the channels of the bacterial nutrient supply from the environment, accelerating apoptosis in the bacteria. On the other hand, the nanocomposites could generate singlet oxygen that damages DNA/RNA, proteins, and phospholipids in the bacterial cells. Also, the presence of CDs in the nanocomposites resulted in strengthened electrostatic interaction between the nanocomposites and the bacteria, and increased singlet oxygen production, which enhanced the bacterial elimination effect of the nanocomposite. Lastly, CD/metal sulfide nanocomposites have been constructed for combating microorganisms. Gao et al. synthesized carbon quantum dots (CQDs) through aldol polymerization reaction using acetone as the carbon source (Figure 3C) [71]. Then $CQDs/Ag_2S/CS$ nanocomposites were prepared through an in situ growth method using polyvinylpyrrolidone (PVP) as the crosslinking agent. The $CQDs/Ag_2S/CS$ nanocomposites exhibited excellent antibacterial property against *E. coli* and *S. aureus* with an MIC of 0.1 mg/mL, and against MRSA with an MIC of 0.25 mg/mL. The $CQDs/Ag_2S/CS$ nanocomposites strongly bound to the surface of the bacteria, leading to the destruction of cell wall and cell membrane and inducing the bacterial cell death (Figure 3C). Notably, no drug resistance was observed for the $CQDs/Ag_2S/CS$ nanocomposites.

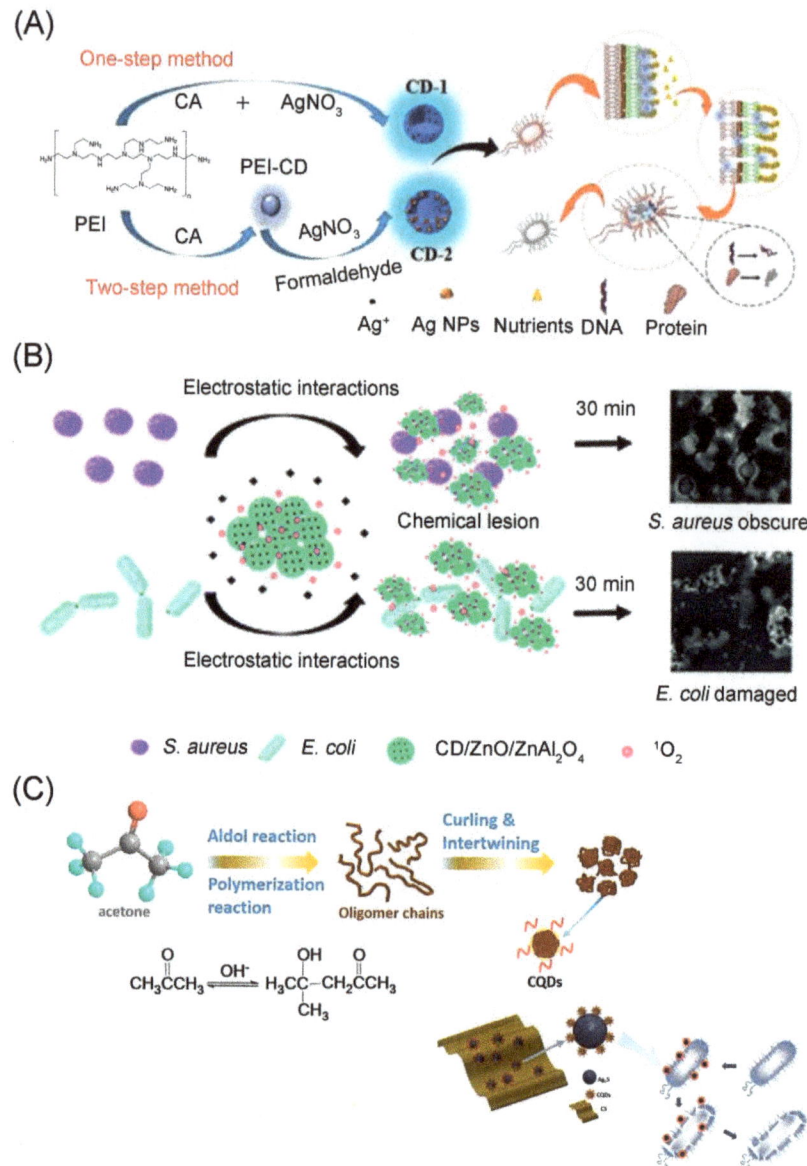

Figure 3. Metal-containing CDs for killing microorganims. (**A**) Scheme depicting the synthesis of CD-1 and CD-2, and their antibacterial properties. Reproduced with permission from [60]. Copyright 2021, Elsevier Inc. (**B**) Antibacterial mechanism of CD/ZnO/ZnAl$_2$O$_4$. Reproduced with permission from [68]. Copyright 2021, The Royal Society of Chemistry. (**C**) Scheme illustrating the formation of CQDs/Ag$_2$S/CS and its antibacterial effect. Reproduced with permission from [71]. Copyright 2021, Elsevier Inc.

2.3. CDs Derived from Antibacterial Compounds (Including Antibiotics)

Antimicrobial CDs have also been developed by using traditional antibiotics and common antibacterial compounds as the raw materials to trade the old for the new, such as kanamycin sulfate [72], levofloxacin hydrochloride [73,74], quaternary ammonium compounds [22,48,49,75,76], and 2,4-dihydroxybenzoic acid [77]. Luo et al. prepared CDs (termed CDs-Kan) from kanamycin sulfate by a one-step hydrothermal method [72]. CDs-Kan were demonstrated to preserve the main bactericidal functional groups of kanamycin like the amino sugar and amino cyclic alcohol, which ensured their good antibacterial activity. Specifically, CDs-Kan inhibited the growth of *E. coli* and *S. aureus* with good biocompatibility. Wu et al. reported cationic levofloxacin-derived CDs (LCDs) with enhanced antibacterial activities and low drug resistance (Figure 4A) [74]. LCDs were synthesized from levofloxacin hydrochloride through a simple one-pot hydrothermal method. With the preservation of the active groups from levofloxacin, LCDs featured notable bactericidal activity against *S. aureus* and *E. coli* with an MIC of 0.125 µg/mL, which was lower than that of levofloxacin hydrochloride. It is worth noting that LCDs displayed low drug resistance, good aqueous dispersity, and outstanding biosafety, while retaining the broad-spectrum antibacterial activity of levofloxacin. In addition to *S. aureus* and *E. coli*, LCDs could kill other microorganisms including MRSA, *Enterococcus faecalis* (*E. faecalis*), *S. epidermidis*, *Listera monocytogenes* (*L. monocytogenes*), *P. aeruginosa*, and *Serratia marcescens*. LCDs could enter the bacterial cells via electrostatic interaction between the positively-charged LCDs and the negatively-charged bacteria. Once entering the bacterial cells, LCDs produced ROS to destroy cell membrane and the normal state of bacteria, causing cell death. LCDs were deployed for the treatment of bacteria-infected wounds and pneumonia in mice with enhanced therapeutic efficacy without harming normal tissues as compared to levofloxacin. Zhao et al. fabricated quaternary ammonium carbon quantum dots (QCQD) from 2,3-epoxypropyltrimethylammonium chloride and diallyldimethylammonium chloride via the hydrothermal reaction (Figure 4B) [75]. QCQD displayed admirable bactericidal effect toward Gram-positive bacteria, including *S. aureus*, MRSA, *E. faecalis*, *L. monocytogenes*, and *S. epidermidis*. QCQD also featured satisfactory biocompatibility as demonstrated by in vivo and in vitro toxicity assays. Thus, QCQD were successfully utilized in the treatment of MRSA-infected pneumonia in mice, prompting the regression of pulmonary inflammation in the mouse lung. As revealed by the quantitative proteomics, the antibacterial ability of QCQD could be attributed to the fact that QCQD might mainly act on ribosomes and upregulate the proteins involved in RNA degradation, causing interference to the protein translation, posttranslational modification, and protein turnover in bacterial cells.

Meanwhile, CDs can be employed as the carriers of existing antibiotics to realize the controlled release of these antibiotics. Saravanan et al. prepared N@CDs by hydrothermal treatment of *m*-phenylenediamine (Figure 4C) [57]. N@CDs exhibited antibacterial activities against *E. coli* and *S. aureus* with an MIC of 1 and 0.75 mg/mL, respectively. Additionally, N@CDs were applied as nanovehicles for sustained time-dependent release of the traditional antibiotic ciprofloxacin in the physiological condition.

2.4. Photoresponsive CDs
2.4.1. Photodynamic Therapy (PDT)

In PDT, photosensitive agents (photosensitizers) are sensitized by light in the presence of oxygen to generate ROS such as free radicals and singlet oxygen [78,79]. The produced ROS can break DNA, inactivate enzymes, and oxidize amino acids, resulting in cell necrosis/apoptosis. PDT represents a promising alternative to antibiotics in killing microorganisms, because of its fascinating advantages such as high spatial controllability, antibiotic resistance independence, and low cumulative toxicity. The photo-generated electrons and holes of CDs that are related with PDT action mechanism entail various catalytic processes [80]. Meanwhile, CDs have a relatively wide visible spectral region. These properties enable CDs to be promising antibacterial photosensitizers. After irradiation with light of a given wavelength, some bare CDs can produce ROS that are capable of inactivating microorganisms, and these CDs can thus serve as potent antimicrobial photosensitizers.

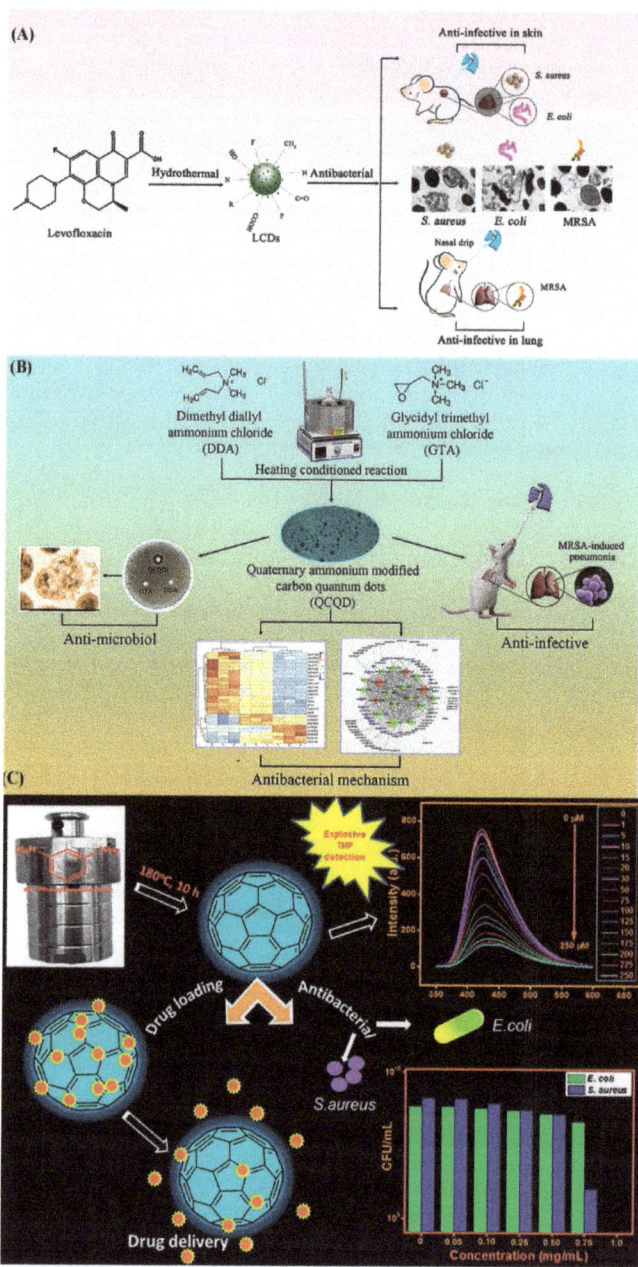

Figure 4. Antibacterial compound-derived CDs for killing microorganisms. (**A**) Construction and antibacterial characteristics of LCDs. Reproduced with permission from [74]. Copyright 2022 Elsevier Ltd. (**B**) Preparation of QCQDs from two quaternary ammonium compounds and their bactericidal properties and antibacterial mechanism. Reproduced with permission from [75]. Copyright 2020, Elsevier Ltd. (**C**) Scheme showing the preparation of N@CDs for delivering the traditional antibiotic ciprofloxacin to kill bacteria. Reproduced with permission from [57]. Copyright 2020, American Chemical Society.

Beyond bare CDs fabricated from mushroom [81], graphite rods [82], and the poloxamer Pluronic F-68 [83], as summarized in our previous review [16], more bare CDs have been reported to possess intrinsic photodynamic characteristics, including graphitic carbon nitride quantum dots (g-CNQDs) [84], red-emitting CDs (R-CDs) [77], CQDs constructed from citric acid and 1,5-diaminonaphthalene by solvothermal reaction [85], nitrogen- and/or sulfur-doped CDs derived from amino acids [43], nitrogen and iodine co-doped CDs (N/I-CD) prepared by hydrothermal treatment of iohexol [86], metal-doped CDs such as zinc-doped CDs [87], copper-doped CDs [88], and terbium-doped CDs [89]. Yadav et al. constructed green-fluorescent g-CNQDs from melamine and ethylene diamine tetraacetic acid (EDTA) sodium salt via a thermal polymerization method [84]. g-CNQDs could effectively produce superoxide and hydroxyl radicals with the irradiation of visible light, eradicating ~99% *E. coli* and ~90% *S. aureus* at a concentration of 0.1 mg/mL. Moreover, g-CNQDs featured low cytotoxicity—3.2 mg/mL g-CNQDs were nontoxic to fibroblast cells. Liu et al. synthesized R-CDs through solvothermal treatment of 2,4-dihydroxybenzoic acid (an organic bactericide) and 6-bromo-2-naphthol [77]. R-CDs possessed both intrinsic antibacterial activity and antibacterial photodynamic activity toward multidrug-resistant *Acinetobacter baumannii* (MRAB). R-CDs could effectively enter bacterial cells and bacterial biofilms with few side effects on animal cells. Therefore, R-CDs were successfully utilized for MRAB biofilm prevention and elimination as well as the treatment of MRAB-induced infected wounds. No microbial drug resistance was observed when using R-CDs to kill MRAB and MRSA. These findings demonstrated that R-CDs represent a potent antibacterial agent for fighting against drug-resistant bacteria. Liu et al. fabricated zinc-doped CDs (Zn-CDs) from citric acid, ethylenediamine, and zinc acetate by a one-step hydrothermal method [87]. Zn-CDs produced ROS under blue light irradiation, showing bactericidal effect toward *S. aureus* and *Streptococcus mutans* with negligible animal cell toxicity. Collectively, the different types of CDs mentioned above can act as new types of photosensitizers for photodynamic antibacterial treatment.

Besides being used as photosensitizers, CDs can be integrated with other photosensitizers such as curcumin [90], black phosphorus (BP) nanosheets [91], TiO_2 [92,93], and ZnO [68,94,95] to afford photoresponsive nanocomposites. In these photosensitive nanocomposites, CDs play different rols in obtaining improved antimicrobial PDT efficacy, such as being used as drug carriers [90], enhancing the interaction of photosensitizers with microorganisms [91,92], preventing the agglomeration of photosensitizers [92], or increasing the light absorption and suppressing photogenerated electron–hole's recombination [95]. For instance, Yan et al. developed a nano-PS system using CDs to deliver the traditional photosensitizer curcumin (Cur) for enhanced antibacterial performance [90]. Zhang et al. decorated BP nanosheets with cationic CDs through in situ growth of CDs from chlorhexidine gluconate on the surface of BPs, resulting in the formation of the nanocomposite BPs@CDs (Figure 5A) [91]. Without light irradiation, BPs@CDs exhibited antibacterial ability due to the electrostatic attraction between the bacteria and the CDs on the surface of BPs@CDs. Under 660 nm laser irradiation, BPs@CDs produced singlet oxygen, exhibiting outstanding photodynamic antibacterial capacity. Under 808 nm laser irradiation, BPs@CDs displayed photothermal antibacterial activity. Accordingly, the BPs@CDs exhibited synergistic intrinsic antibacterial activity, antibacterial PDT activity, and antibacterial PTT activity toward both *E. coli* and *S. aureus*. In addition, BPs@CDs were degradable with no noticeable cytotoxicity. Owing to their triple-mode antibacterial capability, BPs@CDs were successfully deployed for the treatment of bacteria-associated wounds with shortened wound healing time.

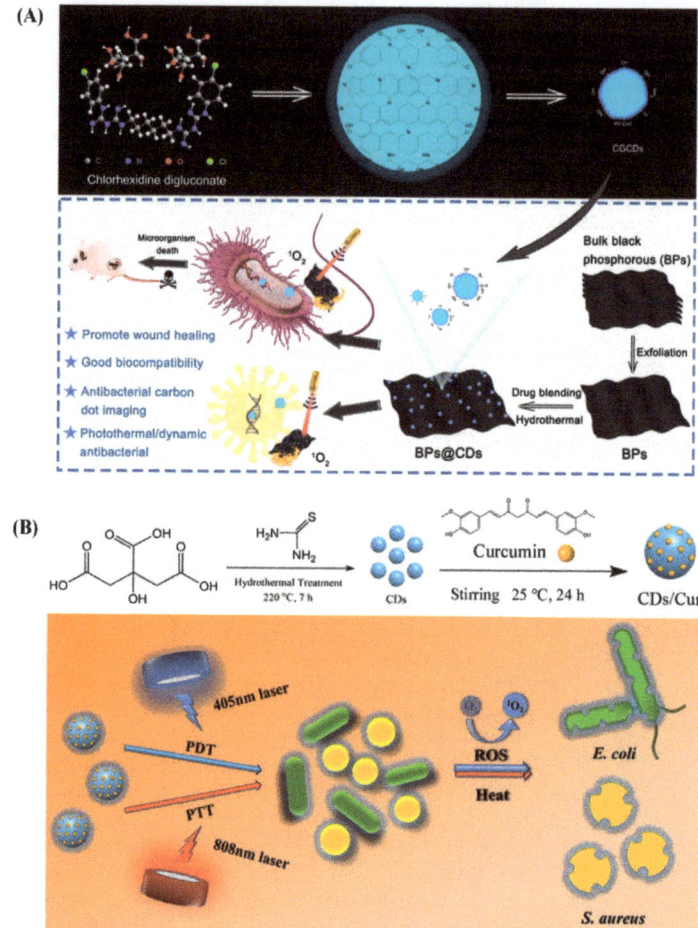

Figure 5. (**A**) Schematic diagram illustrating the preparation of BPs@CDs for eliminating bacteria. Reproduced with permission from [91]. Copyright 2021, Springer Nature. (**B**) Scheme illustrating the preparation of the CDs/Cur nanocomposite and its PDT- and PTT-mediated antimicrobial performance. Reproduced with permission from [90]. Copyright 2021, American Chemical Society.

The CD-involved antimicrobial PDT systems have been developed in the forms of different nanocomposites. Hydrophobic carbon quantum dots (hCQDs) with photodynamic property made from the poloxamer Pluronic F-68 [83] were encapsulated in polymers such as polydimethylsiloxane (PDMS) [96,97], polycaprolactone [98], and polyurethane [99], and to construct light-triggered antibacterial nanocomposites in the form of slide [96], nanofiber [98], and film [99]. For example, hCQDs were embedded into the PDMS polymer matrix to generate hydrophobic CQDs/PDMS surface by a swelling-encapsulation-shrink method [96]. The nanocomposite surface exhibited bactericidal activities against *S. aureus*, *E. coli*, and *K. pneumoniae* by producing ROS upon the excitation at 470 nm. More importantly, the nanocomposite surface showed no toxicity towards NIH/3T3 cells.

2.4.2. Photothermal Therapy (PTT)

PTT eliminates microorganisms by hyperthermia generated from photothermal agents when they absorb light. Only one kind of antimicrobial CDs with PTT capacity had been reported when we prepared the previous review. Since 2019, more CDs have been reported for antimicrobial PTT [76,100,101]. Belkahla et al. produced CDs from glucose as the precursor through the alkali-assisted ultrasonic irradiation approach [100]. The generated CDs possessed the heat-producing capability under illumination at 680 or 808 nm and were employed for photothermal treatment of *E. coli*. Yan et al. constructed a nanosystem termed CDs-Tb-TMPDPA, which consisted of TMPDPA (4-(2,4,6-trimethoxyphenyl)-pyridine-2,6-dicarboxylic acid, a two-photon ligand)-sensitized Tb^{3+} as a temperature-sensitive module and CDs (prepared by microwave heating of citric acid and formamide) as a photothermal antibacterial component [101]. The CDs were coordinated to Tb^{3+} that was further linked with TMPDPA through coordination interaction. The authors demonstrated that the nanosystem could be used for temperature detection based on the temperature-dependent fluorescence intensity (I) ratio ($I(Tb^{3+})/I(CDs)$) and the fluorescence lifetime of CDs-Tb-TMPDPA. Besides, the authors also realized *E. coli* growth inhibition by utilizing the photothermal conversion property of CDs in CDs-Tb-TMPDPA via two-photon excitation (660 nm). This work develops a multifunctional probe for dual-mode temperature detection and antibacterial PTT under two-photon excitation.

Moreover, CDs-involved PTT can be integrated with other antimicrobial strategies such as PDT [90,93] and chemodynamic therapy (CDT) [102] to achieve combined antimicrobial therapies. For instance, N, S-doped CDs with strong fluorescence were first prepared from citric acid and thiourea by a one-step hydrothermal route, and was combined with curcumin (Cur) to afford CDs/Cur (Figure 5B) [90]. In CDs/Cur, the ROS yield of Cur could be enhanced through fluorescence resonance energy transfer (FRET), while the high photothermal conversion efficiency of the CDs due to their strong light absorption was preserved. As a result, upon 405 nm visible light and near-infrared light irradiation, CDs/Cur could yield ROS and a moderate temperature increase, which seriously damaged bacterial cell surface, leading to synergistic PDT- and PTT-promoted antibacterial effects against *E. coli* and *S. aureus*. Furthermore, CDs/Cur displayed low cytotoxicity and negligible hemolytic activity, which ensured their practical application. In another example, Yan et al. developed a nanocomposite termed FeOCl@PEG@CDs by coating poly(ethylene glycol) (PEG) and CDs on iron oxychloride nanosheets (FeOCl NSs) [102]. The hydroxyl radical (•OH) was generated from H_2O_2 activation by the redox cycle of ions on FeOCl NSs, and the heat was produced from the CDs upon the irradiation at 808 nm, inducing the death of *S. aureus* and *E. coli*. Further, the FeOCl@PEG@CDs were successfully applied in synergistic chemodynamic and photothermal treatment of infected wounds.

As far as we know, because only several CDs have been reported for photo-assisted antimicrobial uses, the lethal route was not carefully investigated. Commonly, researchers just reported that bacteria treated with CDs can produce ROS or heat upon light irradiation which can cause cell wall/membrane damage to kill bacteria. Thus, more studies should be performed in the future to thoroughly investigate the underlying lethal mechanism of CD-based antibacterial phototherapy.

3. Applications of Antimicrobial CDs in Medical and Industry Fields

CDs-involved antimicrobial strategies have been deployed in both medical and industry fields. In the medical field, antimicrobial CDs have been leveraged for coating the surface of orthopedic implant materials [66], delivering drugs [24,57,103,104], and repairing infected bone defects [47]. Moradlou et al. grew a thin film of CDs-incorporated hematite (CQDs@α-Fe_2O_3) on a titanium substrate to yield Ti/CQDs@α-Fe_2O_3 [66]. CQDs were prepared from graphite rods via an electrochemical method and used as nano-scaffolds for the growth of CQD@α-Fe_2O_3 nanoparticles as core@shell nanostructures. The Ti/CQDs@α-Fe_2O_3 samples exhibited sustainable antibacterial activity against *S. aureus* but not *E. coli*, offering a way of using CDs to prepare antimicrobial materials for medical devices. In another work, Geng et al. synthesized positively-charged CQDs (p-CQDs) through microwave reaction of spermidine trihydrochloride, and prepared negatively-charged CQDs (n-CQDs) via microwave reaction of 1,3,6-trinitropyrene (TNP) and sodium sulfite (Figure 6A) [47].

The p-CQDs displayed effective antibacterial activity against multidrug-resistant (MDR) bacteria and could realize the inhibition of biofilm formation, while n-CQDs notably promoted bone regeneration. The nearly neutral p-CQD/WS$_2$ hybrids were first fabricated by depositing p-CQDs on WS$_2$ nanosheets, and then coencapsulated with n-CQDs into the gelatin/methacrylate anhydride (GelMA) hydrogel to obtain p-CQD/WS$_2$/n-CQD/GelMA hydrogel scaffold (Figure 6A). The implantation of p-CQD/WS$_2$/n-CQD/GelMA hydrogel scaffold in an MRSA-infected craniotomy defect model induced almost complete repair of an infected bone defect with the new bone area of 97.0 ± 1.6% at 60 days. This work proposes a CD-based strategy for developing biomaterials with both antibacterial and osteogenic activities for the treatment of infected bone defects.

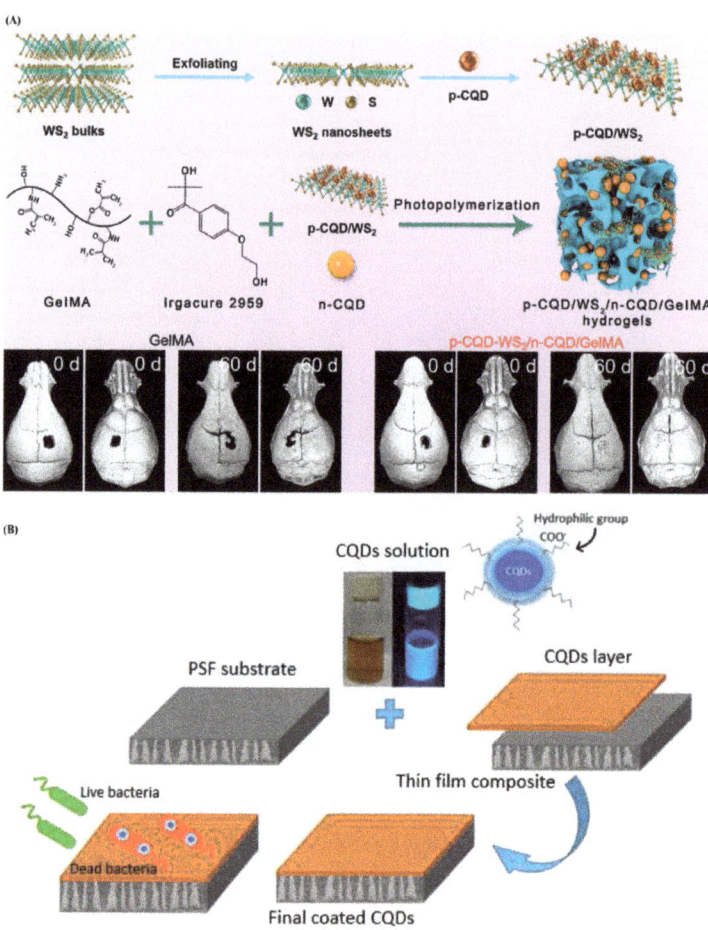

Figure 6. (**A**) Schematic representation of the synthesis of a p-CQD/WS$_2$/n-CQD/GelMA hydrogel for regeneration of bone defects. Micro computed tomography (micro-CT) images of calvarial defects treated with GelMA and p-CQD/WS$_2$/n-CQD/GelMA hydrogels for 0 or 60 days were presented. Reproduced with permission from [47]. Copyright 2021, Elsevier Ltd. (**B**) Scheme showing the fabrication of a CQDs-coated PSF membrane for realizing the antibacterial action in the process of forward osmosis. Reproduced with permission from [105]. Copyright 2020, Elsevier B.V.

In the industry field, antimicrobial CDs have been utilized to construct thin-film composite membranes for forward osmosis [105] and nanofiller [106], packaging materials [107,108], and lubricant additives [53]. Mahat et al. developed thin-film composite membranes for forward osmosis by embedding CQDs derived from oil palm biomass into polysulfone-selective layers, which were denoted as CQDs-PSF (PSF: polysulfone) [105]. The authors proved that the addition of CQDs into PSF membranes increased water flux and improved antibacterial performance. In another study, Koulivand et al. constructed antifouling and antibacterial nanofiltration membranes for efficient salt and dye rejection by incorporating nitrogen-doped CDs (NCDs) to polyethersulfone (PES) using a phase inversion technique [106]. The antibacterial NCDs were synthesized via hydrothermal treatment of ammonium citrate dibasic. The obtained membrane exhibited improved pure water flux and enhanced antifouling property. In addition, Kousheh et al. constructed a nanocellulose film with antimicrobial/antioxidant and ultraviolet (UV) protective activities for food packaging by introducing water-dispersible and photoluminescent CDs [107]. The antimicrobial CDs were synthesized from cell-free supernatant of *Lactobacillus acidophilus* via a hydrothermal method. The as-synthesized CDs were embedded into bacterial nanocellulose (BNC) film due to the hydrogen bonding interaction between CDs and the carboxyl, hydroxyl, and carbonyl groups of BNC, leading to the formation of the CD-BNC film. The CD-BNC film displayed a higher inhibitory activity toward *Listeria monocytogenes* than *E. coli*. In addition to antibacterial activity, the introduction of CDs into the BNC film also endowed the CD-BNC film with UV-blocking activity, fluorescence appearance, and improved flexibility. The CD-BNC film could be used to fabricate nanopaper for wrapping of food commodity and fabrication of forgery-proof packaging. In addition to thin-film composite membranes and packaging materials, antimicrobial CDs have been implemented as lubricant additives. Tang et al. fabricated CDs from PEG and PEI through a hydrothermal approach [53]. The MICs of the CDs toward *E. coli* and *S. aureus* were 62.5 and 15.56 µg mL^{-1}, respectively. Besides the antibacterial activity, the CDs featured anti-friction property. The addition of 0.2% (wt) CDs reduced the mean friction coefficient and wear volume of water-based lubrication by 59.77% and 57.97%, respectively. This example suggests that CDs with antibacterial and anti-friction functions can be utilized as an advanced lubricating additive, thus broadening the practical application of CDs.

4. Conclusions

As reviewed here, CDs are potent antimicrobial nanomaterials and represent promising alternatives to conventional antibiotics for the treatment of infectious diseases caused by microorganisms. Nevertheless, several challenges still exist. First, the potential antibacterial capability and specificity of CDs are difficult to predict from their raw materials, since it is unknown whether the antibacterial groups of raw materials can be retained or there are newly formed antibacterial structures during the complicated CD formation process. The CDs' functional and biological characteristics are directly associated with their core and particularly their surface's functional groups, which are largely dependent on the precursors and synthetic methods. Therefore, the structural analysis of CDs and the clarification of the reaction mechanism of CDs are helpful to better predict the antibacterial activities of CDs. Second, the integration of CDs with other compounds such as antibiotics, metal ions, hydrogels, and photosensitive materials to prepare new composite materials with synergistic antibacterial effect represents an important research direction in this field, which is definitely worth exploring in the future. Third, most antimicrobial CDs possess high MICs, and usually can only kill certain types of bacteria such as Gram-positive bacteria. Thus, it is highly desired to develop CDs with low MICs and broad-spectrum antibacterial ability. Fourth, the current antibacterial research of CDs mainly focuses on the killing of planktonic bacteria. In the future, it is necessary to explore the application potential of CDs in combating bacterial biofilms, eliminating intracellular bacteria, and killing bacteria in tumor. Fifth, despite the extensive investigations on the use of CDs for killing bacteria, few examples have been reported on using CDs to eliminate fungi and viruses which can also cause severe infections, diseases, and even death to humans. Sixth, the reproducible, large-scale, and cost-effective fabrication of CDs still limits the practical antimicrobial applications of CDs. Seventh, studies regarding the interaction of CDs with microbial cells, the distribution of CDs in microbial cells, the antimicrobial mechanisms of CDs, and the possible antimicrobial resistance development of CDs are still lacking, which will benefit the development of CDs with broad-spectrum antimicrobial activities, low MICs, and negligible drug resistance. Finally, although CDs are generally shown to be safe by cytotoxicity assays, the in vivo safety analyses of CDs remain largely unexplored, which is definitely worthy of evaluation in future studies. It is hoped that the current review will further promote the future design of functional CDs and CDs-incorporated advanced materials for combating the microbial infection-caused diseases.

Author Contributions: F.L.: Investigation, Analysis, Writing—original draft & editing. Z.W.: Investigation, Writing—original draft. F.-G.W.: Supervision, Analysis, Writing—original draft & editing. All authors have read and agreed to the published version of the manuscript.

Funding: This work was supported by grants from the Natural Science Foundation of Jiangsu Province (BK20211510) and National Natural Science Foundation of China (32170072).

Institutional Review Board Statement: Not applicable.

Informed Consent Statement: Not applicable.

Data Availability Statement: Data sharing not applicable.

Conflicts of Interest: The authors declare no conflict of interest.

References

1. Hoseinnejad, M.; Jafari, S.M.; Katouzian, I. Inorganic and metal nanoparticles and their antimicrobial activity in food packaging applications. *Crit. Rev. Microbiol.* **2018**, *44*, 161–181. [CrossRef] [PubMed]
2. Raghunath, A.; Perumal, E. Metal oxide nanoparticles as antimicrobial agents: A promise for the future. *Int. J. Antimicrob. Agents* **2017**, *49*, 137–152. [CrossRef] [PubMed]
3. Wang, D.; Lin, Z.; Wang, T.; Yao, Z.; Qin, M.; Zheng, S.; Lu, W. Where does the toxicity of metal oxide nanoparticles come from: The nanoparticles, the ions, or a combination of both? *J. Hazard. Mater.* **2016**, *308*, 328–334. [CrossRef] [PubMed]
4. Soenen, S.J.; Parak, W.J.; Rejman, J.; Manshian, B. (Intra)cellular stability of inorganic nanoparticles: Effects on cytotoxicity, particle functionality, and biomedical applications. *Chem. Rev.* **2015**, *115*, 2109–2135. [CrossRef] [PubMed]
5. Liu, J.; Li, R.; Yang, B. Carbon dots: A new type of carbon-based nanomaterial with wide applications. *ACS Cent. Sci.* **2020**, *6*, 2179–2195. [CrossRef] [PubMed]
6. Wu, H.; Lu, S.; Yang, B. Carbon-dot-enhanced electrocatalytic hydrogen evolution. *Acc. Mater. Res.* **2022**, *3*, 319–330. [CrossRef]
7. Wang, B.; Song, H.; Qu, X.; Chang, J.; Yang, B.; Lu, S. Carbon dots as a new class of nanomedicines: Opportunities and challenges. *Coord. Chem. Rev.* **2021**, *442*, 214010. [CrossRef]
8. Wang, B.; Lu, S. The light of carbon dots: From mechanism to applications. *Matter* **2022**, *5*, 110–149. [CrossRef]
9. Li, B.; Zhao, S.; Li, H.; Wang, Q.; Xiao, J.; Lan, M. Recent advances and prospects of carbon dots in phototherapy. *Chem. Eng. J.* **2021**, *408*, 127245. [CrossRef]
10. Đorđević, L.; Arcudi, F.; Cacioppo, M.; Prato, M. A multifunctional chemical toolbox to engineer carbon dots for biomedical and energy applications. *Nat. Nanotechnol.* **2022**, *17*, 112–130. [CrossRef]
11. Xu, Y.; Wang, B.; Zhang, M.; Zhang, J.; Li, Y.; Jia, P.; Zhang, H.; Duan, L.; Li, Y.; Li, Y.; et al. Carbon dots as a potential therapeutic agent for the treatment of cancer related anemia. *Adv. Mater.* **2022**, *34*, 2200905. [CrossRef]
12. Dhenadhayalan, N.; Lin, K.C.; Saleh, T.A. Recent advances in functionalized carbon dots toward the design of efficient materials for sensing and catalysis applications. *Small* **2020**, *16*, 1905767. [CrossRef] [PubMed]
13. Hua, X.W.; Bao, Y.W.; Chen, Z.; Wu, F.G. Carbon quantum dots with intrinsic mitochondrial targeting ability for mitochondria-based theranostics. *Nanoscale* **2017**, *9*, 10948–10960. [CrossRef] [PubMed]
14. Li, M.; Chen, T.; Gooding, J.J.; Liu, J. Review of carbon and graphene quantum dots for sensing. *ACS Sens.* **2019**, *4*, 1732–1748. [CrossRef] [PubMed]
15. Lin, F.; Jia, C.; Wu, F.G. Carbon dots for intracellular sensing. *Small Struct.* **2022**, *3*, 2200033. [CrossRef]
16. Lin, F.; Bao, Y.W.; Wu, F.G. Carbon dots for sensing and killing microorganisms. *C* **2019**, *5*, 33. [CrossRef]
17. Feng, T.; Tao, S.; Yue, D.; Zeng, Q.; Chen, W.; Yang, B. Recent advances in energy conversion applications of carbon dots: From optoelectronic devices to electrocatalysis. *Small* **2020**, *16*, 2001295. [CrossRef]
18. Hu, C.; Li, M.; Qiu, J.; Sun, Y.P. Design and fabrication of carbon dots for energy conversion and storage. *Chem. Soc. Rev.* **2019**, *48*, 2315–2337. [CrossRef]

19. Hutton, G.A.M.; Martindale, B.C.M.; Reisner, E. Carbon dots as photosensitisers for solar-driven catalysis. *Chem. Soc. Rev.* **2017**, *46*, 6111–6123. [CrossRef] [PubMed]
20. Hassan, M.; Gomes, V.G.; Dehghani, A.; Ardekani, S.M. Engineering carbon quantum dots for photomediated theranostics. *Nano Res.* **2018**, *11*, 1–41. [CrossRef]
21. Yang, J.; Zhang, X.; Ma, Y.H.; Gao, G.; Chen, X.; Jia, H.R.; Li, Y.H.; Chen, Z.; Wu, F.G. Carbon dot-based platform for simultaneous bacterial distinguishment and antibacterial applications. *ACS Appl. Mater. Interfaces* **2016**, *8*, 32170–32181. [CrossRef]
22. Ran, H.H.; Cheng, X.; Bao, Y.W.; Hua, X.W.; Gao, G.; Zhang, X.; Jiang, Y.W.; Zhu, Y.X.; Wu, F.G. Multifunctional quaternized carbon dots with enhanced biofilm penetration and eradication efficiencies. *J. Mater. Chem. B* **2019**, *7*, 5104–5114. [CrossRef]
23. Li, P.; Sun, L.; Xue, S.; Qu, D.; An, L.; Wang, X.; Sun, Z. Recent advances of carbon dots as new antimicrobial agents. *SmartMat* **2022**, *3*, 226–248. [CrossRef]
24. Wang, Z.X.; Wang, Z.; Wu, F.G. Carbon dots as drug delivery vehicles for antimicrobial applications: A minireview. *ChemMedChem* **2022**, *17*, e202200003. [CrossRef]
25. Xu, X.; Ray, R.; Gu, Y.; Ploehn, H.J.; Gearheart, L.; Raker, K.; Scrivens, W.A. Electrophoretic analysis and purification of fluorescent single-walled carbon nanotube fragments. *J. Am. Chem. Soc.* **2004**, *126*, 12736–12737. [CrossRef]
26. Alavi, M.; Jabari, E.; Jabbari, E. Functionalized carbon-based nanomaterials and quantum dots with antibacterial activity: A review. *Expert Rev. Anti. Infect. Ther.* **2021**, *19*, 35–44. [CrossRef]
27. Shahshahanipour, M.; Rezaei, B.; Ensafi, A.A.; Etemadifar, Z. An ancient plant for the synthesis of a novel carbon dot and its applications as an antibacterial agent and probe for sensing of an anti-cancer drug. *Mater. Sci. Eng. C* **2019**, *98*, 826–833. [CrossRef]
28. Boobalan, T.; Sethupathi, M.; Sengottuvelan, N.; Kumar, P.; Balaji, P.; Gulyás, B.; Padmanabhan, P.; Selvan, S.T.; Arun, A. Mushroom-derived carbon dots for toxic metal ion detection and as antibacterial and anticancer agents. *ACS Appl. Nano Mater.* **2020**, *3*, 5910–5919. [CrossRef]
29. Ma, Y.; Zhang, M.; Wang, H.; Wang, B.; Huang, H.; Liu, Y.; Kang, Z. N-doped carbon dots derived from leaves with low toxicity via damaging cytomembrane for broad-spectrum antibacterial activity. *Mater. Today Commun.* **2020**, *24*, 101222. [CrossRef]
30. Surendran, P.; Lakshmanan, A.; Priya, S.S.; Geetha, P.; Rameshkumar, P.; Kannan, K.; Hegde, T.A.; Vinitha, G. Fluorescent carbon quantum dots from Ananas comosus waste peels: A promising material for NLO behaviour, antibacterial, and antioxidant activities. *Inorg. Chem. Commun.* **2021**, *124*, 108397. [CrossRef]
31. Liu, S.; Quan, T.; Yang, L.; Deng, L.; Kang, X.; Gao, M.; Xia, Z.; Li, X.; Gao, D. N,Cl-codoped carbon dots from *Impatiens balsamina* L. stems and a deep eutectic solvent and their applications for Gram-positive bacteria identification, antibacterial activity, cell imaging, and ClO$^-$ sensing. *ACS Omega* **2021**, *6*, 29022–29036. [CrossRef]
32. Genc, M.T.; Yanalak, G.; Aksoy, I.; Aslan, E.; Patır, I.H. Green carbon dots (GCDs) for photocatalytic hydrogen evolution and antibacterial applications. *ChemistrySelect* **2021**, *6*, 7317–7322. [CrossRef]
33. Saravanan, A.; Maruthapandi, M.; Das, P.; Luong, J.H.T.; Gedanken, A. Green synthesis of multifunctional carbon dots with antibacterial activities. *Nanomaterials* **2021**, *11*, 369. [CrossRef]
34. Eskalen, H.; Çeşme, M.; Kerli, S.; Özğan, Ş. Green synthesis of water-soluble fluorescent carbon dots from rosemary leaves: Applications in food storage capacity, fingerprint detection, and antibacterial activity. *J. Chem. Res.* **2021**, *45*, 428–435. [CrossRef]
35. Pandiyan, S.; Arumugam, L.; Srirengan, S.P.; Pitchan, R.; Sevugan, P.; Kannan, K.; Pitchan, G.; Hegde, T.A.; Gandhirajan, V. Biocompatible carbon quantum dots derived from sugarcane industrial wastes for effective nonlinear optical behavior and antimicrobial activity applications. *ACS Omega* **2020**, *5*, 30363–30372. [CrossRef]
36. Qing, W.; Chen, K.; Yang, Y.; Wang, Y.; Liu, X. Cu^{2+}-doped carbon dots as fluorescence probe for specific recognition of Cr(VI) and its antimicrobial activity. *Microchem. J.* **2020**, *152*, 104262. [CrossRef]
37. Das, P.; Maruthapandi, M.; Saravanan, A.; Natan, M.; Jacobi, G.; Banin, E.; Gedanken, A. Carbon dots for heavy-metal sensing, pH-sensitive cargo delivery, and antibacterial applications. *ACS Appl. Nano Mater.* **2020**, *3*, 11777–11790. [CrossRef]
38. Zhao, X.; Wang, L.; Ren, S.; Hu, Z.; Wang, Y. One-pot synthesis of Forsythia@carbon quantum dots with natural anti-wood rot fungus activity. *Mater. Des.* **2021**, *206*, 109800. [CrossRef]
39. Wang, H.; Zhang, M.; Ma, Y.; Wang, B.; Shao, M.; Huang, H.; Liu, Y.; Kang, Z. Selective inactivation of Gram-negative bacteria by carbon dots derived from natural biomass: *Artemisia argyi* leaves. *J. Mater. Chem. B* **2020**, *8*, 2666–2672. [CrossRef]
40. Zhao, D.; Liu, X.; Zhang, R.; Huang, X.; Xiao, X. Facile one-pot synthesis of multifunctional protamine sulfate-derived carbon dots for antibacterial applications and fluorescence imaging of bacteria. *New J. Chem.* **2021**, *45*, 1010–1019. [CrossRef]
41. Suner, S.S.; Sahiner, M.; Ayyala, R.S.; Bhethanabotle, V.R.; Sahiner, N. Nitrogen-doped arginine carbon dots and its metal nanoparticle composites as antibacterial agent. *C* **2020**, *6*, 58. [CrossRef]
42. Pandey, A.; Devkota, A.; Yadegari, Z.; Dumenyo, K.; Taheri, A. Antibacterial properties of citric acid/β-alanine carbon dots against Gram-negative bacteria. *Nanomaterials* **2021**, *11*, 2012. [CrossRef] [PubMed]
43. Kang, J.W.; Kang, D.H. Effect of amino acid-derived nitrogen and/or sulfur doping on the visible-light-driven antimicrobial activity of carbon quantum dots: A comparative study. *Chem. Eng. J.* **2021**, *420*, 129990. [CrossRef]
44. Suner, S.S.; Sahiner, M.; Ayyala, R.S.; Bhethanabotla, V.R.; Sahiner, N. Versatile Fluorescent carbon dots from citric acid and cysteine with antimicrobial, anti-biofilm, antioxidant, and AChE enzyme inhibition capabilities. *J. Fluoresc.* **2021**, *31*, 1705–1717. [CrossRef] [PubMed]

45. Gagic, M.; Kociova, S.; Smerkova, K.; Michalkova, H.; Setka, M.; Svec, P.; Pribyl, J.; Masilko, J.; Balkova, R.; Heger, Z.; et al. One-pot synthesis of natural amine-modified biocompatible carbon quantum dots with antibacterial activity. *J. Colloid Interf. Sci.* **2020**, *580*, 30–48. [CrossRef]
46. Hao, X.; Huang, L.; Zhao, C.; Chen, S.; Lin, W.; Lin, Y.; Zhang, L.; Sun, A.; Miao, C.; Lin, X.; et al. Antibacterial activity of positively charged carbon quantum dots without detectable resistance for wound healing with mixed bacteria infection. *Mater. Sci. Eng. C* **2021**, *123*, 111971. [CrossRef]
47. Geng, B.; Li, P.; Fang, F.; Shi, W.; Glowacki, J.; Pan, D.; Shen, L. Antibacterial and osteogenic carbon quantum dots for regeneration of bone defects infected with multidrug-resistant bacteria. *Carbon* **2021**, *184*, 375–385. [CrossRef]
48. Yang, J.; Gao, G.; Zhang, X.; Ma, Y.H.; Chen, X.; Wu, F.G. One-step synthesis of carbon dots with bacterial contact-enhanced fluorescence emission: Fast Gram-type identification and selective Gram-positive bacterial inactivation. *Carbon* **2019**, *146*, 827–839. [CrossRef]
49. Wang, H.; Song, Z.; Gu, J.; Li, S.; Wu, Y.; Han, H. Nitrogen-doped carbon quantum dots for preventing biofilm formation and eradicating drug-resistant bacteria infection. *ACS Biomater. Sci. Eng.* **2019**, *5*, 4739–4749. [CrossRef]
50. Li, P.; Yang, X.; Zhang, X.; Pan, J.; Tang, W.; Cao, W.; Zhou, J.; Gong, X.; Xing, X. Surface chemistry-dependent antibacterial and antibiofilm activities of polyamine-functionalized carbon quantum dots. *J. Mater. Sci.* **2020**, *55*, 16744–16757. [CrossRef]
51. Zhao, D.; Zhang, Z.; Liu, X.; Zhang, R.; Xiao, X. Rapid and low-temperature synthesis of N, P co-doped yellow emitting carbon dots and their applications as antibacterial agent and detection probe to Sudan Red I. *Mater. Sci. Eng. C* **2021**, *119*, 111468. [CrossRef] [PubMed]
52. Zhao, D.; Zhang, R.; Liu, S.; Huang, X.; Xiao, X.; Yuan, L. One-step synthesis of blue-green luminescent carbon dots by a low-temperature rapid method and their high-performance antibacterial effect and bacterial imaging. *Nanotechnology* **2021**, *32*, 155101. [CrossRef] [PubMed]
53. Tang, W.; Li, P.; Zhang, G.; Yang, X.; Yu, M.; Lu, H.; Xing, X. Antibacterial carbon dots derived from polyethylene glycol/polyethyleneimine with potent anti-friction performance as water-based lubrication additives. *J. Appl. Polym. Sci.* **2021**, *138*, e50620. [CrossRef]
54. Zhao, C.; Wang, X.; Wu, L.; Wu, W.; Zheng, Y.; Lin, L.; Weng, S.; Lin, X. Nitrogen-doped carbon quantum dots as an antimicrobial agent against *Staphylococcus* for the treatment of infected wounds. *Colloids Surf. B* **2019**, *179*, 17–27. [CrossRef] [PubMed]
55. Anand, S.R.; Bhati, A.; Saini, D.; Gunture; Chauhan, N.; Khare, P.; Sonkar, S.K. Antibacterial nitrogen-doped carbon dots as a reversible "fluorescent nanoswitch" and fluorescent ink. *ACS Omega* **2019**, *4*, 1581–1591. [CrossRef]
56. Ye, Z.; Li, G.; Lei, J.; Liu, M.; Jin, Y.; Li, B. One-step and one-precursor hydrothermal synthesis of carbon dots with superior antibacterial activity. *ACS Appl. Bio Mater.* **2020**, *3*, 7095–7102. [CrossRef]
57. Saravanan, A.; Maruthapandi, M.; Das, P.; Ganguly, S.; Margel, S.; Luong, J.H.T.; Gedanken, A. Applications of N-doped carbon dots as antimicrobial agents, antibiotic carriers, and selective fluorescent probes for nitro explosives. *ACS Appl. Bio Mater.* **2020**, *3*, 8023–8031. [CrossRef]
58. Chu, X.; Wu, F.; Sun, B.; Zhang, M.; Song, S.; Zhang, P.; Wang, Y.; Zhang, Q.; Zhou, N.; Shen, J. Genipin cross-linked carbon dots for antimicrobial, bioimaging and bacterial discrimination. *Colloids Surf. B* **2020**, *190*, 110930. [CrossRef]
59. Li, P.; Yu, M.; Ke, X.; Gong, X.; Li, Z.; Xing, X. Cytocompatible amphipathic carbon quantum dots as potent membrane-active antibacterial agents with low drug resistance and effective inhibition of biofilm formation. *ACS Appl. Bio Mater.* **2022**, *5*, 3290–3299. [CrossRef]
60. Zhao, D.; Liu, X.; Zhang, R.; Xiao, X.; Li, J. Preparation of two types of silver-doped fluorescent carbon dots and determination of their antibacterial properties. *J. Inorg. Biochem.* **2021**, *214*, 111306. [CrossRef]
61. Liu, Y.; Xu, B.; Lu, M.; Li, S.; Guo, J.; Chen, F.; Xiong, X.; Yin, Z.; Liu, H.; Zhou, D. Ultrasmall Fe-doped carbon dots nanozymes for photoenhanced antibacterial therapy and wound healing. *Bioact. Mater.* **2022**, *12*, 246–256. [CrossRef] [PubMed]
62. Liu, M.; Huang, L.; Xu, X.; Wei, X.; Yang, X.; Li, X.; Wang, B.; Xu, Y.; Li, L.; Yang, Z. Copper doped carbon dots for addressing bacterial biofilm formation, wound infection, and tooth staining. *ACS Nano* **2022**, *16*, 9479–9497. [CrossRef] [PubMed]
63. Xu, Z.; He, H.; Zhang, S.; Wang, B.; Jin, J.; Li, C.; Chen, X.; Jiang, B.; Liu, Y. Mechanistic studies on the antibacterial behavior of Ag nanoparticles decorated with carbon dots having different oxidation degrees. *Environ. Sci. Nano* **2019**, *6*, 1168–1179. [CrossRef]
64. Raina, S.; Thakur, A.; Sharma, A.; Pooja, D.; Minhas, A.P. Bactericidal activity of *Cannabis sativa* phytochemicals from leaf extract and their derived carbon dots and Ag@carbon dots. *Mater. Lett.* **2020**, *262*, 127122. [CrossRef]
65. Wei, X.; Cheng, F.; Yao, Y.; Yi, X.; Wei, B.; Li, H.; Wu, Y.; He, J. Facile synthesis of a carbon dots and silver nanoparticles (CDs/AgNPs) composite for antibacterial application. *RSC Adv.* **2021**, *11*, 18417–18422. [CrossRef]
66. Moradlou, O.; Rabiei, Z.; Delavari, N. Antibacterial effects of carbon quantum dots@hematite nanostructures deposited on titanium against Gram-positive and Gram-negative bacteria. *J. Potoch. Photobio. A* **2019**, *379*, 144–149. [CrossRef]
67. Abd Elkodous, M.; El-Sayyad, G.S.; Youssry, S.M.; Nada, H.G.; Gobara, M.; Elsayed, M.A.; El-Khawaga, A.M.; Kawamura, G.; Tan, W.K.; El-Batal, A.I.; et al. Carbon-dot-loaded $Co_xNi_{1-x}Fe_2O_4$; $x = 0.9/SiO_2/TiO_2$ nanocomposite with enhanced photocatalytic and antimicrobial potential: An engineered nanocomposite for wastewater treatment. *Sci. Rep.* **2020**, *10*, 11534. [CrossRef]
68. Gao, X.; Ma, X.; Han, X.; Wang, X.; Li, S.; Yao, J.; Shi, W. Synthesis of carbon dot-ZnO-based nanomaterials for antibacterial application. *New J. Chem.* **2021**, *45*, 4496–4505. [CrossRef]

69. Chen, Y.; Cheng, X.; Wang, W.; Jin, Z.; Liu, Q.; Yang, H.; Cao, Y.; Li, W.; Fakhri, A.; Gupta, V.K. Preparation of carbon dots-hematite quantum dots-loaded hydroxypropyl cellulose-chitosan nanocomposites for drug delivery, sunlight catalytic and antimicrobial application. *J. Potoch. Photobio. B* **2021**, *219*, 112201. [CrossRef] [PubMed]
70. Nemera, D.J.; Etefa, H.F.; Kumar, V.; Dejene, F.B. Hybridization of nickel oxide nanoparticles with carbon dots and its application for antibacterial activities. *Luminescence* **2022**, *37*, 965–970. [CrossRef] [PubMed]
71. Gao, X.; Li, H.; Niu, X.; Zhang, D.; Wang, Y.; Fan, H.; Wang, K. Carbon quantum dots modified Ag_2S/CS nanocomposite as effective antibacterial agents. *J. Inorg. Biochem.* **2021**, *220*, 111456. [CrossRef] [PubMed]
72. Luo, Q.; Qin, K.; Liu, F.; Zheng, X.; Ding, Y.; Zhang, C.; Xu, M.; Liu, X.; Wei, Y. Carbon dots derived from kanamycin sulfate with antibacterial activity and selectivity for Cr^{6+} detection. *Analyst* **2021**, *146*, 1965–1972. [CrossRef] [PubMed]
73. Liang, J.; Li, W.; Chen, J.; Huang, X.; Liu, Y.; Zhang, X.; Shu, W.; Lei, B.; Zhang, H. Antibacterial activity and synergetic mechanism of carbon dots against Gram-positive and -negative bacteria. *ACS Appl. Bio Mater.* **2021**, *4*, 6937–6945. [CrossRef]
74. Wu, L.N.; Yang, Y.J.; Huang, L.X.; Zhong, Y.; Chen, Y.; Gao, Y.R.; Lin, L.Q.; Lei, Y.; Liu, A.L. Levofloxacin-based carbon dots to enhance antibacterial activities and combat antibiotic resistance. *Carbon* **2022**, *186*, 452–464. [CrossRef]
75. Zhao, C.; Wu, L.; Wang, X.; Weng, S.; Ruan, Z.; Liu, Q.; Lin, L.; Lin, X. Quaternary ammonium carbon quantum dots as an antimicrobial agent against Gram-positive bacteria for the treatment of MRSA-infected pneumonia in mice. *Carbon* **2020**, *163*, 70–84. [CrossRef]
76. Chu, X.; Zhang, P.; Liu, Y.; Sun, B.; Huang, X.; Zhou, N.; Shen, J.; Meng, N. A multifunctional carbon dot-based nanoplatform for bioimaging and quaternary ammonium salt/photothermal synergistic antibacterial therapy. *J. Mater. Chem. B* **2022**, *10*, 2865–2874. [CrossRef] [PubMed]
77. Liu, W.; Gu, H.; Ran, B.; Liu, W.; Sun, W.; Wang, D.; Du, J.; Fan, J.; Peng, X. Accelerated antibacterial red-carbon dots with photodynamic therapy against multidrug-resistant *Acinetobacter baumannii*. *Sci. China Mater.* **2022**, *65*, 845–854. [CrossRef] [PubMed]
78. Li, C.; Lin, F.; Sun, W.; Wu, F.G.; Yang, H.; Lv, R.; Zhu, Y.X.; Jia, H.R.; Wang, C.; Gao, G.; et al. Self-assembled rose bengal-exopolysaccharide nanoparticles for improved photodynamic inactivation of bacteria by enhancing singlet oxygen generation directly in the solution. *ACS Appl. Mater. Interfaces* **2018**, *10*, 16715–16722. [CrossRef] [PubMed]
79. Lin, F.; Bao, Y.W.; Wu, F.G. Improving the phototherapeutic efficiencies of molecular and nanoscale materials by targeting mitochondria. *Molecules* **2018**, *23*, 3016. [CrossRef] [PubMed]
80. LeCroy, G.E.; Yang, S.T.; Yang, F.; Liu, Y.; Shiral Fernando, K.A.; Bunker, C.E.; Hu, Y.; Luo, P.G.; Sun, Y.P. Functionalized carbon nanoparticles: Syntheses and applications in optical bioimaging and energy conversion. *Coord. Chem. Rev.* **2016**, *320–321*, 66–81. [CrossRef]
81. Venkateswarlu, S.; Viswanath, B.; Reddy, A.S.; Yoon, M. Fungus-derived photoluminescent carbon nanodots for ultrasensitive detection of Hg^{2+} ions and photoinduced bactericidal activity. *Sens. Actuators B Chem.* **2018**, *258*, 172–183. [CrossRef]
82. Ristic, B.Z.; Milenkovic, M.M.; Dakic, I.R.; Todorovic-Markovic, B.M.; Milosavljevic, M.S.; Budimir, M.D.; Paunovic, V.G.; Dramicanin, M.D.; Markovic, Z.M.; Trajkovic, V.S. Photodynamic antibacterial effect of graphene quantum dots. *Biomaterials* **2014**, *35*, 4428–4435. [CrossRef] [PubMed]
83. Stanković, N.K.; Bodik, M.; Šiffalovič, P.; Kotlar, M.; Mičušik, M.; Špitalsky, Z.; Danko, M.; Milivojević, D.D.; Kleinova, A.; Kubat, P.; et al. Antibacterial and antibiofouling properties of light triggered fluorescent hydrophobic carbon quantum dots Langmuir–Blodgett thin films. *ACS Sustain. Chem. Eng.* **2018**, *6*, 4154–4163. [CrossRef]
84. Yadav, P.; Nishanthi, S.T.; Purohit, B.; Shanavas, A.; Kailasam, K. Metal-free visible light photocatalytic carbon nitride quantum dots as efficient antibacterial agents: An insight study. *Carbon* **2019**, *152*, 587–597. [CrossRef]
85. Nie, X.; Jiang, C.; Wu, S.; Chen, W.; Lv, P.; Wang, Q.; Liu, J.; Narh, C.; Cao, X.; Ghiladi, R.A.; et al. Carbon quantum dots: A bright future as photosensitizers for in vitro antibacterial photodynamic inactivation. *J. Potoch. Photobio. B* **2020**, *206*, 111864. [CrossRef]
86. Wang, X.; Lu, Y.; Hua, K.; Yang, D.; Yang, Y. Iodine-doped carbon dots with inherent peroxidase catalytic activity for photocatalytic antibacterial and wound disinfection. *Anal. Bioanal. Chem.* **2021**, *413*, 1373–1382. [CrossRef] [PubMed]
87. Liu, D.; Yang, M.; Liu, X.; Yu, W. Zinc-doped carbon dots as effective blue-light-activated antibacterial agent. *Nano* **2021**, *16*, 2150031. [CrossRef]
88. Nichols, F.; Lu, J.E.; Mecado, R.; Rojas-Andrade, M.D.; Ning, S.; Azhar, Z.; Sandhu, J.; Cazares, R.; Saltikov, C.; Chen, S. Antibacterial activity of nitrogen-doped carbon dots enhanced by atomic aispersion of copper. *Langmuir* **2020**, *36*, 11629–11636. [CrossRef]
89. Li, H.; Tian, H.; Yu, L.; Gao, Z.; Yang, Y.; Li, W. Preparation of terbium doped carbon dots and their antibacterial capacity against *Escherichia coli*. *Acta Agric. Zhejiangensis* **2020**, *32*, 291–298.
90. Yan, H.; Zhang, B.; Zhang, Y.; Su, R.; Li, P.; Su, W. Fluorescent carbon dot-curcumin nanocomposites for remarkable antibacterial activity with synergistic photodynamic and photothermal abilities. *ACS Appl. Bio Mater.* **2021**, *4*, 6703–6718. [CrossRef] [PubMed]
91. Zhang, P.; Sun, B.; Wu, F.; Zhang, Q.; Chu, X.; Ge, M.; Zhou, N.; Shen, J. Wound healing acceleration by antibacterial biodegradable black phosphorus nanosheets loaded with cationic carbon dots. *J. Mater. Sci.* **2021**, *56*, 6411–6426. [CrossRef]
92. Yan, Y.; Kuang, W.; Shi, L.; Ye, X.; Yang, Y.; Xie, X.; Shi, Q.; Tan, S. Carbon quantum dot-decorated TiO_2 for fast and sustainable antibacterial properties under visible-light. *J. Alloys Compd.* **2019**, *777*, 234–243. [CrossRef]

93. Jin, C.; Su, K.; Tan, L.; Liu, X.; Cui, Z.; Yang, X.; Li, Z.; Liang, Y.; Zhu, S.; Yeung, K.W.K. Near-infrared light photocatalysis and photothermy of carbon quantum dots and Au nanoparticles loaded titania nanotube array. *Mater. Des.* **2019**, *177*, 107845. [CrossRef]
94. Kuang, W.; Zhong, Q.; Ye, X.; Yan, Y.; Yang, Y.; Zhang, J.; Huang, L.; Tan, S.; Shi, Q. Antibacterial nanorods made of carbon quantum dots-ZnO under visible light irradiation. *J. Nanosci. Nanotechno.* **2019**, *19*, 3982–3990. [CrossRef]
95. Gao, D.; Zhao, P.; Lyu, B.; Li, Y.; Hou, Y.; Ma, J. Carbon quantum dots decorated on ZnO nanoparticles: An efficient visible-light responsive antibacterial agents. *Appl. Organomet. Chem.* **2020**, *34*, e5665. [CrossRef]
96. Marković, Z.M.; Kováčová, M.; Humpolíček, P.; Budimir, M.D.; Vajd'ák, J.; Kubát, P.; Mičušík, M.; Švajdlenková, H.; Danko, M.; Capáková, Z.; et al. Antibacterial photodynamic activity of carbon quantum dots/polydimethylsiloxane nanocomposites against *Staphylococcus aureus*, *Escherichia coli* and *Klebsiella pneumoniae*. *Photodiagnosis Photodyn. Ther.* **2019**, *26*, 342–349. [CrossRef]
97. Kováčová, M.; Bodík, M.; Mičušík, M.; Humpolíček, P.; Šiffalovič, P.; Špitalský, Z. Increasing the effectivity of the antimicrobial surface of carbon quantum dots-based nanocomposite by atmospheric pressure plasma. *Clin. Plasma Med.* **2020**, *19–20*, 100111. [CrossRef]
98. Ghosal, K.; Kováčová, M.; Humpolíček, P.; Vajd'ák, J.; Bodík, M.; Špitalský, Z. Antibacterial photodynamic activity of hydrophobic carbon quantum dots and polycaprolactone based nanocomposite processed via both electrospinning and solvent casting method. *Photodiagnosis Photodyn. Ther.* **2021**, *35*, 102455. [CrossRef]
99. Budimir, M.; Marković, Z.; Vajdak, J.; Jovanović, S.; Kubat, P.; Humpolíček, P.; Mičušik, M.; Danko, M.; Barras, A.; Milivojević, D.; et al. Enhanced visible light-triggered antibacterial activity of carbon quantum dots/polyurethane nanocomposites by gamma rays induced pre-treatment. *Radiat. Phys. Chem.* **2021**, *185*, 109499. [CrossRef]
100. Belkahla, H.; Boudjemaa, R.; Caosi, V.; Pineau, D.; Curcio, A.; Lomas, J.S.; Decorse, P.; Chevillot-Biraud, A.; Azaïs, T.; Wilhelm, C.; et al. Carbon dots, a powerful non-toxic support for bioimaging by fluorescence nanoscopy and eradication of bacteria by photothermia. *Nanoscale Adv.* **2019**, *1*, 2571. [CrossRef] [PubMed]
101. Yan, H.; Ni, H.; Yang, Y.; Shan, C.; Yang, X.; Li, X.; Cao, J.; Wu, W.; Liu, W.; Tang, Y. Smart nanoprobe based on two-photon sensitized terbium-carbon dots for dual-mode fluorescence thermometer and antibacterial. *Chin. Chem. Lett.* **2020**, *31*, 1792–1796. [CrossRef]
102. Yan, X.; Yang, J.; Wu, F.; Su, H.; Sun, G.; Ni, Y.; Sun, W. Antibacterial carbon dots/iron oxychloride nanoplatform for chemodynamic and photothermal therapy. *Colloids Interface Sci. Commun.* **2021**, *45*, 100552. [CrossRef]
103. Mazumdar, A.; Haddad, Y.; Milosavljevic, V.; Michalkova, H. Guran, R.; Bhowmick, S.; Moulick, A. Peptide-carbon quantum dots conjugate, derived from human retinoic acid receptor responder protein 2, against antibiotic-resistant Gram positive and Gram negative pathogenic bacteria. *Nanomaterials* **2020**, *10*, 325. [CrossRef] [PubMed]
104. Liu, S.; Lv, K.; Chen, Z.; Li, C.; Chen, T.; Ma, D. Fluorescent carbon dots with a high nitric oxide payload for effective antibacterial activity and bacterial imaging. *Biomater. Sci.* **2021**, *9*, 6486–6500. [CrossRef] [PubMed]
105. Mahat, N.A.; Shamsudin, S.A.; Jullok, N.; Ma'Radzi, A.H. Carbon quantum dots embedded polysulfone membranes for antibacterial performance in the process of forward osmosis. *Desalination* **2020**, *493*, 114618. [CrossRef]
106. Koulivand, H.; Shahbazi, A.; Vatanpour, V.; Rahmandoost, M. Novel antifouling and antibacterial polyethersulfone membrane prepared by embedding nitrogen-doped carbon dots for efficient salt and dye rejection. *Mater. Sci. Eng. C* **2020**, *111*, 110787. [CrossRef]
107. Kousheh, S.A.; Moradi, M.; Tajik, H.; Molaei, R. Preparation of antimicrobial/ultraviolet protective bacterial nanocellulose film with carbon dots synthesized from lactic acid bacteria. *Int. J. Biol. Macromol.* **2020**, *155*, 216–225.
108. Riahi, Z.; Rhim, J.W.; Bagheri, R.; Pircheraghi, G.; Lotfali, E. Carboxymethyl cellulose-based functional film integrated with chitosan-based carbon quantum dots for active food packaging applications. *Prog. Org. Coat.* **2022**, *166*, 106794. [CrossRef]

Review

Photodynamic Anti-Bacteria by Carbon Dots and Their Nano-Composites

Xiaoyan Wu [1], Khurram Abbas [1], Yuxiang Yang [1], Zijian Li [2], Antonio Claudio Tedesco [1,3] and Hong Bi [1,2,*]

1. School of Chemistry and Chemical Engineering, Anhui University, Hefei 230601, China; wuxiaoyan1213@163.com (X.W.); abbaskhurram93@gmail.com (K.A.); yyx18715093366@163.com (Y.Y.); atedesco@usp.br (A.C.T.)
2. School of Materials Science and Engineering, Anhui University, Hefei 230601, China; 22018@ahu.edu.cn
3. Department of Chemistry, Center of Nanotechnology and Tissue Engineering-Photobiology and Photomedicine Research Group, Faculty of Philosophy, Sciences and Letters of Ribeirão Preto, University of São Paulo, Ribeirão Preto, São Paulo 14040-901, Brazil
* Correspondence: bihong@ahu.edu.cn; Tel.: +86-551-63861279

Abstract: The misuse of many types of broad-spectrum antibiotics leads to increased antimicrobial resistance. As a result, the development of a novel antibacterial agent is essential. Photodynamic antimicrobial chemotherapy (PACT) is becoming more popular due to its advantages in eliminating drug-resistant strains and providing broad-spectrum antibacterial resistance. Carbon dots (CDs), zero-dimensional nanomaterials with diameters smaller than 10 nm, offer a green and cost-effective alternative to PACT photosensitizers. This article reviewed the synthesis methods of antibacterial CDs as well as the recent progress of CDs and their nanocomposites in photodynamic sterilization, focusing on maximizing the bactericidal impact of CDs photosensitizers. This review establishes the base for future CDs development in the PACT field.

Keywords: carbon dots; antimicrobial; light activation; photodynamic effect; reactive oxygen species

1. Introduction

Infections caused by fungi, bacteria, parasites, or viruses cause many severe diseases. Our healthcare systems face substantial problems, from treatment needs to prevention in hospital settings and routine work dealing with many critical pathologies, food and water environments and sources protection, and worldwide public health impact [1,2]. Antibiotics have historically been the primary weapon in the fight against infectious diseases. However, due to the high cost and long pathways to new drugs discovery, clinical testing, and scaling up the production process, approval of the development of next-generation antibiotics takes longer [3]. Additionally, bacteria have several ways of rapidly acquiring resistance, which can endanger the health of patients and delay wound recovery after treatment [4]. As a result of multidrug resistance (MDR), many of these diseases will become more challenging to treat and result in higher medical costs and mortality rates [5–7]. Since the appearance of multidrug resistance in pathogenic bacteria, traditional antibiotics/antimicrobials cannot meet the expectations of today's society and the urgent needs to efficiently prevent and treat a considerable spectrum of bacterial infections [8]. It is imperative to find and develop alternative antibacterial techniques to combat MDR effectively and to prevent and treat diseases and their undesirable side effects. In environmental contamination, the consequences and expenses to eliminate the impact could be worse and could take a year and in some cases decades [9].

Photodynamic inactivation of bacteria mediated by photoactive compounds, more precisely photosensitizer molecules (PSs), is one of the most promising techniques in the fight against MDR pathogens [10], such that as used and developed in remote ancient Egypt approximately 4000 years ago, when a skin disease such as vitiligo was treated by a combination of plants orally administered and exposure of the patients to sunlight. The successes of the treatment were a result of photodynamic reactions mediated by a natural product present in the extract of *Ammimajus*, a furanocoumarin and a psoralen. Photodynamic antimicrobial chemotherapy (PACT) is a fast, intense and challenging field that has been developed to address the growing antibiotic resistance among harmful bacteria [11]. It was developed in response to the need for better treatment and prevention of bacterial infectious diseases.

Several issues are involved in PACT, including the design and choice of the nanostructured photoactive molecules and their isolation or synthetic route to make it feasible to penetrate the cellular cytoplasm or induce specific damage to the cellular organelles in the target tissue [12]. Fungi are eukaryotic microorganisms, similar to mammalian cells, and the development of new antifungal drugs remains challenging due to a number of reasons, such as the presence of the nucleus and the structure of the cell wall. Bacteria are prokaryote microorganisms that can be easily distinguished from mammalian cells. Conversely, fungal diseases are usually caused by pathogens (fungi), and these fungal diseases feature a variety of symptoms that are commonly related to the attack of the skin and respiratory systems. Fungi and bacteria can form biofilms versus staying in their planktonic forms, increasing their drug resistance [13].

Carbon dots (CDs) have been proposed as a potential fluorescent nanomaterial for identifying and inactivating different types of bacterial species among a wide variety of PSs, already used in the past [14,15]. CDs are carbon-based nanomaterials that are quasi-spherical in shape and have a typical size of less than 10 nm. They have good photoelectric properties, high water solubility, and chemical durability. CDs also present low toxicity and have good biocompatibility, making them ideal for bioimaging [16–20], drug delivery [21], gene delivery [22], biosensors [21] and fluorescent-labeling applications [23,24]. CDs are well known to undergo optical absorption via π-plasmon transitions [9]. In contrast, fluorescence emission occurs in the visible to the near-infrared spectral range due to photogenerated holes and electrons trapped at different surface sites and associated radiative recombination [25]. CDs exhibit powerful photodynamic effects due to their optical properties [26], which have been exploited to kill bacterial and cancer cells under visible light irradiation [27].

In this review, we summarize the most common synthetic methods for producing CDs and CDs nanocomposites, their application in photodynamic antimicrobial applications in recent years, and the factors and improvements affecting the antimicrobial effectiveness of CDs.

2. Synthesis Techniques of Carbon Dots Employed for Antimicrobials

The properties of CDs are closely related to their preparation methods [28]. Top-down and bottom-up approaches are two commonly used approaches for preparing CDs [29,30]. Carbon quantum dots' final physicochemical and functional properties, including their photophysical behaviors, biocompatibility and antibacterial activity, are influenced by the method employed and the carbon source used during the synthetic process [31].

Using a top-down technique, large-sized carbon materials, such as carbon nanotubes and graphite ash, are decomposed into small CDs, from the macro to the nanoscale. Different carbon sources are exposed to laser ablation, arc discharge, plasma treatment, chemical oxidation, electrochemical oxidation, and others [32–36]. Using different types of acid treatment, the concentration of the oxygen-containing groups attached to the CDs structure can be easily changed. However, doping additional materials onto CDs is tricky and is a powerful option to potentialize the CDs' nanomaterial. Furthermore, the strong acid may cause CDs to lose their conjugated structure, changing their photophysical properties, resulting in lower absorption and emission wavelengths [37].

Chemical processes such as hydrothermal, pyrolysis, combustion, ultrasonic, microwave irradiation, thermal, and biogenic procedures, conversely, are used in the bottom-up approach [38,39]. CDs can also be prepared using non-graphite carbon sources such as tiny polymers and monomers as carbon precursors [40,41]. Using this method, a wide range of elements, including N, P, S, B, and even metal ions, can be doped into the CDs' structure [42]. The addition of heteroatoms to the CDs structure improves the fluorescent properties of these nanomaterials by changing the absorption and emission peak positions and boosting the fluorescent quantum yield. Wu et al. found a link between CDs' photo-oxidation activity, phosphorescent quantum yield, and N content, underlining the importance of N-doping in boosting CDs' photosensitization performance [43]. Marković et al. found that the photodynamic antibacterial properties directly impact the ROS production by the CDs doping process with F and Cl compared with undoped nanoparticles [44]. It is essential to optimize the photodynamic antibacterial effect of CDs by choosing suitable precursors and the proper selection of doping elements. However, doping sites and better concentrations are still challenging to manage and archive.

Unfortunately, many of the current methods require toxic chemicals and solvents, high temperatures, long reaction times and complex processing steps. Therefore, the development of green chemistry concepts to manufacturing fluorescent CDs through simple, economical and sustainable pathways represents a meaningful topic [45]. More recently, efforts have been devoted to utilization of green carbon sources and the development of green synthesis processes. For the former, biomass, which is renewable organic material that comes from plants and animals, represents a typical green carbon source. The use of biomass for CDs synthesis, especially in large-scale production, is attracting increasing attention among researchers. For the green synthesis process, toxic chemicals that are harmful to people's health and the environment should be avoided. In addition, the preparation procedure, reaction time and conditions should be optimized to increase economic efficiency. Currently, the main green synthesis processes for producing CDs include ultrasonication, microwave irradiation, hydrothermal carbonization, self-exothermic synthesis, and ozone/hydrogen peroxide oxidation [46].

3. Carbon Dots in Antimicrobial Photodynamic Therapy

The currently known antibacterial mechanism of CDs is shown in Figure 1. CDs with positively charged surfaces interact electrostatically with negatively charged bacteria, facilitating CDs internalization and killing bacteria [47,48] This CDs behavior is critical for the success of the PACT. As observed in the past, bacteria and biofilm frameworks present a tremendous challenge in the treatment protocol design, considering the difficulty of the photoactive compounds penetrating through the exopolysaccharide matrix [49]. Neutral or negatively charged photoactive compounds have been proposed to treat Gram-negative bacteria without success. The low permeability of the Gram-negative outer membrane avoids effective incorporation of molecules, reducing the light activation effect. Previous examples of treatment of the tissue with biological or chemical agents, such as $CaCl_2$ or Tris-EDTA, which are expected to increase the release of the molecules by up to 50% of the outer membrane lipopolysaccharide present as desired, but they showed undesirable side effects in a clinical trial [50,51].

Bacteria can also be killed by disintegrating bacterial cell walls, resulting in cytoplasmic material leaking [52], which could induce secondary side effects. Furthermore, the higher temperature caused by the photothermal therapy (PTT) effect or the release of ROS [53] by the PACT effect can directly damage the bacterial DNA and proteins, leading to a bluster effect. CDs are also valued for their capacity to produce highly active ROS. CDs' present a visible and near-infrared spectrum of light absorbance, destroying bacteria through classical mechanisms' photoinduced production of ROS.

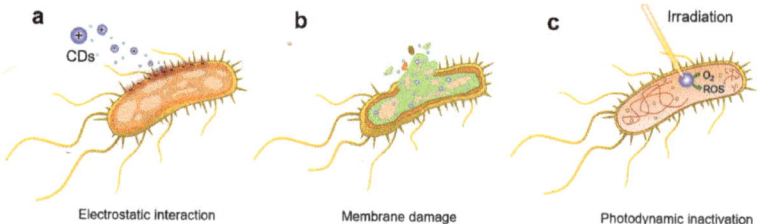

Figure 1. General bactericidal mechanisms of action of CDs. (**a**) Schematic representation of the initial electrostatic interaction between CDs and the bacterial cell wall. (**b**) CDs internalization, intercalation in the bacterial membrane, and irreversible disruption with a leak of cytoplasmatic material. (**c**) CDs-promoted bacterial photodynamic inactivation with ROS production and DNA damage. Reproduced with permission from Ref. [10]. Copyright © 2021 Nanomaterials MDPI.

3.1. Photosensitization Mechanisms

PDT was discovered in the 20th century, has received much attention, and has established a background in cancer treatment worldwide because of its advantages, such as fewer side effects, less invasive surgery, and repeatable treatment [54–56]. It has also been used as adjunctive therapy and works synergistically with the classical approaches for cancer. The classical photophysical and photochemical steps involved in PDT are summarized by Jablonski Diagram. Basically, the PSs in the ground non-excited state (S_0) can absorb visible light (photons), moving to the first electronically excited singlet state (S_1), and then by intersystem crossing (ISC), produce, by spin inversion, the first excited triplet state (T_1). The excited triplet state (T_1) has a long lifetime (around milliseconds)—enough to move forward by energy and electron transfer processes. The deactivation of the first excited state was sometimes a fast and helpful process, defined by a fluorescence emission used in many cases for diagnostic. The excited triplet state (T_1) deactivation is a phosphorescence process with long-time light emission. It can also be deactivated by a thermal decay that can also be used to disperse the energy of the excited PS. Three classical photoreductions govern their photoreaction on PDT and photon energy transfer mechanisms (Type I, Type II, and Type III (Figure 2)) found today [57–59]. Photoactive compounds acting by the Type I mechanism are primarily based on hydrogen atoms or electrons that transfer from the excited photosensitizer molecules *($T1$) states, react with oxygen, and produce ROS products such as $HO^\bullet$, $O_2^{\bullet -}$, and H_2O_2 [60–63].

Photosensitization mechanism Type I-Redox reactions (1–4) with biomolecules [64].

$$^0S \rightarrow {}^1S \rightarrow {}^3S \qquad (1)$$

$$^3S + Sub \rightarrow Sub^{+\bullet} + Sub^{-\bullet} \text{ (electron transfer)}, \qquad (2)$$

$$S^{-\bullet} + {}^3O_2 \rightarrow Sub + O_2^{\bullet -} \rightarrow HO^\bullet + HO^- \qquad (3)$$

$$S = \text{photosensitizer molecule and Sub = organic substrate.} \qquad (4)$$

Conversely, photoactive compounds operating by Type II are propelled by electron spin exchange between the photosensitizer *(T_1) and triplet oxygen ($^3O_2$), which results in the T_1–T_1 annihilation process and the formation of non-radical but highly reactive singlet oxygen ($^1O_2$).

Mechanism photosensitizing Type II-mediated production $^1O_2$, such as lipid peroxidation, is based on the above reaction (1) [64] and then follows reactions (5–6).

$$^3S + {}^3O_2 \rightarrow {}^0S + {}^1O_2 \text{ (energy transfer)}, \qquad (5)$$

$$^1O_2 + Sub \rightarrow Sub\text{-OOH(peroxides, etc.)}, \qquad (6)$$

The reactive oxygen species produced by the photosensitization Type I mechanism are constantly made in the living organisms at a low level [65–67]. Since these species are highly reactive, the microorganisms need protective systems to neutralize the action of these radicals. This protective system includes enzymes such as superoxide dismutase (SOD) that have its activity over the superoxide anion radical, as well as the catalase and the peroxidase, which control the harmful effects of hydrogen peroxide:

$$O_2^{\bullet-} + O_2^{\bullet-} + 2H^+ \xrightarrow{SOD} H_2O_2 \quad (7)$$

$$2H_2O_2 \xrightarrow{catalase} H_2O + O_2 \quad (8)$$

$$H_2O_2 + 2H^+ \xrightarrow{peroxidase} 2H_2O \quad (9)$$

In the Type II mechanism, energy is transferred directly to the molecular oxygen. The PSs return to the ground state after an absorption cycle and active oxygen generation. The photodynamic effect is directly affected by the single-linear oxygen yield in the Type II mechanism [68–71].

PSs should also be selectively targeted biomolecules such as nucleic acids, proteins, and other macromolecules by a Type III mechanism. When paired with Type III photosensitizers, the PSs can directly and effectively destroy the exciting biological target molecule [59].

Few PSs can be employed in Type III PDT due to the stringent requirements for photosensitizers. CDs have not yet been proven to work by Type III PDT but remain an open challenge. Type II PDT may be the most studied of the three mechanisms, and $^1O_2$ QY is a valuable metric of photophysical properties for assessing photodynamic performance [72,73]. The better the photodynamic effect, the higher the $^1O_2$ QY. Several methods for detecting $^1O_2$ include electron spin resonance (ESR), fluorescence, UV–visible absorption indirect method (classical Donors–Acceptor quenching process) and by time-resolved methods such as flash photolysis and near-infrared (NIR) $^1O_2$ detection. Fluorescence and UV–visible absorption can be used to calculate the QY of $^1O_2$ [74,75].

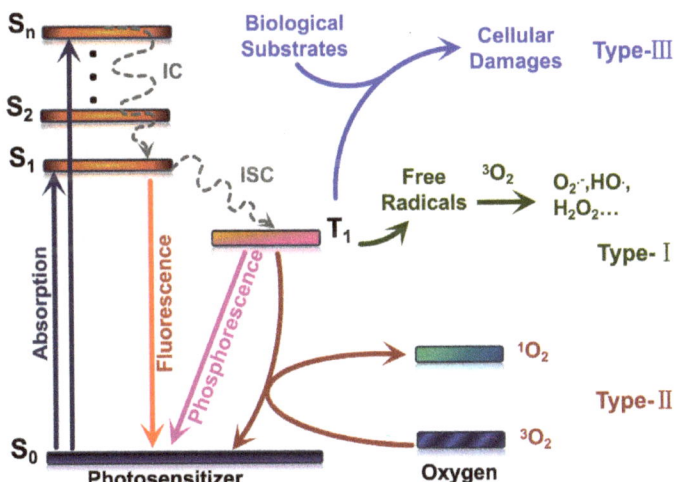

Figure 2. Schematic illustration of three types of mechanisms in PDT. Reproduced with permission from Ref. [59]. Copyright © 2022 Chem Elsevier B.V.

3.2. Photodynamic Anti-Bacteria by Carbon Dots

PACT can damage DNA, oxidize amino acids, inactivate enzymes, and kill bacteria by releasing ROS. Because of its features of low cumulative toxicity, high spatial targeting processes, and drug resistance independence, PACT is considered a practical approach for antibacterial applications. Under visible and NIR light irradiation, exposed CDs produce ROS by electrons or by an energy transfer process, leading to ROS, employed as a new photosensitive nano-agent in PACT against microorganisms [76–79].

Some CDs, such as graphene [76] and mushroom CDs [78], have intrinsic photodynamic characteristics that do not require extra surface changes or doping processes. Trajkovic et al. used an electrochemical approach to make graphene quantum dots that can produce reactive oxygen species under photoexcitation (470 nm, 1 W) and kill Gram-negative bacteria such as *Escherichia coli* and methicillin-resistant *Staphylococcus aureus* (MRSA) [76]. As shown in Figure 3a, Yoon et al. used mushrooms as a raw material to make carbon dots (MCDs) with high blue fluorescence, with the most significant emission at 456 nm under the excitation of 360 nm UV light. Under LED, visible light illumination (2.70 mW cm^{-2}), MCDs can produce ROS such as hydroxyl radicals and superoxide radicals, which can directly adhere to the surface of *Escherichia coli* (*E. coli*) cells and induce cell membrane damage [78]. Conversely, CDs have an effective photodynamic action with the advantages of photostability, non-toxicity, and high quantum yield of fluorescence, and use citric acid as the sole carbon source in the production of CDs. The entire process is considered a green and inexpensive environmental safety process and material production with many medical applications [78]. Bagnato et al. used citric acid as a raw material to prepare CDs that displayed an antibacterial photodynamic effect. It emits its maximum light at 530 nm when excited by light at 450 nm. CDs-mediated PACT removes *Staphylococcus aureus* (*S. aureus*) suspensions and biofilms, as shown in Figure 3b, and is an effective, affordable, and simple PACT reagent that may be utilized in both in vitro and in vivo studies [80].

However, UV–vis light can induce side effects and damage to the human body and has a low penetrability to tissue. Still, near-infrared light (NIR, 780–1700 nm) has an advantage in PACT because of its longer wavelength, less scattering, tissue absorption, and more importantly, higher penetration efficiency of biological tissue [81,82]. Pu et al. obtained graphene oxide (GO) sheets. As shown in Figure 3c, they prepared two-photon GQDs capable of producing ROS through ultrasonic shearing, with high two-photon absorption, a large two-photon excitation absolute cross-section (TPE), and two-photon solid luminescence of the NIR. It is possible to perform two-photon bioimaging and two-photon photodynamic therapy on both Gram-negative and Gram-positive bacterial [83].

Although some organic photosensitizers have a high singlet oxygen yield (i.e., rose benga (RB, 75%), methylene blue (MB, 52%), indocyanine green (ICG, 0.80%)) [55,84–86], CDs have outstanding advantages such as biocompatibility, water solubility, targeting specificity, NIR absorption, fast and reliable synthetic processes and many others features. Hence, CDs photosensitizers continue to have promising antibacterial uses. Some organic PSs, such as curcumin and riboflavin, have been under evaluation for decades and present remarkable photostability but poor water solubility, reducing their photodynamic effect [87,88]. Su et al. used a hydrothermal approach to create curcumin carbon dots (Cur-NRCDs) with imaging and antibacterial properties. They used curcumin (Cur), neutral red (NR), and citrate (CA) as raw materials. Under 405 nm excitation, Cur-NRCDs fluoresced brightly red. Cur-NRCDs had better photosensitivity than Cur. Cur-NRCDs have outstanding antibacterial activity, cytocompatibility, photostability, and ROS efficiency. Under xenon lamp irradiation, Cur-NRCDs can inactivate 100% *E. coli* and *S. aureus* at 15 and 10 mM concentrations, respectively [88].

Several investigations have been conducted in recent years to improve the antibacterial efficacy of CDs photosensitizers [9]. As previously stated, increasing the nitrogen concentration in CDs helps to strengthen the PACT effect. Kuo et al. prepared graphene oxide sheets using a modified Hummers process and produced N-GQDs using an ultrasonic shear reaction method. N-GQDs have a bright PL emission spectrum in the near-infrared region, at 728 nm. Their superior luminescence properties and photostability make them a viable contrast agent for bacteria tracking in bioimaging techniques. Under 670 nm (0.10 W cm^{-2}) laser irradiation, a considerable amount of $^1O_2$ and $O_2^{\bullet-}$ may be created simultaneously, and the effect of killing *E. coli* can approach 100% efficacy in only three minutes of light exposure. They also discovered that components with higher nitrogen content in graphene quantum dots could perform photodynamic therapy more effectively after the same treatment than components with lower nitrogen bonding content, indicating that future clinical applications, particularly for multidrug-resistant bacteria, are possible [89]. Probably, the mechanism of action works by synergic production of ROS and RNO (reactive nitrogen species).

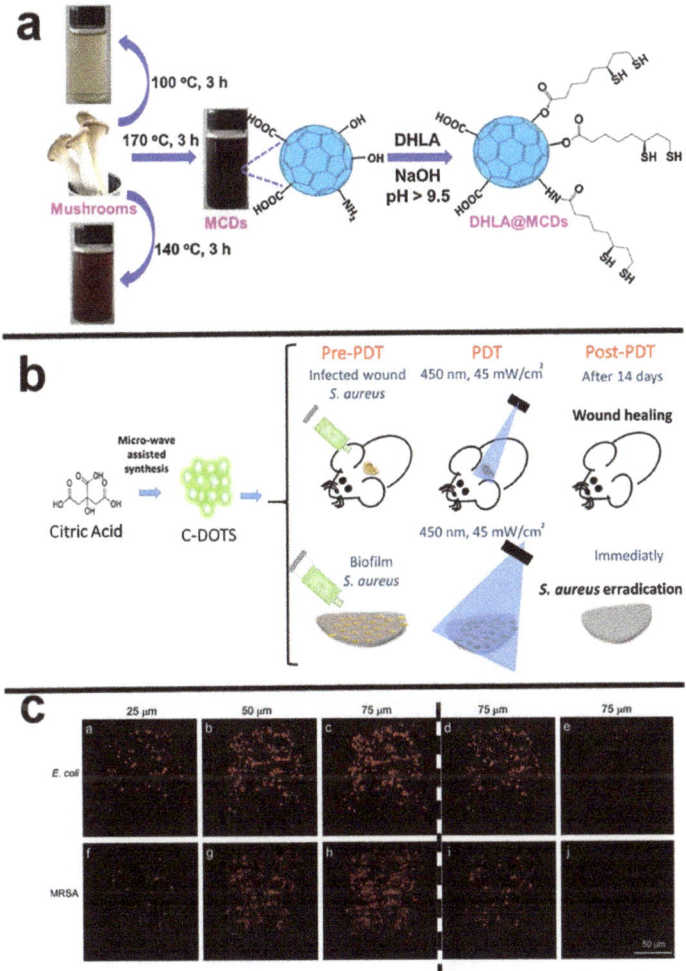

Figure 3. (a) Schematic illustration of synthesis of MCDs by thermal treatment at various temperatures and surface functionalization with DHLA. Reproduced with permission from Ref. [78]. Copyright © 2017, *Sensors and Actuators B: Chemical*, Elsevier B.V. (b) Brief description of the procedures carried out in this study. In vivo and in vitro PACT studies evaluated PACT mediated by CDs and blue LED light against *S. aureus*. Reproduced with permission from Ref. [80]. Copyright © 2021 *Frontiers in Microbiology*, Romero, Alves, Stringasci, Buzzá, Ciol, Inada and Bagnato. (c) Two-photon luminescence images of *E. coli* and MRSA Excitation wavelength: 800 nm. Delivered dose: $OD_{600} \sim 0.05$ of bacteria and 0.50 µg mL^{-1} GQD-Ab$_{LPS}$ and 0.50 µg mL^{-1} GQD-Ab$_{protein\ A}$, respectively. Reproduced with permission from Ref. [83]. Copyright © 2022 *ACS Applied Materials & Interfaces*, American Chemical Society.

Optimizing the photophysical characteristics of CDs molecules has been an essential strategy to improve the efficiency of PACT, as seen by increased fluorescence and phosphorescence quantum yields. Wu et al. used citric acid and ethylenediamine as raw ingredients to make a variety of nitrogen-doped carbon dots (CDs). The phosphorescent quantum yield was positively associated with the CDs' photosensitization ability. From the standpoint of material design, a better photosensitizer can be obtained from the cross between systems from an excited singlet state to an excited triplet state (Figure 4). When the designed CDs were applied to photodynamic antibiotics, the inhibition rates of *Salmonella* and *E. coli* (92% and 86%) were much higher than the growth inhibition rates of phloxine B (40% and 55%), demonstrating the excellent photodynamic antibacterial effect of these CDs [43].

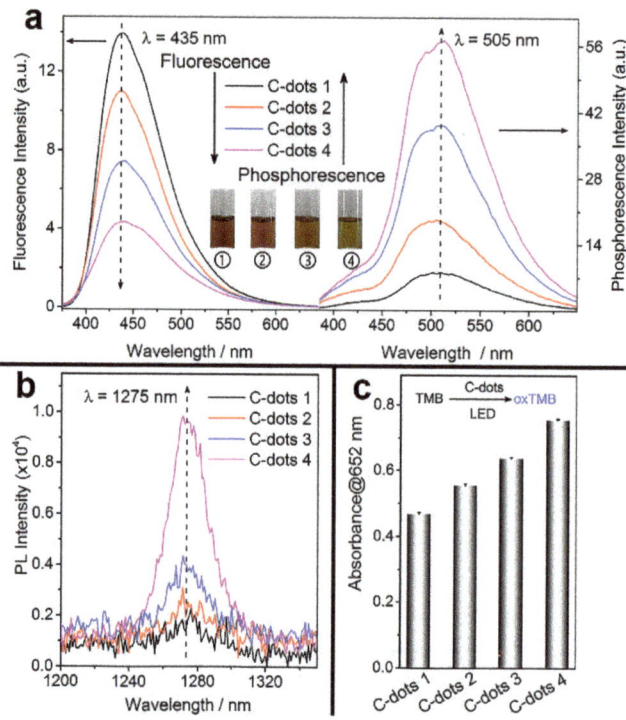

Figure 4. Relationship between phosphorescence (delay time of 1 ms) and the oxygen activation performance of CDs. (**a**) Fluorescence and phosphorescence (in PVA matrix) spectra of four types of CDs. Inset shows the solution of CDs at different temperatures of synthesis. (**b**) $^1O_2$ phosphorescence emission spectrum of the four types of CDs in a CD_3CN-D_2O mixed solvent (v/v = 15/1). (**c**) TMB photo-oxidation efficiencies of the four CDs. Reproduced with permission from Ref. [43]. Copyright © 2022 ACS Applied Materials & Interfaces, American Chemical Society.

In addition, using the "heavy atom effect" can also be demonstrated to improve the advantages of intersystem crossing (ISC), leading to a phosphorescent deactivation process, energy and electron transfer and more effectively antibacterial PSs [90]. The work of Knoblauch et al. showed that the intersystem crossing could be optimized by doping bromine on the CDs produced [91]. The "heavy atom effect," which involves incorporating elements such as bromine into tiny molecules, has long been an approach for obtaining better phosphorescence from fluorophores [92]. ROS generation benefits from the triplet interaction between the triplet-excited state and molecular oxygen at the ground state. It has been demonstrated that adding bromine to CDs can result in excellent spin-orbit coupling and subsequent phosphorescence detection. Under UV-A irradiation, the prepared brominated carbon dots (BrCDs) produced HO• (Type I) and $^1O_2$ (Type II) and displayed considerable antibacterial activity. The potential of hydrogen-departing reactive nitrogen species was discovered in the synthesized CDs, which impeded colony formation even without photodynamic processes [91].

The surface charge of CDs has a significant impact on their PACT effect. Yang et al. performed three studies in which they prepared CDs and then surface functionalized them with 2,2-(ethylenedioxy)bis(ethylamine) (EDA) and 3-ethoxypropylamine (EPA) to produce EDA-CDs and EPA-CDs, respectively [93]. EDA is protonated and becomes positively charged at the neutral potential of hydrogen, but EPA is not. The researchers discovered that EDA-CDs have considerably stronger antibacterial activity than EPA-CDs, implying that the positive charge on EDA-CDs can aid CDs to electrostatically attach a negatively charged bacterial film, improving the PACT effect. According to the authors, a high fluorescence quantum yield is also favorable to the PACT effect. Furthermore, the antibacterial activity of CDs is enhanced by a thin polymer passivation layer.

It is also required to prepare CDs with targeting ability to optimize the photodynamic effect of CDs. Galan et al. found that green fluorescent FCDs made by microwave using glucosamine hydrochloride and m-phenylenediamine as precursors can label *E. coli*, *Klebsiella pneumoniae*, *Pseudomonas aeruginosa*, and Gram-positive (*S. aureus*) pathogens in less than 10 min. FCDs paired with LED irradiation successfully kill Gram-negative and Gram-positive bacteria in the visible range [94].

4. Carbon Dots-Based Nanocomposites in PACT

While basic research into the mechanism and optimization of CDs as photosensitizers is still ongoing, researchers have begun to investigate how these particles might be integrated into hybrid systems to improve antimicrobial efficiency. In other cases, the CDs' intrinsic antibacterial capabilities have been exploited to increase the overall system's efficiency.

4.1. Antibiotic-Modified Carbon Dots

CDs not only have high antibacterial and antimicrobial membrane activity, but they also include a variety of hydrophilic groups on the surface, including $-NH_2$, $-OH$, and $-COOH$, and their unique nanostructure allows them to be used in many drug delivery applications. The higher cellular internalization of these CDs is used to boost drug absorption. At the same time, it can bring active molecules attached to the main structure by functionalization groups. This property has been discovered to be directly employed in the administration of antibiotic molecules such as vancomycin [95], ampicillin [96], penicillin [97], ciprofloxacin [98,99], and antiparasitic creams [97] to destroy microorganisms in bacterial cells. Bacteria can be killed more effectively by a synergistic combination of the PACT action with CDs antibiotics.

Boukherrub et al. used a hydrothermal approach to prepare amine-functionalized carbon dots (CDs-NH_2) using citric acid and ethylenediamine as primary materials. To make CDs-AMP nanostructures, the main amine groups on the surface of CDs-NH_2 were employed to be covalently connected to ampicillin (AMP), a classical β-lactam antibiotic. The lowest dose of CDs-AMP conjugate(14 g mL^{-1})-inhibited *E. coli* cells was higher than free AMP (25 g mL^{-1}), confirming the superiority of the CDs-AMP conjugate. As shown in Figure 5a, the authors also confirmed that exposing CDs-AMP to visible light irradiation increased its bactericidal action. When compared to free AMP, the results of this investigation demonstrated that AMP placed on CDs had improved stability and antibacterial activity when exposed to visible light [96].

Mandal et al. used a solvothermal approach to make 1,5-dihydroxyanthraquinone-based CDs that emit green fluorescence. BSA was coated on the surface of the CDs via an amidation reaction to boost ROS activity. As given in Figure 5b, Ciprofloxacin reacted non-covalently with BSA-CDs conjugates to generate drug nanocomplexes. At the same time, PACT functioned synergistically with antibiotic drug release to kill 95% of *E. coli* and *S. aureus* at concentrations as low as 1.47 g mL^{-1} in their complexes [100].

Using CDs conjugated with antibiotics can improve their antimicrobial activity in many cases. Combining photodynamic sterilization with antibiotic sterilization can maximize bacteria-killing because CDs improve the internalization efficiency and targeting of the material and have a photodynamic effect [96].

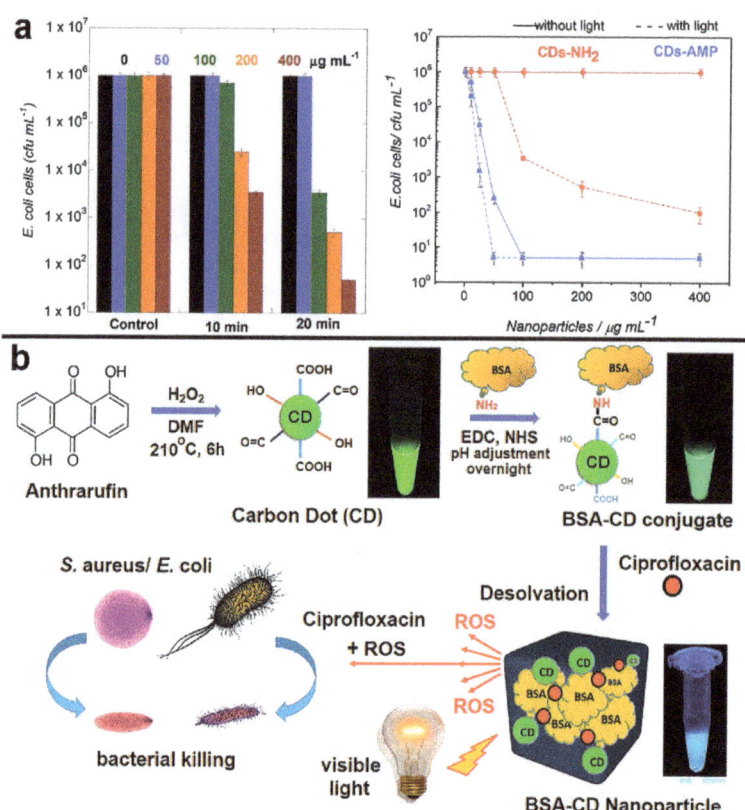

Figure 5. (a) Photodynamic efficiency of CDs-NH$_2$ for the inactivation of *E. coli* K12-MG 1655 upon irradiation at 0.30 W for 10 and 20 min and influence of the CDs-NH$_2$ and CDs-AMP concentration on the treatment efficiency of *E. coli* without (solid lines) and with (dash lines) visible light illumination (20 min, 0.30 W). Reproduced with permission from Ref. [96]. Copyright © 2018 *Colloids and Surfaces B: Biointerfaces*, Elsevier B.V. (b) Scheme for Synthesis of CDs, conjugation of the CDs to BSA, and subsequent creation of BSA-CDs nanoparticles for visible-light-induced ROS generation and simultaneous release of ciprofloxacin for antibacterial activity. Reproduced with permission from Ref. [100]. Copyright © 2022 *ACS Applied Materials & Interfaces*, American Chemical Society.

4.2. Carbon Dots as Nanocarriers for Photosensitizers

PSs such as MB and curcumin can also be effectively bound to CDs through covalent coupling or supramolecular interactions [101] (including π–π stacking, electrostatic interactions, etc.) using the same rational approaches to the development of CDs-attached antibiotics.

CDs' intrinsic antibacterial properties can sometimes boost antimicrobial efficacy. In recent years, mixing CDs with photosensitizers molecules has been discovered to boost overall antimicrobial activity. Dong et al. for example increased the antibacterial action of CDs by combining them with the photosensitizers MB and toluidine blue (TB). According to the scientists, CDs (5 g mL^{-1}) alone had no antibacterial effect, and MB (1 g mL^{-1}) had little antimicrobial activity, but mixing CDs (5 g mL^{-1}) with MB (1 g mL^{-1}) considerably increased their antimicrobial effect and almost entirely inhibited bacterial growth, according to the scientists. In *E. coli* cells, the combination of CDs and TB showed a similar synergistic impact. The authors speculate that this is due to (1) increased cellular penetration by small-molecule photosensitizers, (2) improved solubility of their small-molecule counterparts by CDs and thus improved uptake/localization and target delivery, or (3) increased overall intracellular ROS by the combination of both photosensitizers, for example through a fluorescence resonance energy transfer (FRET) mechanism [25].

Similarly, Kholikov et al. synthesized biocompatible and photostable GQDs that produced more monoclinic oxygen when mixed with MB. The population of excited triplet-state photosensitizers created by the inter-systemic crossover (ISC) of excited singlet-state photosensitizers is related to the oxygen generation efficiency of singlet photosensitizers. As a result, GQDs may lengthen the duration of the MB triplet state, boosting (ISC) efficiency from the singlet excited state to the triplet excited state of MB. Within 5 min of irradiation with visible light at the appropriate wavelength, Gram-positive and Gram-negative bacteria treated with the MB-GQD were inactivated (Figure 6a). The lower doses of MB provided higher antimicrobial activity when sulfur-doped GQDs were combined with MB [102]. Yameen et al. synthesized a compound (cur-GQDs) by loading curcumin onto GQDs to increase the solubility and biocompatibility of curcumin. It was shown that loading curcumin onto GQDs solved the problem of curcumin's poor water solubility and increased its ROS production by three-fold, effectively being an inhibitory effect on *Pseudomonas aeruginosa*, MRSA, *E. coli* and *Candida albicans*. The results were observed when the samples were exposed to 405 nm blue light at 30 J cm^{-2}. As shown in Figure 6b, *Pseudomonas aeruginosa*, MRSA, *E. coli* and *Candida albicans* were significantly inhibited [103].

Other types of CDs have excellent photothermal properties but are not directly associated with photodynamic properties. This material class could be under near-infrared light, transforming light energy into heat, and inducing desaturation of enzymes on bacteria's surfaces by raising their temperature and destroying the cell membrane and biofilms framework. Based on this foundation, a nanocomposite system with synergistic PDT and PTT therapy can be built. The antimicrobial effect can be considerably increased by mixing CDs with photosensitizers molecules via FRET. Su et al. developed a carbon dot composite system CDs/Cur that uses CDs as a carrier for curcumin, which improves curcumin biocompatibility and ROS yield and has an excellent photothermal impact. CDs/Cur may create both heat and generate ROS under dual-wavelength irradiation at 405 and 808 nm, enhancing the antibacterial efficacy by combining PDT/PTT (Figure 6c). The killing effect on *E. coli* was up to 100% of the final concentration of 1 M CDs/Cur concentration. In contrast, the lethal effect on *S. aureus* was even stronger, with a mortality rate of 100% at the lowest concentration of 0.10 nM [104].

CDs have excellent fluorescence properties, including broad-spectrum absorption, tunable photothermal effects, upconversion luminescence, and visible light absorption, allowing them to be used as FRET donors in various applications. Traditional photosensitizers such as PpIX have low solubility and are prone to aggregation-induced bursts. Das et al. developed the CD-DNA-PpIX hybrid hydrogel using protoporphyrin as the acceptor. DNA functions as a linker to join CDs and PpIX, combining the preceding advantages of CDs. As shown in Figure 6d, CDs serve two purposes: a cross-linker to disseminate PpIX and a FRET donor to stimulate PpIX. The photodynamic effect of PpIX in visible light and CDs to produce FRET can work together to generate additional ROS, which can considerably improve the photodynamic outcome. The length of the DNA's sequence impacts the distance between the CDs and PpIX, as well as the efficiency of FRET, which can minimize PpIX's self-burst and ensure its delayed release. The hydrogel was entirely over at 10–11 days in the experiment, while PpIX produced ROS slowly and consistently, killing Gram-positive bacteria (*S. aureus*) continually [105].

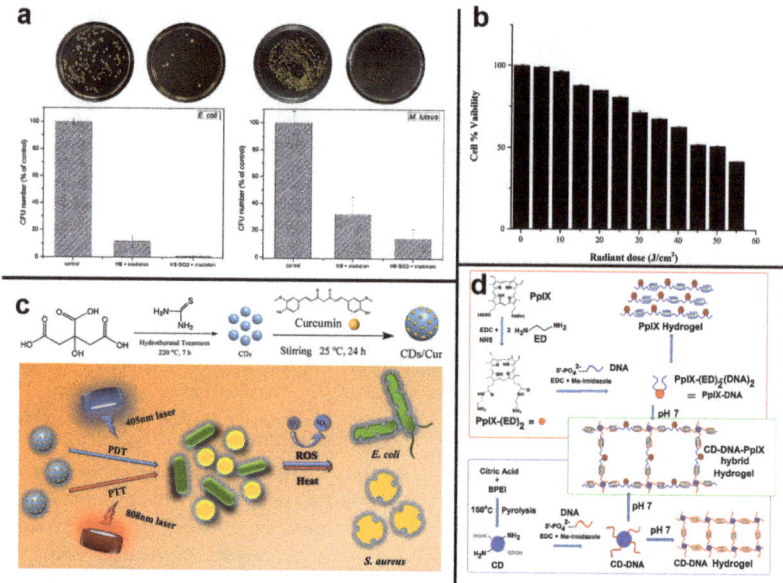

Figure 6. (**a**) (left) *E. coli* (5 min irradiation) and (right) M. Luteus colonies (2 min irradiation) in PBS before and after irradiation and their percentage irradiation of MBGQD with 660 nm light. Reproduced with permission from Ref. [102]. Copyright 2018 *Photodiagnosis and Photodynamic Therapy*, Elsevier B.V. (**b**) Effects of optimized blue light and optimized cur-GQD concentration on NIH/3t3 cells, indicating nontoxic effect of light dose at 30 J cm^{-2} with cur-GQD. Reproduced with permission from Ref. [103]. Copyright © 1999–2022 *Photochemistry and Photobiology*, John Wiley & Sons. (**c**) Synthesis of the CDs/Cur nanocomposite photosensitizer, and bactericidal activities of CDs/Cur upon dual-wavelength (405 + 808 nm) illumination. Reproduced with permission from Ref. [104]. Copyright 2022 *ACS Applied Bio Materials*, American Chemical Society. (**d**) Scheme for conjugation of cytosine, rich single-stranded DNA to CDs and PpIX for hydrogel formation. The blue and red color-coded DNA sequence is the same. Reproduced with permission from Ref. [105]. Copyright © 2019 *Journal of Colloid and Interface Science*, Elsevier Inc.

4.3. Carbon Dots/Metal Oxide Nanocomposites

CDs/metal oxide nanocomposites, such as ZnO/GQDs [106], CDs/$Na_2W_4O_{13}$/WO_3 [107], CDs/TiO_2 [108] and CDs/Cu_2O [109], are also under evaluation for its antimicrobial activity. When these materials are exposed to UV–visible light, these nanocomposites release ROS, which kill microorganisms by some previous mechanisms presented. For example, Chen et al. used a hydrothermal technique to make ZnO/GQDs nanocomposites that could form reactive oxygen species to kill *E. coli* when exposed to UV light. Their bactericidal activity was much higher than ZnO and GQDs [106].

4.4. Other Hybrid Carbon Dots

Many experiments have been conducted to improve the antibacterial photodynamic properties of CDs. The embedding of light-responsive CDs into soft hyaluronic acid hydrogels is frequently used as a photoactive antibacterial technique. Infectious bacteria in the target tissue can dissolve hydrogels structure and liberate CDs because of the action of hyaluronidase present on infectious bacteria naturally. Park et al. created a light-responsive carbon dot-embedding soft hyaluronic acid hydrogel (CDgel) to be used as a photodynamic antibacterial agent in vivo and in vitro by embedding CDs into a hyaluronic acid cross-linked hydrogel. As previously stated, the hyaluronic acid backbone of CDgel is broken by the bacterial hyaluronidase enzyme when applied to the bacterial site. CDgel degrades at this moment, transitioning from a gel to a liquid state, and CDs are released as a result. CDs release vast levels of $^1O_2$ when exposed to white LED light, which can kill up to 99% of E. coli and 97% of S. aureus (Figure 7a) [110].

In addition, the development of a synergistic antibacterial platform can also greatly enhance the antibacterial performance. As shown in Figure 7b, Shen et al. developed silicon-based near-infrared CDs (QPCuRC@MSiO$_2$) and a bicarbonate (BC) nanoplatform (BC/QPCuRC@MSiO$_2$@PDA). It has triple synergistic antibacterial properties such as PDT, PTT and quaternary ammonium compounds (QACs). In vitro and in vivo experiments showed that BC/QPCuRC@MSiO$_2$@PDA had excellent antibacterial properties, and the antibacterial rates against S. aureus and E. coli could reach 99.99% and 99.60%, respectively [111].

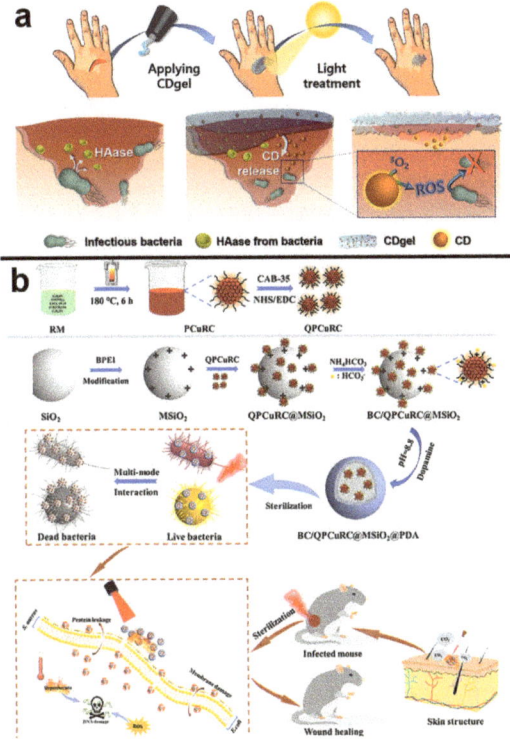

Figure 7. (a) Diagram of a CDs-embedded hyaluronic acid-based hydrogel (CDgel) for photoablation of infectious bacteria. Infectious bacteria such as S. aureus and E. coli produce hyaluronidases (HAase) when they increase. The CD gel's hyaluronic acid is decomposed by hyaluronidases, resulting in the release of the CDs. The released CDs use $^1O_2$ production to photodynamically ablate infectious bacteria, allowing for the selective and targeted removal of malignant germs from infected wounds. Reproduced with permission from Ref. [110]. Copyright © 2022 ACS Applied Bio Materials, American Chemical Society. (b) Schematic illustration of the synthesis of BC/QPCuRC@MSiO$_2$@PDA and their related biological applications. Reproduced with permission from Ref. [111]. Copyright © 2021 Journal of Colloid and Interface Science, Elsevier Inc.

The application of carbon dots-based nanocomposites in PACT is summarized in this section. To boost the effect of PACT, CDs can be employed as an antibiotic carrier, mixed with a photosensitizer, combined with a metal oxide, or formed into various hybrid materials.

In Table 1, we have listed some representative CDs structure presenting high potential activity on PACT, based on the material discussed in Sections 3 and 4.

Table 1. Representative CDs for killing microorganisms; from Sections 3 and 4.

CDs Label [a]	The Precursor of CDs	Excitation Wavelength	Emission Wavelength	QY	Light Wavelength	Light Power	ROS Sensitization Yields	Microorganism	Reduction of Bacteria	Ref.
GQD	graphite rods	328 nm	494 nm	—	blue light (470 nm)	1 W	—	S. aureus and E. coli	80% E. coli and 90–95% S. aureus were eliminated after 15 min *	[76]
MCDs	edible mushroom	360 nm	456 nm	25%	visible LED light	2.70 mW cm^{-2}	—	E. coli	>90% elimination of E. coli in 12 h	[78]
CDs	citric acid	370 nm	450 nm	—	blue light (450 nm)	40 J cm^{-2}	—	S. aureus	total elimination of S. aureus suspension was achieved (CDs: 6.90 mg/mL) and total elimination of the biofilm cultures was achieved (CDs: 13.80 mg/mL)	[80]
GQD	graphite	480/740 nm	618–647 nm	18.50%	800 nm	2.64 mW	QY = 0.51 ($^1O_2$)	E. coli and MRSA	all E. coli and MRSA to be dead after the 15 s laser photoexcitation	[83]
Cur-NRCDs	curcumin, neutral red and citric acid	540 nm	635 nm	—	xenon light (400–450 nm)	—	—	S. aureus and E. coli	after 10 min of xenon irradiation, 10 mM and 15 mM of Cur-NRCDs can kill 100% of S. aureus and E. coli, respectively	[88]
N-GQD (5.1%)	graphite	365 nm	624 nm	25.90%	670 nm laser	0.10 W cm^{-2}	QY = 0.64 ($^1O_2$)	E. coli	100% was eliminated by N-GQDs (5.1%) after a 3-min exposure	[89]
CDs	citric acid and ethylenediamine	350 nm *	450 nm	20%	LED light (365 nm)	3 V/3 W.	QY = 0.82 ($^1O_2$)	E. coli and Salmonella	bacteria growth inhibition efficiencies of 92% and 86% were obtained for E. coli and Salmonella in the presence of 5 μM CDs with light in 1 h, respectively	[43]
BrCDs	natural gas, HBr	302 nm	>355 nm	—	Ultraviolet lamp (365 nm)	3 mW	—	Listeria monocytogenes, S. aureus and E. coli.	with 10 min of UV exposure the growth of each bacterium is further decreased, achieving minimal to no colony formation visible for each	[91]
EDA-CDs/ EPA-CDs	carbon nano-powders	—	—	20%	400–800 nm light bulb	36 W, 12 V	—	Bacillus subtilis	1 h of EDA-CDs and EPA-CDs treatment resulted in a reduction of approximately 5.80 log and 0.84 log, respectively	[93]
FCDs	glucosamine hydrochloride and m-phenylenediamine	—	—	—	blue–LED strip lights (460 nm)	24 W, 12 V	—	Klebsiella pneumoniae, Pseudomonas aeruginosa, E. coli and S. aureus	complete killing of each bacterium was reproducibly observed after treatment with 200 μg/mL FCDs with 4 h of irradiation, and significant killing (>95%) could be observed after only 90 min LED irradiation	[94]
Antibiotic-Modified CDs									
CDs-AMP	citric acid and ethylenediamine	350 nm	450 nm	19%	visible light	0.30 W	—	E. coli	>4 log$_{10}$ inhibition of E. coli by CDs-AMP after 20 min of irradiation *	[96]
BSA-CDs NP	1,5-dihydroxyanthraquinone	395 nm	525 nm	75% (CDs)	Tungsten bulb (300–900 nm)	100 W	—	S. aureus and E. coli	99.97% and 99.53% elimination of E. coli and S. aureus in 1h	[100]

Table 1. *Cont.*

CDs Label [a]	The Precursor of CDs	Excitation Wavelength	Emission Wavelength	QY	Light Wavelength	Light Power	ROS Sensitization Yields	Microorganism	Reduction of Bacteria	Ref.
CDs as nanocarriers for photosensitizers											
CDs/MB or CDs/TB	carbon nanopowders	400 nm	–	12% (CDs)	white light bulb	36 W	–	E. coli	5 µg/mL CDs combined with 1 µg/mL MB completely inhibited bacteria growth, resulting in 6.20 log viable cell number reduction	[25]
GQDs	sulfur and nickel (II) oxide powder and benzene	310 nm	420 nm	–	660 nm red light	12 W	–	E. coli and Micrococcus luteus	10^6 CFU/mL E. coli and Micrococcus luteus can be eradicated entirely in 10 min with MB-GQD irradiation	[102]
cur-GQDs	coal and curcumin	407 nm	525–550 nm *	–	405 nm LEDs	30 J cm^{-2}.	–	Pseudomonas aeruginosa, MRSA, E. coli and Candida albicans.	for S. aureus Pseudomonas aeruginosa, MRSA, E. coli and Candida albicans, cur-GQDs caused 5.68 log$_{10}$, 5.02 log$_{10}$, 5.44 log$_{10}$, 2.26 log$_{10}$ and 3.82 log$_{10}$ CFU reduction, respectively	[103]
CDs/Cur	citric acid and thiourea	420 nm	550–575 nm*	–	405 + 808 nm light	808 nm (500 mW cm^{-2}), 405 nm (200 mW cm^{-2})	–	E. coli and S. aureus	death rate of E. coli and S. aureus increased to 100% for 1 µM and 0.1 nM CD/Cur, respectively	[104]
CD-DNA-PpIX hybrid hydrogel	citric acid and Branched Polyethylenimine	350 nm	625–650 nm *	–	UV lamp (302 nm)	–	–	S. aureus	UV irradiation for 2.50 min followed by incubation for 24 h affected > 4.50 log (>99.99%) reduction of S. aureus cells	[105]
CDs/metal oxide nanocomposites											
ZnO/GQDs	citric acid	365 nm	460 nm	–	UV light (365nm)	100 W, 1000–1500 lumen	–	E. coli	100% was eliminated by ZnO/GQDs after 5 min of UV exposure	[106]
Other hybrid CDs											
CDgel	ammonium citrate and polyethylenimine	390 nm	400–500 nm	–	white light irradiation	5 mW cm^{-2}	–	S. aureus and E. coli	CDgel under light giving approximately 99% and 97% mortality for S. aureus and E. coli, respectively	[110]
BC/QPCuRC@MSiO$_2$@PDA	Citric, urea and CuCl$_2$·2H$_2$O	360 nm	722 nm	–	808 nm	2 W cm^{-2}	–	S. aureus and E. coli	antibacterial rate up to 99.60% and 99.99% to E. coli and S. aureus, respectively	[111]

[a] Labels indicate either additional details regarding the nature of the reported carbon dots or indicate the abbreviation/common label used within the cited study to describe the particle.
* Denotes values extrapolated from relevant in-text details from the specified reference.

5. Toxicology and Safety Profile of Carbon Dots

CDs generally have good cytotoxicity and biocompatibility, which is beneficial for biomedical applications. However, the influence of CDs on host cells must be carefully evaluated before CDs can be widely used in PACT applications, and its possible medical complications and detailed toxicology need to be further studied. The toxicity of CDs is particularly important because of the unavoidable contact between mammalian cells or tissues and photosensitizers during PACT application. Additionally, photosensitizers are prone to transfer from the surface to the biological system, resulting in potential safety risks.

Thus far, much work has been performed on the cytotoxicity of CDs on mammalian cells, and it has been reported that they are non-toxic at proper concentrations both in vitro and in vivo [112–114]. However, the cytotoxicity of the CDs should not be overlooked because the properties of CDs vary greatly between different precursors and synthetic strategies. Although CDs prepared from "bio-safe" precursors such as common glucose are commonly believed to be nontoxic, sometimes they maintain low cytotoxicity only in dark conditions, while cytotoxic substances could be produced upon light irradiation. Therefore, extensive studies must be carried out to fully explore the possible toxic effects of CDs on humans. The used concentration of CDs is also an important factor affecting its toxicity. A high concentration of CDs will exert toxic effects on the central nervous system. Toxicology reports of GQDs indicate that although most existing studies support the safe use of GQDs, their toxicity may vary depending on the concentration and test method used in the synthesis technology. Studies have found that small-sized CDs are more toxic than large-sized CDs [115], and CDs with negative charged are more cytotoxic to mammalian cells [116,117]. ROS production plays an important role in the sterilization process of CDs, but ROS may also cause cell death. In order to solve the above problems, it is necessary to promote safe and controllable CDs synthesis strategies and application methods, and the safe application of CDs in the treatment of infectious diseases requires in-depth research on its possible toxic side effects and complications.

6. Conclusions

As presented, CDs have received much attention in chemical sensing, biological imaging, photocatalysis, phototherapy, and drug administration. The CDs' structure work as a photosensitizer in PACT is discussed in many aspects and applications in this paper. The synthetic process for CDs preparation was evaluated first to propose classical and new routes for synthesizing CDs with high photodynamic antibacterial effects. Then, the key elements that could be impacting the antibacterial effect of CDs were also discussed. Furthermore, more effective antibacterial materials can be created by mixing CDs with other photosensitizers molecules and antibiotics or by creating hybrid materials based on CDs. CDs have been shown to be one of the most promising carbon classes of material to work properly as an antibacterial material because of their excellent physical and chemical properties, optical qualities, and photophysical and photochemical behavior associated with exceptional water solubility.

Conversely, CDs face some problems, limiting their practical application. The exact process of photoluminescence is unknown, and CDs with extended excitation and emission wavelengths are still uncommon, leading to complex tissue and biofilm penetration. Second, relatively few CDs have intrinsic microbe targeting ability, resulting in a significantly reduced antibacterial effect that is essential in developing antibacterial CDs. Finally, CDs' water solubility and biocompatibility influence their microbial therapy usage. To summarize, the creation of high-efficiency antibacterial CDs faces numerous hurdles. As these issues are resolved, CDs may have more good results in microbial therapeutics.

Author Contributions: For this review, X.W. initiated the review concept and designed the manuscript contents; K.A., Y.Y., Z.L. and A.C.T. helped in the revision of the manuscript; H.B. supervised the entire project. All the authors conjointly reviewed and edited the manuscript. All authors have read and agreed to the published version of the manuscript.

Funding: This project was funded by the National Natural Science Foundation of China (grant number 52172033 and 51772001) and the Anhui Province Key Research and Development Plan Project International Science and Technology Cooperation Special Project (no. 202004b11020015).

Institutional Review Board Statement: Not applicable.

Informed Consent Statement: Not applicable.

Data Availability Statement: Data sharing not applicable.

Acknowledgments: The National Natural Science Foundation of China (grant no. 52172033 and 51772001) and the Anhui Province Key Research and Development Plan Project International Science and Technology Cooperation Special Project provided financial support for this research (no. 202004b11020015). We gratefully acknowledge the support of the Ministry of Education's Key Laboratory of Structure and Functional Regulation of Hybrid Materials, Anhui University, Hefei, Anhui 230601, China. We also appreciate the help of the Anhui Province Key Laboratory of Environmentally Friendly Polymer Materials in Hefei, China.

Conflicts of Interest: The authors declare no conflict of interest.

References

1. Mu, H.; Tang, J.; Liu, Q.; Sun, C.; Wang, T.; Duan, J. Potent antibacterial nanoparticles against biofilm and intracellular bacteria. *Sci. Rep.* **2016**, *6*, 18877. [CrossRef] [PubMed]
2. Dong, X.; Al Awak, M.; Tomlinson, N.; Tang, Y.; Sun, Y.P.; Yang, L. Antibacterial effects of carbon dots in combination with other antimicrobial reagents. *PLoS ONE* **2017**, *12*, e0185324. [CrossRef] [PubMed]
3. Varghese, M.; Balachandran, M. Antibacterial efficiency of carbon dots against Gram-positive and Gram-negative bacteria: A review. *J. Environ. Chem. Eng.* **2021**, *9*, 106821. [CrossRef]
4. Durao, P.; Balbontin, R.; Gordo, I. Evolutionary mechanisms shaping the maintenance of antibiotic resistance. *Trends Microbiol.* **2018**, *26*, 677–691. [CrossRef] [PubMed]
5. Liu, Y.; Liu, X.; Xiao, Y.; Chen, F.; Xiao, F. A multifunctional nanoplatform based on mesoporous silica nanoparticles for imaging-guided chemo/photodynamic synergetic therapy. *RSC Adv.* **2017**, *7*, 31133–31141. [CrossRef]
6. Russell, S.P.; Neary, C.; Elwahab, S.A.; Powell, J.; O'Connell, N.; Power, L.; Tormey, S.; Merrigan, B.A.; Lowery, A.J. Breast infections–Microbiology and treatment in an era of antibiotic resistance. *Surgeon* **2020**, *18*, 1–7. [CrossRef]
7. Sattarahmady, N.; Rezaie-Yazdi, M.; Tondro, G.H.; Akbari, N. Bactericidal laser ablation of carbon dots: An in vitro study on wild-type and antibiotic-resistant staphylococcus aureus. *J. Photochem. Photobiol. B-Biol.* **2017**, *166*, 323–332. [CrossRef]
8. Pidot, S.J.; Gao, W.; Buultjens, A.H.; Monk, I.R.; Guerillot, R.; Carter, G.P.; Lee, J.Y.H.; Lam, M.M.C.; Grayson, M.L.; Ballard, S.A.; et al. Increasing tolerance of hospital enterococcus faecium to handwash alcohols. *Sci. Transl. Med.* **2018**, *10*, eaar6115. [CrossRef]
9. Dong, X.; Liang, W.; Meziani, M.; Sun, Y.P.; Yang, L. Carbon dots as potent antimicrobial agents. *Theranostics* **2020**, *10*, 671–686. [CrossRef]
10. Ghirardello, M.; Ramos-Soriano, J.; Galan, M.C. Carbon dots as an emergent class of antimicrobial agents. *Nanomaterials* **2021**, *11*, 1877. [CrossRef]
11. Al Awak, M.M.; Wang, P.; Wang, S.; Tang, Y.; Sun, Y.P.; Yang, L. Correlation of carbon dots' light-activated antimicrobial activities and fluorescence quantum yield. *RSC Adv.* **2017**, *7*, 30177–30184. [CrossRef] [PubMed]
12. Zhou, C.; Peng, C.; Shi, C.; Jiang, M.; Chau, J.H.; Liu, Z.; Bai, H.; Kwok, R.T.; Lam, J.W.; Shi, Y. Mitochondria-specific aggregation-induced emission luminogens for selective photodynamic killing of fungi and efficacious treatment of keratitis. *ACS Nano* **2021**, *15*, 12129–12139. [CrossRef] [PubMed]
13. Cui, J.; Ren, B.; Tong, Y.; Dai, H.; Zhang, L. Synergistic combinations of antifungals and anti-virulence agents to fight against Candida albicans. *Virulence* **2015**, *6*, 362–371. [CrossRef] [PubMed]
14. Dong, X.; Ge, L.; Abu Rabe, D.I.; Mohammed, O.O.; Wang, P.; Tang, Y.; Kathariou, S.; Yang, L.; Sun, Y.P. Photoexcited state properties and antibacterial activities of carbon dots relevant to mechanistic features and implications. *Carbon* **2020**, *170*, 137–145. [CrossRef]
15. Makabenta, J.M.V.; Nabawy, A.; Li, C.H.; Schmidt-Malan, S.; Patel, R.; Rotello, V.M. Nanomaterial-based therapeutics for antibiotic-resistant bacterial infections. *Nat. Rev. Microbiol.* **2021**, *19*, 23–36. [CrossRef]
16. Yang, J.; Gao, G.; Zhang, X.; Ma, Y.H.; Jia, H.R.; Jiang, Y.W.; Wang, Z.; Wu, F.G. Ultrasmall and photostable nanotheranostic agents based on carbon quantum dots passivated with polyamine-containing organosilane molecules. *Nanoscale* **2017**, *9*, 15441–15452. [CrossRef]
17. Hua, X.W.; Bao, Y.W.; Wu, F.G. Fluorescent carbon quantum dots with intrinsic nucleolus-targeting capability for nucleolus imaging and enhanced cytosolic and nuclear drug delivery. *ACS Appl. Mater. Interfaces* **2018**, *10*, 16924. [CrossRef]
18. Jiang, L.; Ding, H.; Xu, M.; Hu, X.; Li, S.; Zhang, M.; Zhang, Q.; Wang, Q.; Lu, S.; Tian, Y.; et al. UV-Vis-NIR full-range responsive carbon dots with large multiphoton absorption cross sections and deep-red fluorescence at nucleoli and in vivo. *Small* **2020**, *16*, 2000680. [CrossRef]

19. Sun, Y.; Qin, H.; Geng, X.; Yang, R.; Qu, L.; Kani, A.N.; Li, Z. Rational design of far-red to near-infrared emitting carbon dots for ultrafast lysosomal polarity imaging. *ACS Appl. Mater. Interfaces* **2020**, *12*, 31738–31744. [CrossRef]
20. Kaminari, A.; Nikoli, E.; Athanasopoulos, A.; Sakellis, E.; Sideratou, Z.; Tsiourvas, D. Engineering mitochondriotropic carbon dots for targeting cancer cells. *Pharmaceuticals* **2021**, *14*, 932. [CrossRef]
21. Wang, L.; Chang, Y.; Feng, Y.; Li, X.; Cheng, Y.; Jian, H.; Ma, X.; Zheng, R.; Wu, X.; Xu, K.; et al. Nitric oxide stimulated programmable drug release of nanosystem for multidrug resistance cancer therapy. *Nano Lett.* **2019**, *19*, 6800–6811. [CrossRef] [PubMed]
22. Dou, Q.; Fang, X.; Jiang, S.; Chee, P.L.; Lee, T.C.; Loh, X.J. Multi-functional fluorescent carbon dots with antibacterial and gene delivery properties. *RSC Adv.* **2015**, *5*, 46817–46822. [CrossRef]
23. Tuerhong, M.; Xu, Y.; Yin, X.B. Review on carbon dots and their applications. *Chin. J. Anal. Chem.* **2017**, *45*, 139–149. [CrossRef]
24. Shen, C.; Dong, C.; Cheng, L.; Shi, X.; Bi, H. Fluorescent carbon dots from Shewanella oneidensis MR-1 for Hg^{2+} and tetracycline detection and selective fluorescence imaging of Gram-positive bacteria. *J. Environ. Chem. Eng.* **2022**, *10*, 107020. [CrossRef]
25. Dong, X.; Bond, A.E.; Pan, N.Y.; Coleman, M.; Tang, Y.; Sun, Y.P.; Yang, L. Synergistic photoactivated antimicrobial effects of carbon dots combined with dye photosensitizers. *Int. J. Nanomed.* **2018**, *13*, 8025–8035. [CrossRef]
26. Luo, P.G.; Yang, F.; Yang, S.T.; Sonkar, S.K.; Yang, L.; Broglie, J.J.; Liu, Y.; Sun, Y.P. Carbon-based quantum dots for fluorescence imaging of cells and tissues. *RSC Adv.* **2014**, *4*, 10791–10807. [CrossRef]
27. Luo, Q.; Ding, H.; Hu, X.; Xu, J.; Sadat, A.; Xu, M.; Primo, F.L.; Tedesco, A.C.; Zhang, H.; Bi, H. Sn^{4+} complexation with sulfonated-carbon dots in pursuit of enhanced fluorescence and singlet oxygen quantum yield. *Dalton Trans.* **2020**, *49*, 6950–6956. [CrossRef]
28. Wang, X.; Feng, Y.; Dong, P.; Huang, J. A mini review on carbon quantum dots: Preparation, properties, and electrocatalytic application. *Front. Chem.* **2019**, *7*, 671. [CrossRef]
29. Zheng, X.; Ananthanarayanan, A.; Luo, K.Q.; Chen, P. Glowing graphene quantum dots and carbon dots: Properties, syntheses, and biological applications. *Small* **2015**, *11*, 1620–1636. [CrossRef]
30. De, B.; Karak, N. Recent progress in carbon dot-metal based nanohybrids for photochemical and electrochemical applications. *J. Mater. Chem. A* **2017**, *5*, 1826–1859. [CrossRef]
31. Wu, Y.; Li, C.; van der Mei, H.C.; Busscher, H.J.; Ren, Y. Carbon quantum dots derived from different carbon sources for antibacterial applications. *Antibiotics* **2021**, *10*, 623. [CrossRef] [PubMed]
32. Zhu, H.; Wang, X.; Li, Y.; Wang, Z.; Yang, F.; Yang, X. Microwave synthesis of fluorescent carbon nanoparticles with electrochemiluminescence properties. *Chem. Commun.* **2009**, *34*, 5118–5120. [CrossRef] [PubMed]
33. Li, X.; Wang, H.Q.; Shimizu, Y.; Pyatenko, A.; Kawaguchi, K.; Koshizaki, N. Preparation of carbon quantum dots with tunable photoluminescence by rapid laser passivation in ordinary organic solvents. *Chem. Commun.* **2011**, *47*, 932–934. [CrossRef] [PubMed]
34. Li, H.; Ming, H.; Liu, Y.; Yu, H.; He, X.; Huang, H.; Pan, K.; Kang, Z.; Lee, S.T. Fluorescent carbon nanoparticles: Electrochemical synthesis and their pH sensitive photoluminescence properties. *New J. Chem.* **2011**, *35*, 2666–2670. [CrossRef]
35. Dong, Y.; Pang, H.; Yang, H.; Guo, C.; Shao, J.; Chi, Y.; Li, C.; Yu, T. Carbon-based dots co-doped with nitrogen and sulfur for high quantum yield and excitation-independent emission. *Angew. Chem.-Int. Ed.* **2013**, *52*, 7800–7804. [CrossRef] [PubMed]
36. Liu, W.; Li, C.; Ren, Y.; Sun, X.; Pan, W.; Li, Y.; Wang, J.; Wang, W. Carbon dots: Surface engineering and applications. *J. Mater. Chem. B* **2016**, *4*, 5772–5788. [CrossRef]
37. Li, B.; Zhao, S.; Huang, L.; Wang, Q.; Xiao, J.; Lan, M. Recent advances and prospects of carbon dots in phototherapy. *Chem. Eng. J.* **2021**, *408*, 127245. [CrossRef]
38. Xiao, D.; Qi, H.X.; Teng, Y.; Pierre, D.; Kutoka, P.T.; Liu, D. Advances and challenges of fluorescent nanomaterials for synthesis and biomedical applications. *Nanoscale Res. Lett.* **2021**, *16*, 167. [CrossRef]
39. Wang, B.; Song, H.; Qu, X.; Chang, J.; Yang, B.; Lu, S. Carbon dots as a new class of nanomedicines: Opportunities and challenges. *Coord. Chem. Rev.* **2021**, *442*, 214010. [CrossRef]
40. Anwar, S.; Ding, H.; Xu, M.; Hu, X.; Li, Z.; Wang, J.; Liu, L.; Jiang, L.; Wang, D.; Dong, C.; et al. Recent advances in synthesis, optical properties, and biomedical applications of carbon dots. *ACS Appl. Bio Mater.* **2019**, *2*, 2317–2338. [CrossRef]
41. Zuo, P.; Lu, X.; Sun, Z.; Guo, Y.; He, H. A review on syntheses, properties, characterization and bioanalytical applications of fluorescent carbon dots. *Microchim. Acta* **2016**, *183*, 519–542. [CrossRef]
42. Miao, S.; Liang, K.; Zhu, J.; Yang, B.; Zhao, D.; Kong, B. Hetero-atom-doped carbon dots: Doping strategies, properties and applications. *Nano Today* **2020**, *33*, 100879. [CrossRef]
43. Zhang, J.; Lu, X.; Tang, D.; Wu, S.; Hou, X.; Liu, J.; Wu, P. Phosphorescent carbon dots for highly efficient oxygen photosensitization and as photo-oxidative nanozymes. *ACS Appl. Mater. Interfaces* **2018**, *10*, 40808–40814. [CrossRef] [PubMed]
44. Markovic, Z.M.; Labudova, M.; Danko, M.; Matijasevic, D.; Micusik, M.; Nadazdy, V.; Kovacova, M.; Kleinova, A.; Spitalsky, Z.; Pavlovic, V.; et al. Highly efficient antioxidant F- and Cl-doped carbon quantum dots for bioimaging. *ACS Sustain. Chem. Eng.* **2020**, *8*, 16327–16338. [CrossRef]
45. Noh, S.-H.; Na, W.; Jang, J.-T.; Lee, J.-H.; Lee, E.J.; Moon, S.H.; Lim, Y.; Shin, J.-S.; Cheon, J. Nanoscale magnetism control via surface and exchange anisotropy for optimized ferrimagnetic hysteresis. *Nano Lett.* **2012**, *12*, 3716–3721. [CrossRef]
46. Feng, Z.; Adolfsson, K.H.; Xu, Y.; Fang, H.; Hakkarainen, M.; Wu, M. Carbon dot/polymer nanocomposites: From green synthesis to energy, environmental and biomedical applications. *Sustain. Mater. Technol.* **2021**, *29*, e00304. [CrossRef]

47. Saravanan, A.; Maruthapandi, M.; Das, P.; Ganguly, S.; Margel, S.; Luong, J.H.T.; Gedanken, A. Applications of N-doped carbon dots as antimicrobial agents, antibiotic carriers, and selective fluorescent probes for nitro explosives. *ACS Appl. Bio Mater.* **2020**, *3*, 8023–8031. [CrossRef]
48. Bing, W.; Sun, H.; Yan, Z.; Ren, J.; Qu, X. Programmed bacteria death induced by carbon dots with different surface charge. *Small* **2016**, *12*, 4713–4718. [CrossRef]
49. Vatansever, F.; de Melo, W.; Avci, P.; Vecchio, D.; Sadasivam, M.; Gupta, A.; Chandran, R.; Karimi, M.; Parizotto, N.A.; Yin, R.; et al. Antimicrobial strategies centered around reactive oxygen species-bactericidal antibiotics, photodynamic therapy, and beyond. *Fems Microbiol. Rev.* **2013**, *37*, 955–989. [CrossRef]
50. Lacey, J.A.; Phillips, D. The photosensitisation of *Escherichia coli* using disulphonated aluminium phthalocyanine. *J. Photochem. Photobio. A Chem.* **2001**, *142*, 145–150. [CrossRef]
51. Reddi, E.; Ceccon, M.; Valduga, G.; Jori, G.; Bommer, J.C.; Elisei, F.; Latterini, L.; Mazzucato, U. Photophysical properties and antibacterial activity of meso-substituted cationic porphyrins. *Photochem. Photobio.* **2002**, *75*, 462–470. [CrossRef]
52. Yang, J.; Zhang, X.; Ma, Y.H.; Gao, G.; Chen, X.; Jia, H.R.; Li, Y.H.; Chen, Z.; Wu, F.G. Carbon dot-based platform for simultaneous bacterial distinguishment and antibacterial applications. *ACS Appl. Mater. Interfaces* **2016**, *8*, 32170–32181. [CrossRef] [PubMed]
53. Xu, N.; Du, J.; Yao, Q.; Ge, H.; Li, H.; Xu, F.; Gao, F.; Xian, L.; Fan, J.; Peng, X. Precise photodynamic therapy: Penetrating the nuclear envelope with photosensitive carbon dots. *Carbon* **2020**, *159*, 74–82. [CrossRef]
54. Chilakamarthi, U.; Giribabu, L. Photodynamic therapy: Past, present and future. *Chem. Rev.* **2017**, *17*, 775–802. [CrossRef]
55. He, H.; Zheng, X.; Liu, S.; Zheng, M.; Xie, Z.; Wang, Y.; Yu, M.; Shuai, X. Diketopyrrolopyrrole-based carbon dots for photodynamic therapy. *Nanoscale* **2018**, *10*, 10991–10998. [CrossRef]
56. Xie, J.; Wang, Y.; Choi, W.; Jangili, P.; Ge, Y.; Xu, Y.; Kang, J.; Liu, L.; Zhang, B.; Xie, Z.; et al. Overcoming barriers in photodynamic therapy harnessing nano-formulation strategies. *Chem. Soc. Rev.* **2021**, *50*, 9152–9201. [CrossRef]
57. Bayona, A.M.D.; Mroz, P.; Thunshelle, C.; Hamblin, M.R. Design features for optimization of tetrapyrrole macrocycles as antimicrobial and anticancer photosensitizers. *Chem. Biol. Drug Des.* **2017**, *89*, 192–206. [CrossRef]
58. Horne, T.K.; Cronje, M.J. Mechanistics and photo-energetics of macrocycles and photodynamic therapy: An overview of aspects to consider for research. *Chem. Biol. Drug Des.* **2017**, *89*, 221–242. [CrossRef]
59. Yao, Q.; Fan, J.; Long, S.; Zhao, X.; Li, H.; Du, J.; Shao, K.; Peng, X. The concept and examples of type-III photosensitizers for cancer photodynamic therapy. *Chem* **2022**, *8*, 197–209. [CrossRef]
60. Nguyen, V.N.; Qi, S.; Kim, S.; Kwon, N.; Kim, G.; Yim, Y.; Park, S.; Yoon, J. An emerging molecular design approach to heavy-atom-free photosensitizers for enhanced photodynamic therapy under hypoxia. *J. Am. Chem. Soc.* **2019**, *141*, 16243–16248. [CrossRef]
61. Shi, Z.; Zhang, K.; Zada, S.; Zhang, C.; Meng, X.; Yang, Z.; Dong, H. Upconversion nanoparticle-induced multimode photodynamic therapy based on a metal-organic framework/titanium dioxide nanocomposite. *ACS Appl. Mater. Interfaces* **2020**, *12*, 12600–12608. [CrossRef] [PubMed]
62. Cheng, J.; Zhao, H.; Yao, L.; Li, Y.; Qi, B.; Wang, J.; Yang, X. Simple and multifunctional natural self-assembled sterols with anticancer activity-mediated supramolecular photosensitizers for enhanced antitumor photodynamic therapy. *ACS Appl. Mater. Interfaces* **2019**, *11*, 29498–29511. [CrossRef] [PubMed]
63. Wu, X.; Xu, M.; Wang, S.; Abbas, K.; Huang, X.; Zhang, R.; Tedesco, A.C.; Bi, H. F,N-doped carbon dots as efficient Type I photosensitizers for photodynamic therapy. *Dalton Trans.* **2022**, *51*, 2296–2303. [CrossRef] [PubMed]
64. Foote, C.S. Definition of type I and type II photosensitized oxidation. *Photochem. Photobio.* **1991**, *54*, 659. [CrossRef] [PubMed]
65. Wainwright, M. Special issue on photoantimicrobials. *J. Photochem. Photobiol. B-Biol.* **2015**, *150*, 1. [CrossRef]
66. Planas, O.; Bresoli-Obach, R.; Nos, J.; Gallavardin, T.; Ruiz-Gonzalez, R.; Agut, M.; Nonell, S. Synthesis, Photophysical characterization, and photoinduced antibacterial activity of methylene blue-loaded amino- and mannose-targeted mesoporous silica nanoparticles. *Molecules* **2015**, *20*, 6284–6298. [CrossRef]
67. DeRosa, M.C.; Crutchley, R.J. Photosensitized singlet oxygen and its applications. *Coord. Chem. Rev.* **2002**, *233*, 351–371. [CrossRef]
68. Baptista, M.S.; Cadet, J.; Di Mascio, P.; Ghogare, A.A.; Greer, A.; Hamblin, M.R.; Lorente, C.; Nunez, S.C.; Ribeiro, M.S.; Thomas, A.H.; et al. Type I and Type II photosensitized oxidation reactions: Guidelines and mechanistic pathways. *Photochem. Photobiol.* **2017**, *93*, 912–919. [CrossRef]
69. Wainwright, M.; Maisch, T.; Nonell, S.; Plaetzer, K.; Almeida, A.; Tegos, G.P.; Hamblin, M.R. Photoantimicrobials-are we afraid of the light? *Lancet Infect. Dis.* **2017**, *17*, E49–E55. [CrossRef]
70. Alves, E.; Faustino, M.A.F.; Neves, M.; Cunha, A.; Tome, J.; Almeida, A. An insight on bacterial cellular targets of photodynamic inactivation. *Future Med. Chem.* **2014**, *6*, 141–164. [CrossRef]
71. Wang, J.; Xu, M.; Wang, D.; Li, Z.; Primo, F.L.; Tedesco, A.C.; Bi, H. Copper-doped carbon dots for optical bioimaging and photodynamic therapy. *Inorg. Chem.* **2019**, *58*, 13394–13402. [CrossRef] [PubMed]
72. Yu, B.; Goel, S.; Ni, D.; Ellison, P.A.; Siamof, C.M.; Jiang, D.; Cheng, L.; Kang, L.; Yu, F.; Liu, Z.; et al. Reassembly of Zr-89-labeled cancer cell membranes into multicompartment membrane-derived liposomes for PET-trackable tumor-targeted theranostics. *Adv. Mater.* **2018**, *30*, 1704934. [CrossRef] [PubMed]
73. Abrahamse, H.; Hamblin, M.R. New photosensitizers for photodynamic therapy. *Biochem. J.* **2016**, *473*, 347–364. [CrossRef] [PubMed]

74. Ge, J.; Lan, M.; Zhou, B.; Liu, W.; Guo, L.; Wang, H.; Jia, Q.; Niu, G.; Huang, X.; Zhou, H.; et al. A graphene quantum dot photodynamic therapy agent with high singlet oxygen generation. *Nat. Commun.* **2014**, *5*, 4596. [CrossRef]
75. Li, Z.; Wang, D.; Xu, M.; Wang, J.; Hu, X.; Anwar, S.; Tedesco, A.C.; Morais, P.C.; Bi, H. Fluorine-containing graphene quantum dots with a high singlet oxygen generation applied for photodynamic therapy. *J. Mat. Chem. B* **2020**, *8*, 2598–2606. [CrossRef]
76. Ristic, B.Z.; Milenkovic, M.M.; Dakic, I.R.; Todorovic-Markovic, B.M.; Milosavljevic, M.S.; Budimir, M.D.; Paunovic, V.G.; Dramicanin, M.D.; Markovic, Z.M.; Trajkovic, V.S. Photodynamic antibacterial effect of graphene quantum dots. *Biomaterials* **2014**, *35*, 4428–4435. [CrossRef]
77. Stankovic, N.K.; Bodik, M.; Siffalovic, P.; Kotlar, M.; Micusik, M.; Spitalsky, Z.; Danko, M.; Milivojevic, D.D.; Kleinova, A.; Kubat, P.; et al. Antibacterial and antibiofouling properties of light triggered fluorescent hydrophobic carbon quantum dots langmuir-blodgett thin films. *ACS Sustain. Chem. Eng.* **2018**, *6*, 4154. [CrossRef]
78. Venkateswarlu, S.; Viswanath, B.; Reddy, A.S.; Yoon, M. Fungus-derived photoluminescent carbon nanodots for ultrasensitive detection of Hg2+ ions and photoinduced bactericidal activity. *Sens. Actuator B. Chem.* **2018**, *258*, 172–183. [CrossRef]
79. Dong, X.; Al Awak, M.; Wang, P.; Sun, Y.P.; Yang, L. Carbon dot incorporated multi-walled carbon nanotube coated filters for bacterial removal and inactivation. *RSC Adv.* **2018**, *8*, 8292–8301. [CrossRef]
80. Romero, M.P.; Alves, F.; Stringasci, M.D.; Buzza, H.H.; Ciol, H.; Inada, N.M.; Bagnato, V.S. One-pot microwave-assisted synthesis of carbon dots and in vivo and in vitro antimicrobial photodynamic applications. *Front. Microbiol.* **2021**, *12*, 662149. [CrossRef]
81. Cai, Y.; Wei, Z.; Song, C.H.; Tang, C.; Han, W.; Dong, X. Optical nano-agents in the second near-infrared window for biomedical applications. *Chem. Soc. Rev.* **2019**, *48*, 22–37. [CrossRef] [PubMed]
82. Tang, Y.; Wang, G. NIR light-responsive nanocarriers for controlled release. *J. Photochem. Photobiol. C-Photochem. Rev.* **2021**, *47*, 100420. [CrossRef]
83. Kuo, W.S.; Chang, C.Y.; Chen, H.H.; Hsu, C.L.L.; Wang, J.Y.; Kao, H.F.; Chou, L.C.S.; Chen, Y.C.; Chen, S.J.; Chang, W.T.; et al. Two-photon photoexcited photodynamic therapy and contrast agent with antimicrobial graphene quantum dots. *ACS Appl. Mater. Interfaces* **2016**, *8*, 30467–30474. [CrossRef] [PubMed]
84. Li, Y.; Zheng, X.; Zhang, X.; Liu, S.; Pei, Q.; Zheng, M.; Xie, Z. Porphyrin-based carbon dots for photodynamic therapy of hepatoma. *Adv. Healthc. Mater.* **2017**, *6*, 1600924. [CrossRef]
85. Jia, Q.; Ge, J.; Liu, W.; Zheng, X.; Chen, S.; Wen, Y.; Zhang, H.; Wang, P. A magnetofluorescent carbon dot assembly as an acidic H$_2$O$_2$-driven oxygenator to regulate tumor hypoxia for simultaneous bimodal imaging and enhanced photodynamic therapy. *Adv. Mater.* **2018**, *30*, 1706090. [CrossRef]
86. Wen, Y.; Jia, Q.; Nan, F.; Zheng, X.; Liu, W.; Wu, J.; Ren, H.; Ge, J.; Wang, P. Pheophytin derived near-infrared-light responsive carbon dot assembly as a new phototheranotic agent for bioimaging and photodynamic therapy. *Chem.-Asian J.* **2019**, *14*, 2162–2168. [CrossRef]
87. Yue, J.; Li, L.; Jiang, C.; Mei, Q.; Dong, W.F.; Yan, R. Riboflavin-based carbon dots with high singlet oxygen generation for photodynamic therapy. *J. Mater. Chem. B* **2021**, *9*, 7972–7978. [CrossRef]
88. Su, R.; Yan, H.; Jiang, X.; Zhang, Y.; Li, P.; Su, W. Orange-red to NIR emissive carbon dots for antimicrobial, bioimaging and bacteria diagnosis. *J. Mater. Chem. B* **2022**, *10*, 1250–1264. [CrossRef]
89. Kuo, W.S.; Chen, H.H.; Chen, S.Y.; Chang, C.Y.; Chen, P.C.; Hou, Y.I.; Shao, Y.T.; Kao, H.F.; Hsu, C.L.L.; Chen, Y.C.; et al. Graphene quantum dots with nitrogen-doped content dependence for highly efficient dual-modality photodynamic antimicrobial therapy and bioimaging. *Biomaterials* **2017**, *120*, 185–194. [CrossRef]
90. Knoblauch, R.; Geddes, C.D. Carbon Nanodots in Photodynamic Antimicrobial Therapy: A Review. *Materials* **2020**, *13*, 4004. [CrossRef]
91. Knoblauch, R.; Harvey, A.; Ra, E.; Greenberg, K.M.; Lau, J.; Hawkins, E.; Geddes, C.D. Antimicrobial carbon nanodots: Photodynamic inactivation and dark antimicrobial effects on bacteria by brominated carbon nanodots. *Nanoscale* **2021**, *13*, 85–99. [CrossRef] [PubMed]
92. Turro, N.J. *Modern Molecular Photochemistry*; University Science Books: Herndon, VA, USA, 1991.
93. Abu Rabe, D.I.; Al Awak, M.M.; Yang, F.; Okonjo, P.A.; Dong, X.; Teisl, L.R.; Wang, P.; Tang, Y.; Pan, N.; Sun, Y.P.; et al. The dominant role of surface functionalization in carbon dots' photo-activated antibacterial activity. *Int. J. Nanomed.* **2019**, *14*, 2655–2665. [CrossRef] [PubMed]
94. Samphire, J.; Takebayashi, Y.; Hill, S.A.; Hill, N.; Heesom, K.J.; Lewis, P.A.; Alibhai, D.; Bragginton, E.C.; Dorh, J.; Dorh, N. Green fluorescent carbon dots as targeting probes for LED-dependent bacterial killing. *Nano Select* **2021**, *3*, 662–672. [CrossRef]
95. Pan, C.L.; Chen, M.H.; Tung, F.I.; Liu, T.Y. A nanovehicle developed for treating deep-seated bacteria using low-dose X-ray. *Acta Biomater.* **2017**, *47*, 159–169. [CrossRef] [PubMed]
96. Jijie, R.; Barras, A.; Bouckaert, J.; Dumitrascu, N.; Szunerits, S.; Boukherroub, R. Enhanced antibacterial activity of carbon dots functionalized with ampicillin combined with visible light triggered photodynamic effects. *Colloid Surf. B-Biointerf.* **2018**, *170*, 347–354. [CrossRef] [PubMed]
97. Sidhu, J.S.; Mayank; Pandiyan, T.; Kaur, N.; Singh, N. The Photochemical degradation of bacterial cell wall using penicillin-based carbon dots: Weapons against multi-drug resistant (MDR) Strains. *Chem. Sel.* **2017**, *2*, 9277–9283. [CrossRef]
98. Thakur, M.; Pandey, S.; Mewada, A.; Patil, V.; Khade, M.; Goshi, E.; Sharon, M. Antibiotic conjugated fluorescent carbon dots as a theranostic agent for controlled drug release, bioimaging, and enhanced antimicrobial activity. *J. Drug Deliv.* **2014**, *2014*, 282193. [CrossRef]

99. Hou, P.; Yang, T.; Liu, H.; Li, Y.; Huang, C. An active structure preservation method for developing functional graphitic carbon dots as an effective antibacterial agent and a sensitive pH and Al (III) nanosensor. *Nanoscale* **2017**, *9*, 17334–17341. [CrossRef]
100. Mandal, S.; Prasad, S.R.; Mandal, D.; Das, P. Bovine serum albumin amplified reactive oxygen species generation from anthrarufin-derived carbon dot and concomitant nanoassembly for combination antibiotic-photodynamic therapy application. *ACS Appl. Mater. Interfaces* **2019**, *11*, 33273–33284. [CrossRef]
101. Cao, Y.; Dong, H.; Yang, Z.; Zhong, X.; Chen, Y.; Dai, W.; Zhang, X. Aptamer-conjugated graphene quantum dots/porphyrin derivative theranostic agent for intracellular cancer-related microRNA detection and fluorescence-guided photothermal/photodynamic synergetic therapy. *ACS Appl. Mater. Interfaces* **2017**, *9*, 159–166. [CrossRef]
102. Kholikov, K.; Ilhom, S.; Sajjad, M.; Smith, M.E.; Monroe, J.D.; San, O.; Er, A.O. Improved singlet oxygen generation and antimicrobial activity of sulphur-doped graphene quantum dots coupled with methylene blue for photodynamic therapy applications. *Photodiagn. Photodyn. Therp.* **2018**, *24*, 7–14. [CrossRef] [PubMed]
103. Mushtaq, S.; Yasin, T.; Saleem, M.; Dai, T.; Yameen, M.A. Potentiation of antimicrobial photodynamic therapy by curcumin-loaded graphene quantum dots. *Photochem. Photobio.* **2022**, *98*, 202–210. [CrossRef] [PubMed]
104. Yan, H.; Zhang, B.; Zhang, Y.; Su, R.; Li, P.; Su, W. Fluorescent carbon dot-curcumin nanocomposites for remarkable antibacterial activity with synergistic photodynamic and photothermal abilities. *ACS Appl. Bio. Mater.* **2021**, *4*, 6703–6718. [CrossRef] [PubMed]
105. Kumari, S.; Prasad, S.R.; Mandal, D.; Das, P. Carbon dot-DNA-protoporphyrin hybrid hydrogel for sustained photoinduced antimicrobial activity. *J. Colloid Interface Sci.* **2019**, *553*, 228–238. [CrossRef] [PubMed]
106. Liu, J.; Rojas-Andrade, M.D.; Chata, G.; Peng, Y.; Roseman, G.; Lu, J.E.; Millhauser, G.L.; Saltikov, C.; Chen, S. Photo-enhanced antibacterial activity of ZnO/graphene quantum dot nanocomposites. *Nanoscale* **2018**, *10*, 158–166. [CrossRef]
107. Zhang, J.; Liu, X.; Wang, X.; Mu, L.; Yuan, M.; Liu, B.; Shi, H. Carbon dots-decorated $Na_2W_4O_{13}$ composite with WO_3 for highly efficient photocatalytic antibacterial activity. *J. Hazard. Mater.* **2018**, *359*, 1–8. [CrossRef]
108. Yan, Y.; Kuang, W.; Shi, L.; Ye, X.; Yang, Y.; Xie, X.; Shi, Q.; Tan, S. Carbon quantum dot-decorated TiO_2 for fast and sustainable antibacterial properties under visible-light. *J. Alloy. Compd.* **2019**, *777*, 234–243. [CrossRef]
109. De, B.; Gupta, K.; Mandal, M.; Karak, N. Biocide immobilized OMMT-carbon dot reduced Cu_2O nanohybrid/hyperbranched epoxy nanocomposites: Mechanical, thermal, antimicrobial and optical properties. *Mater. Sci. Eng. C-Mater. Biol. Appl.* **2015**, *56*, 74–83. [CrossRef]
110. Lee, C.H.; Song, S.Y.; Chung, Y.J.; Choi, E.K.; Jang, J.; Lee, D.H.; Kim, H.D.; Kim, D.-U.; Park, C.B. Light-stimulated carbon dot hydrogel: Targeting and clearing infectious bacteria in vivo. *ACS Appl. Bio Mater.* **2022**, *5*, 761–770. [CrossRef]
111. Chu, X.; Liu, Y.; Zhang, P.; Li, K.; Feng, W.; Sun, B.; Zhou, N.; Shen, J. Silica-supported near-infrared carbon dots and bicarbonate nanoplatform for triple synergistic sterilization and wound healing promotion therapy. *J. Colloid Interface Sci.* **2022**, *608*, 1308–1322. [CrossRef]
112. Wang, J.; Qiu, J. A review of carbon dots in biological applications. *J. Mater. Sci.* **2016**, *51*, 4728–4738. [CrossRef]
113. Tao, H.; Yang, K.; Ma, Z.; Wan, J.; Zhang, Y.; Kang, Z.; Liu, Z. In vivo NIR fluorescence imaging, biodistribution, and toxicology of photoluminescent carbon dots produced from carbon nanotubes and graphite. *Small* **2012**, *8*, 281–290. [CrossRef] [PubMed]
114. Li, Q.; Ohulchanskyy, T.Y.; Liu, R.; Koynov, K.; Wu, D.; Best, A.; Kumar, R.; Bonoiu, A.; Prasad, P.N. Photoluminescent carbon dots as biocompatible nanoprobes for targeting cancer cells in vitro. *J. Phys. Chem. C* **2010**, *114*, 12062–12068. [CrossRef]
115. Wang, S.; Cole, I.S.; Li, Q. The toxicity of graphene quantum dots. *RSC Adv.* **2016**, *6*, 89867–89878. [CrossRef]
116. Li, Y.J.; Harroun, S.G.; Su, Y.C.; Huang, C.F.; Unnikrishnan, B.; Lin, H.J.; Lin, C.H.; Huang, C.C. Synthesis of self-assembled spermidine-carbon quantum dots effective against multidrug-resistant bacteria. *Adv. Healthcare Mater.* **2016**, *5*, 2545–2554. [CrossRef] [PubMed]
117. Jian, H.J.; Wu, R.S.; Lin, T.Y.; Li, Y.J.; Lin, H.J.; Harroun, S.G.; Lai, J.Y.; Huang, C.C. Super-cationic carbon quantum dots synthesized from spermidine as an eye drop formulation for topical treatment of bacterial keratitis. *ACS Nano* **2017**, *11*, 6703–6716. [CrossRef]

Combatting Antibiotic Resistance Using Supramolecular Assemblies

Shuwen Guo [1,*], Yuling He [2], Yuanyuan Zhu [1], Yanli Tang [1,*] and Bingran Yu [3,*]

1. Key Laboratory of Analytical Chemistry for Life Science of Shaanxi Province, School of Chemistry and Chemical Engineering, Shaanxi Normal University, Xi'an 710100, China; zyyzyy304@snnu.edu.cn
2. Institute of Basic and Translational Medicine, Xi'an Medical University, No. 1 Xinwang Road, Xi'an 710021, China; heyuling0925@163.com
3. State Key Laboratory of Chemical Resource Engineering, Key Lab of Biomedical Materials of Natural Macromolecules (Beijing University of Chemical Technology, Ministry of Education), Beijing 100029, China
* Correspondence: guo-shuwen15@foxmail.com (S.G.); yltang@snnu.edu.cn (Y.T.); yubr@mail.buct.edu.cn (B.Y.)

Abstract: Antibiotic resistance has posed a great threat to human health. The emergence of antibiotic resistance has always outpaced the development of new antibiotics, and the investment in the development of new antibiotics is diminishing. Supramolecular self-assembly of the conventional antibacterial agents has been proved to be a promising and versatile strategy to tackle the serious problem of antibiotic resistance. In this review, the recent development of antibacterial agents based on supramolecular self-assembly strategies will be introduced.

Keywords: supramolecular assembly; antibacteria; antibiotic resistance

1. Introduction

Antibiotic resistance is a growing problem that causes 700,000 deaths per year worldwide, and a recent prediction indicates that bacterial infections will cause 10 million deaths by 2050 [1–6]. The rapidly growing antibiotic resistance is mainly driven by misuse of antibiotics in human medicine and abuse of antibiotics in animal medicine and husbandry [7–10]. The emergence of multi-drug-resistant infections makes the situation worse, as multi-drug-resistant infections require prolonged and high-cost antibiotic therapy [11–14]. However, the investment in research and development of new antibiotics is diminishing, as the emergence of resistance has outpaced the development of new antibiotics [15–18]. Therefore, developing novel approaches to tackle the serious antibiotic resistance crisis is urgently needed.

Researchers in relevant fields are making great efforts to discover new drugs and develop new antibacterial strategies and technologies to address the serious issues caused by antibiotic resistance [19–26]. Supramolecular self-assembly focuses on the autonomous organization of components into multi-component systems through intermolecular noncovalent interactions, including electrostatic, hydrophobic, hydrogen-bonding, metal-ligand coordination, charge–transfer, van der Waals, and π-π stacking interactions [27–32]. The dynamic properties and integration features of supramolecular self-assembly endow the assembly with extraordinary functions, which cannot be empowered by traditional covalent modification strategies. Therefore, supramolecular self-assembly strategies have demonstrated great potential applications in the biomaterial field. In particular, supramolecular self-assembly strategies have been applied in combating bacteria and bacterial biofilms.

In this review, we briefly highlight and discuss the recent development of antibacterial agents based on the supramolecular self-assembly strategies, which involve the above-mentioned noncovalent interactions. The typical and novel supramolecular self-assembly strategies, which could improve the therapeutic performance of antibacterial agents and overcome their shortcomings, will be focused on in this review. The review is divided into six sections according to the type of conventional antibacterial agents constituting the supramolecular self-assembly, including antibiotics, antimicrobial peptides, cationic surfactants and polymers, antibacterial photodynamic therapy (aPDT) agents, antibacterial photothermal therapy (aPTT) agents, and macrophages.

2. Supramolecular Self-Assembly of Antibiotics

Since their introduction in the 1940s, antibiotics have been heavily applied to treat infectious diseases. After the "Golden Age" of antibiotics, semisynthetic and chemically modified antibiotics have been deemed as the bright way to develop new antibiotics [33–37]. However, these innovative compounds cannot address the antibiotic-resistant issues as bacteria could rapidly acquire tolerance after a period of gene response. Thus, developing a novel antibacterial strategy based on the existing antibiotics to combat resistance is greatly needed. Supramolecular self-assembly of antibiotics is a promising strategy to improve therapeutic efficiency, reduce nonspecific cytotoxicity, and depress drug resistance to antibiotics. Bhosale and coworkers have summarized the supramolecular self-assembly of amino-glycoside antibiotics [38], and for readers interested in this aspect, please refer to the review.

The host–guest complexations between antibiotics and macrocyclic hosts are highly potential approaches to enhance antibiotics activity. Sinisterra et al. reported that the complexation of doxycycline (Dox) and β-cyclodextrin (β-CD) showed greater antimicrobial activity than free doxycycline (Figure 1) [39]. Isothermal titration calorimetry (ITC) data indicated that the free doxycycline and the Dox/β-CD complexation bind with the cell membrane through different interactions. The interaction of free doxycycline with the cell membrane is an ion-paring and hydrogen bond, while the interaction of Dox/β-CD complexation with the cell membrane is a much stronger hydrogen bond. The Dox/β-CD complexation could serve as a local and sustained source to improve treatment efficacy. Dend et al. also reported the complexation of per-6(4-methoxylbenzyl)-amino-6-deoxy-β-cyclodextrin (pMBA-βCD) and methicillin (Met) to increase the water solubility of Met and its antibacterial activities against methicillin-resistant Staphylococcus aureus (MRSA) [40], which was presumed that the Met/pMBA-βCD complex improved the affinity between the active β-lactam moiety and the narrow active site groove of MRSA PBP2a, and facilitate the acylation [41,42] (Figure 1). The solubility and antibacterial activity of ciprofloxacin were also enhanced by Mono-6-deoxy-6-aminoethylamino-β-cyclodextrin (mET-βCD, Figure 1) [43]. NOESY NMR indicated that the ethylenediamine moiety of mET-βCD induces stable hydrogen bonding with primary hydroxyls of β CD, leading to the formation of this oval-shaped cavity, which could encapsulate quinolone and the cyclopropyl groups of ciprofloxacin. D-mannose and D-glucose-grafted cyclodextrins were also developed as "Trojan Horse"-like nanocarriers for loading hydrophobic antibiotics (ciprofloxacin, erythromycin, and rifampicin) and potentiate their activity, as the functionalized sugars on CD are chemoattractant for the bacteria and could promote uptake (Figure 1) [44].

Kumari et al. also fabricated the host–guest complexation between resorcin[4]arene and gatifloxacin to improve antibacterial activity [45]. These host–guest complexations indicated that strategies based on host–guest interactions are highly potential approaches to constructing novel antimicrobial formulations, which would improve antibacterial activity, water-solubility, and biocompatibility of traditional antibiotics.

Microbial biofilms showed restricted drug accessibility and antibiotic resistance, and most cells in the biofilms changed their metabolic mode to a dormant state. All of these factors result in great challenges in disrupting biofilms [46,47]. Supramolecular self-assembly of antibiotics is also a potential strategy to disrupt biofilms. Wang and coworkers also developed host–guest complexation between guanidinium per-functionalized pillar[5]arene (GP5) and the antibiotic cefazolin sodium to synergistically eradicate *Escherichia coli* (*E. coli*) biofilm [48]. In vitro experiment showed that GP5 could disrupt biofilm, as guanidinium per-functionalized pillar[5]arene could penetrate through biofilm barriers and destroy biofilm-enclosed bacteria. After GP5 formed the host–guest complex with cefazolin sodium, the host–guest complex penetrated through biofilm due to the high penetrability of pre-organized multiple guanidiniums on pillar[5]arene skeleton (Figure 2). The GP5 and cefazolin sodium, therefore, could effectively breach the bacterial membranes and work synergistically to kill the bacteria within the biofilm matrix.

Encapsulation and delivery of antibiotics through the self-assembly of lipids is a promising way to enhance the bioavailability of antibiotics [49]. Macrophage membrane encapsulated antibiotic was construed to selectively enter into the infected macrophages and efficiently kill intracellular bacteria [50]. The hydrophobic triclosan and hydrophilic ciprofloxacin were covalently conjugated together, and the obtained amphiphilic pro-drug self-assembled into nanoparticles. With further encapsulation with macrophage membranes, the nanoparticles showed similar Toll-like receptor expression and negative surface charge as their precursor murine macrophage/human monocyte cell lines. Such features allowed uptake of the infected macrophages/monocytes through positively charged, lysozyme-rich membrane scars created during staphylococcal engulfment.

Combined antibiotic administration based on co-assembly is also an effective way to eliminate resistant infections. Lehr et al. developed a co-assembled nanoparticle of synthesized amphiphilic lipid (squalenyl hydrogen sulfate), hydrophilic antibiotic tobramycin, and a quorum sensing inhibitor (QS1) to synergistically eradicate *Pseudomonas aeruginosa* (*P. aeruginosa*) biofilm [51].

Wang and coworkers fabricated cascade-targeting poly(amino acid) nanoparticles to sequentially target macrophages and intracellular bacteria and fulfill on-site antibiotic delivery. The mannose-decorated poly(α-N-acryloylphenylalanine)-block-poly(β-N-acryloyl-D-aminoalanine) could self-assemble into nanoparticles and efficiently load rifampicin due to abundant noncovalent interactions between the drug and polymer backbone (Figure 3). The mannose groups from the surface of the nanoparticles promote macrophage-specific uptake and intracellular accumulation. After exposing the D-aminoalanine moieties in the acidic phagolysosome, the nanoparticles escape from lysosomes and target intracellular bacteria through peptidoglycan-specific binding. Subsequently, rifampicin was precisely released regardless of the states and locations of the intracellular MRSA [52].

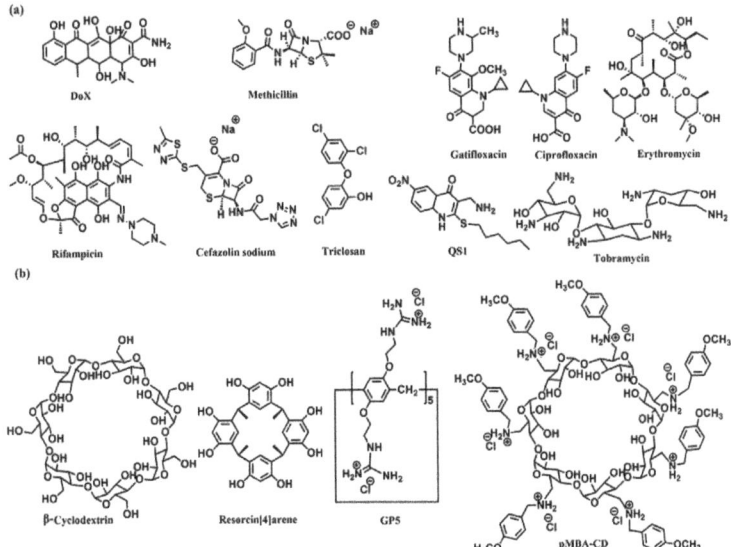

Figure 1. Chemical structures of (a) the antibiotics, and (b) the co-assembly materials used for fabricating supramolecular self-assembly of antibiotics.

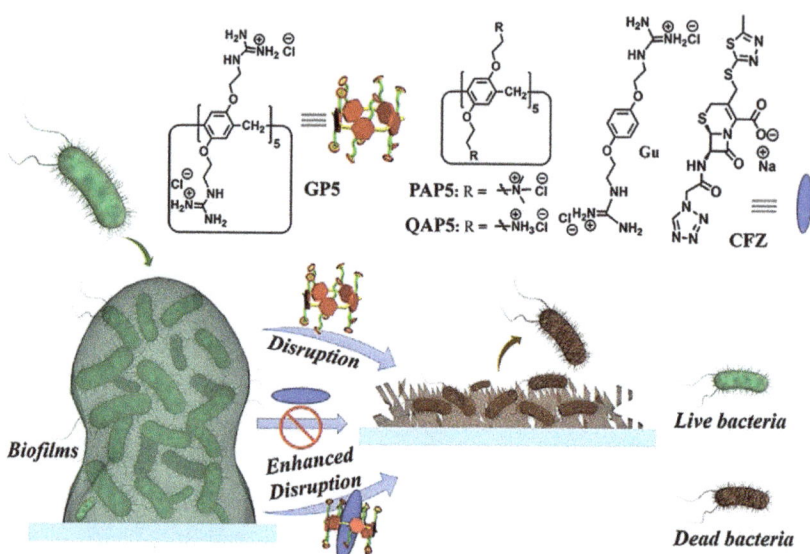

Figure 2. Disruption of biofilms by GP5 and GP5⊃ ∩CFZ. Reprinted with permission from Ref. [48]. Copyright 2021 Wiley-VCH.

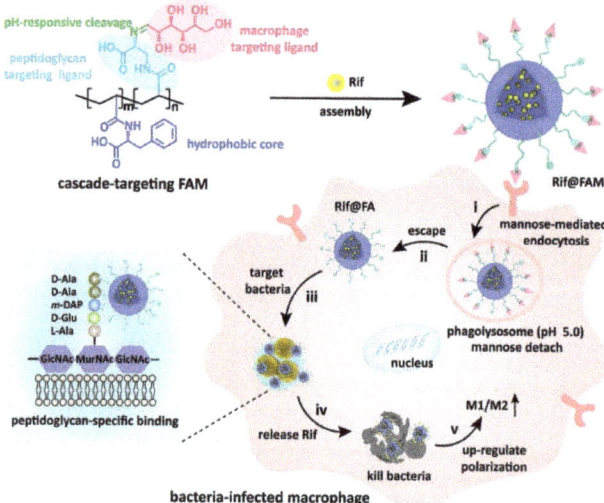

Figure 3. The cascade-targeting DDS eliminates intracellular MRSA. Reprinted with permission from Ref. [52]. Copyright 2022 Wiley-VCH.

3. Supramolecular Self-Assembly of AMPs

Antimicrobial peptides (AMPs), a group of functional peptides that display antimicrobial activities, are potential therapeutic agents due to their broad-spectrum antibacterial activities and low degrees of antimicrobial resistance [53–55]. As peptides are outstanding and feasible building blocks for self-assembly, the artificial peptides with self-assembling capacities have been applied in antibiotic delivery, antibacterial surface defense, bio-sensing of bacteria, and as bacteriostatic agents [56–58]. The self-assembly of AMPs could not only enhance antimicrobial efficacy by locally enhancing the antimicrobial peptide concentration on the bacterial surface but also promote the biological stability of the fragile peptide.

AMPs are drawing great attention in modern medicine due to their high biological activity. Their low stability leads to undesirable bioavailability, as these agents could be degraded by proteases. Further, systemic toxicity is also a major issue for some classes of AMPs. Recently, Li et al. exploited the host–guest complexion between and pexiganan (PXG) to decrease hemolysis and improve the stability of PXG [59]. The two large-sized macrocycles showed strong interactions towards PXG, giving association constants in the magnitude of 10^4 and 10^5 M^{-1}. The host–guest complexation remarkably improves metabolic stability under endogenous proteases conditions and decreases hemolysis of PXG.

Lin et al. developed a supramolecular trap to boost the low-dose antibacterial activity of colistin against *E. coli* by preventing the interaction between colistin and free LPS [60]. The molecular trap was fabricated from a subnanometer gold nanosheet with methyl motifs (SAuM) and could directly target and capture free LPS by binding to and compressing the packing density of lipid A, thus reducing the risk of endotoxin-induced sepsis and preventing its interaction with colistin (Figure 4). The binding specificity assay results showed that SAuM is a strong competitor of colistin for binding with free LPS. The higher driving force of SAuM to bind with LPS is mainly due to the fact that the sheet-like structure of SAuM can precisely bind the lipid A of LPS to form a steric wall that efficiently inhibits colistin binding. This supramolecular trap allows the therapeutic window of colistin to be expanded to low-dose concentrations for the treatment of Gram-negative bacteria infections while also minimizing the risk of endotoxemia.

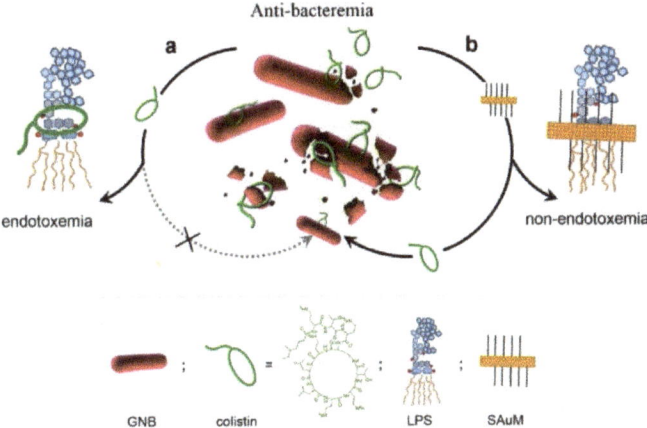

Figure 4. A Supramolecular Trap to Increase the Antibacterial Activity of Colistin. In Path a, free LPS can block the function of colistin. And the free LPS can be sequestered by SAuM in the circulating blood (in Path b), thus promoting the killing efficiency of colistin against GNB while also minimizing endotoxemia. Reprinted with permission from Ref. [60]. Copyright 2020, Wiley-VCH.

Tang and coworkers developed a novel peptide-based co-assembled hydrogel to effectively and specifically capture and kill MRSA bacteria. The supramolecular self-assembly of 9-fluorenylmethyloxycarbonyl-L-phenylalanin (Fmoc-L-Phe) and amino-acid-modified conjugated oligomer OTE-D-Phe formed a new and biocompatible low-molecular weight hydrogel [61]. The hydrogel was composed of a thick and rough fibrous network, which could spontaneously capture MRSA and *E. coli* efficiently. The hydrogel exhibited efficient and specific antibacterial activity against MRSA due to specific interaction with lipid domains of the MRSA membrane.

4. Supramolecular Self-Assembly of Cationic Surfactants and Polymers

Cationic surfactants have been extensively applied as antimicrobial agents in various fields due to their excellent bactericidal activities [62–64]. The cationic surfactants exert biological activity mainly by disrupting bacterial membranes via electrostatic and hydrophobic interaction, which further leads to cell lysis and bacterial death [65,66]. The cationic surfactants were once supposed to be impervious to resistance. However, after huge accumulation in the environment over the past decades, bacteria have acquired surfactant resistance [67–69]. Moreover, it is a great challenge to design highly antibacterial surfactants with high selectivity, as the essential factors for antibacterial activity also predominate their cytotoxicity. Recently, the application of the supramolecular self-assembly strategy for cationic surfactants has shown encouraging results in improving their antimicrobial activity, bacterial resistance, selectivity, etc.

Wang and coworkers developed supramolecular antimicrobial agents with switchable activities, which could reduce the pressure on bacteria and prevent the emergence of bacterial resistance. The supramolecular antibacterial switches were constructed based on the host–guest complexation of cucurbit(7)uril (CB[7]) and cationic surfactants (DDBAC, DDBAB, BCDAC, and DTAC) [70]. The antimicrobial ability was switched off when the alkyl chain of surfactants was encapsulated by CB[7], which concealed the active sites. The antimicrobial ability was switched on when the active sites recovered upon the addition of competitive guest amantadine hydrochloride (AD). As the exposure frequency of bacteria to active surfactants could be reduced by regulating the antimicrobial switch, the occurrence of drug resistance was prevented by the antimicrobial switch.

Wang et al. also fabricated the host–guest complexation of biocompatible CDs and cationic surfactant (tri(dodecyldimethylammonioacetoxy)-diethyltriamine trichloride, DTAD) to improve the mildness of cationic surfactant [71]. The outstanding antimicrobial DTAD showed a strong skin irritation effect due to its larger numbers of cationic charges and multiple hydrophobic chains. The host–guest complexation of DTAD and three types of CDs (α-, β-, γ-CD) could self-assemble to nanostructure with different morphology, and the antibacterial activity and mildness of DTAT were all improved by the host–guest complexation and further self-assembly.

Tang and coworkers fabricated cationic nanoparticles through the co-assembly of cationic surfactant cetyltrimethylammonium bromide (CTAB) and conjugated polymers (PFVBT), which could enhance the antibacterial effect and lower the cytotoxicity of CTAB [72]. When mixing CTAB and PFVBT together, the hydrophobic chain of the polymers was entwined with the hydrophobic part of CTAB to form the nucleus, and the hydrophilic quaternary ammonium groups formed shells on the surface, resulting in the spherical cationic nanoparticles. Instead of light source requirement, which is vital for the antibacterial activity of most conjugated polymers and nanoparticles, the cationic nanoparticles showed efficient and broad-spectrum biocidal activities in the dark. This work provides a new way to combine cationic surfactant and conjugated polymers together to combat bacteria.

The commonly used antibacterial quaternary ammonium surfactants (QAS) generally lack long-lasting performances. The frequent usage of these agent cause accumulation in the environment, which would exert a huge burden on the ecosystem and selective pressure on bacteria, triggering the emergence of antimicrobial resistance. Supramolecular self-assembly is a potential strategy to fabricate long-lasting antimicrobial agents. For example, Wang et al. fabricated long-lasting antimicrobial aggregate by co-assembling QAS and biocompatible gallic acid through multiple synergistic noncovalent interactions, including electrostatic interaction, hydrogen bonding, hydrophobic interaction, and π-π stacking [73]. Three QASs, including dodecyltrimethylammonium bromide (DTAB), trimethylene-1,3-bis-(dodecyldimethyl-ammonium bromide) and methyldodecylbis[3-(dimethyldodecylammonio) propane] ammonium tribromide were chosen to co-assemble with gallic acid (GA) (Figure 5). GA could counterbalance the charge repulsion of the surfactant headgroups after electrostatic binding with the surfactant, and GA could serve as a linker to adhere surfactant micelles together through intermolecular hydrogen bonding of the phenolic hydroxyl groups. The parallelly stacked micelles formed hexagonal columns, and the hexagonal columns interpenetrated spheres (HCISs) were finally generated when these hexagonal columns intersperse at different orientations. The strong dehydration of HCISs and their adhesion force to substrates endowed the HCISs with anti-water washing properties and long-lasting antimicrobial performance.

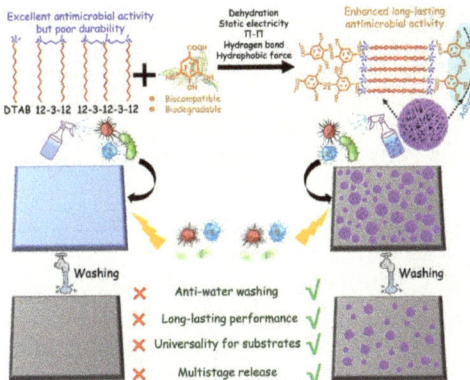

Figure 5. Construction of substrate-adhesive aggregates (hexagonal columns interpenetrated spheres, HCISs) based on noncovalent assembly of gallic acid (GA) and three quaternary ammonium surfactants (QASs). Reprinted with permission from Ref. [73]. Copyright 2021, Wiley-VCH.

Traditional antibiotic abuse has caused a lot of health problems, including antimicrobial resistance. In order to tackle these problems, Lei and coworkers investigated two self-assembling modes between berberine (BBR) and flavonoid glycosides and compared their antibacterial potency (Figure 1). Based on electrostatic and hydrophobic interactions, the mixture of BBR and baicalin formed nanoparticles in an aqueous solution, while BBR and wogonoside generated nanofibers [74]. The nanoparticles exhibited significantly enhanced bacteriostatic activity, whereas nanofibers showed a weaker effect than BBR. The enhanced bacteriostatic activity of nanoparticles is likely attributed to their stronger affinity to bacteria as hydrophilic glucuronic acid was distributed on the surface of nanoparticles. Using a similar strategy, they also developed nanoparticles by the self-assembly of BBR and cinnamic acid to combat MRSA and biofilm [75]. The self-assembly of the nanoparticles was governed by hydrogen bonds and π-π stacking interactions.

A precise pesticide delivery nanoplatform was fabricated by iron mineralization after electrostatic self-assembly between sodium lignosulphonate (SL) and dodecyl dimethyl benzyl ammonium chloride (DDBAC). The nanoplatform was a core–shell structure, which was excellently stabilized by iron mineralization. The outer layer of the shell was constructed by in situ mineralization with iron ions, while the endothecium of the shell was formed by the electrostatic self-assembly of SL and DDBAC. The nanoplatform was suitable for encapsulating agrochemicals with hydrophobicity and low volatility. The nanoplatform showed a specific response to alkalescency and laccase in the alimentary tract of insects and also improved the photostability and environmental security of agricultural chemicals [76].

Shen et al. designed an electrochemical redox-controlled bacterial inhibition agent [(12-ferrocenyl) benzalkonium bromide (FBZK)] containing ferrocene quaternary ammonium salt based on supramolecular self-assembly and disassembly [77]. They demonstrated that using a redox-active cationic antibiotic to regulate antibacterial activity with an alternating electric field takes the critical micelle concentration (CMC) as the boundary, inhibition of antimicrobial activity by redox-induced changes in hydrophilicity and hydrophobicity. At a concentration over the CMC, self-assembly into micelles led to a decrease in antibacterial activity; on the contrary, the disassembly of micelles can enhance the bactericidal effect. When the concentration was below the CMC, REFBZK with a hydrophobic tail penetrated more easily into the membrane.

Yang et al. exploited the supramolecular self-assemblies of cationic polyaspartamides and anionic carboxylatopillar[5]arene (CP[5]A) to target Gram-positive bacteria and mitigate antimicrobial resistance (Figure 6) [78]. The cationic polyaspartamides derivatives were functionalized with quaternary ammonium on the side chains, which contribute to the insertion of the catiomers into the negatively charged bacterial membranes and promotion of membranolysis and cell death. The anionic carboxylatopillar[5]arene (CP[5]A) host was further introduced to complex with the catiomers, which increased the biocompatibility of catiomers. The host–guest interaction endowed the systems with pH-responsiveness, as CP[5]A could depart from cationic quaternary ammonium compounds under acid conditions. Therefore, the biocomposite could selectively eliminate Gram-positive bacteria. In vivo MRSA-infected wound treatment further demonstrated that the system possessed potential values in practical application.

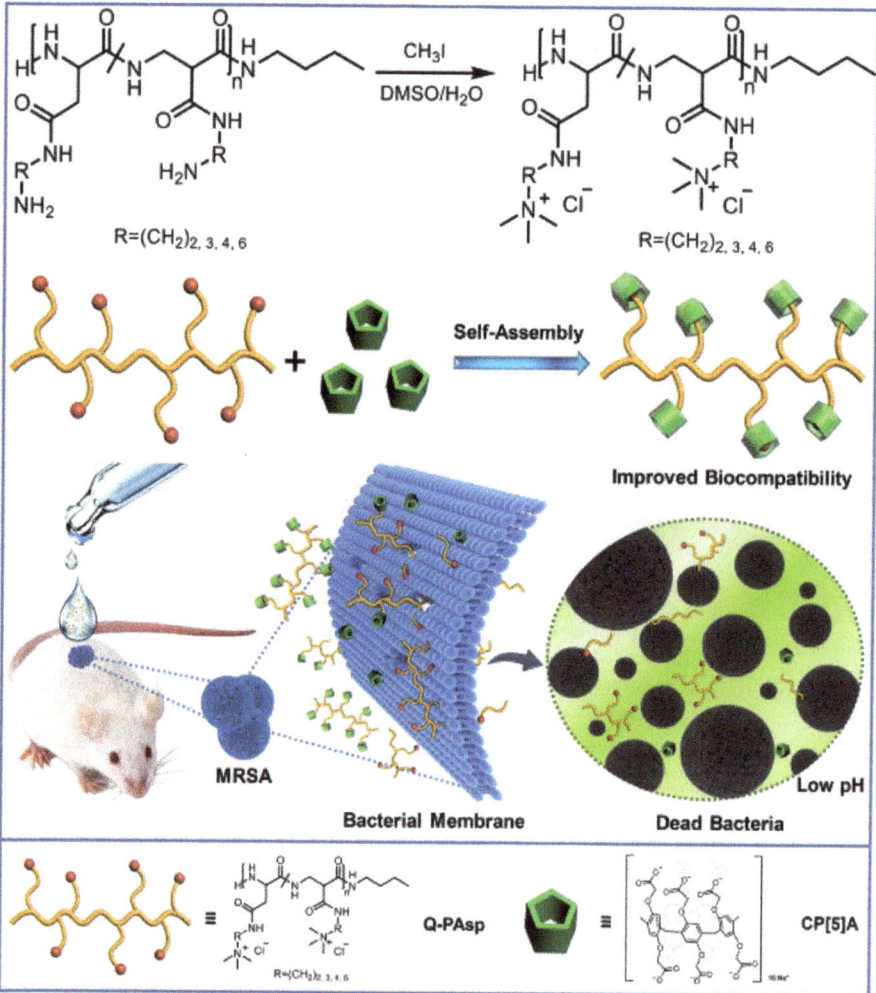

Figure 6. Schematic illustration for construction of supramolecular antimicrobial system and its antibacterial property for healing MRSA-infected wound. Reprinted with permission from Ref. [78]. Copyright 2019 Wiley-VCH.

Chan-Park and coworkers also developed block copolymer nanoparticles to remove biofilms of drug-resistant Gram-positive bacteria using nanoscale bacterial debridement [79]. The block copolymer DA95B5, dextran-block–poly((3-5 acrylamidopropyl) trimethylammonium chloride (AMPTMA)-co-butyl methacrylate (BMA)), could self-assemble into core–shell nanoparticles with a non-fouling dextran shell and a cationic core. The nanoparticles could diffuse through Gram-positive biofilm and attach to bacterial surfaces. The dextran corona enhanced the bacterial/nanoparticle interactions to promote bacterial detachment from the biofilm, which reduced biofilm biomass. The bacterial debridement mechanism is orthogonal to antibiotic resistance so that it removes biofilms of drug-resistant strains as effectively as those of drug-sensitive strains.

5. Supramolecular Self-Assembly for Improvement of aPDT

Antibacterial Photodynamic therapy (aPDT) is a promising way to eliminate bacteria and biofilm infections. In the aPDT process, the reactive oxygen species (ROS) destroy biomolecules in bacterial cells, leading to cell death [80–82]. The antibacterial efficiency of aPDT generally is very high, and it also does not induce drug resistance; thus, aPDT has been applied in antibacterial therapy for body infections. However, more excellent aPDT systems are still needed to overcome the drawbacks of aPDT. For instance, the ROS has limited diffusion distances and short lifetimes, and it is inefficient for Gram-negative bacteria due to its complicated cell membrane. The light employed for activation of photosensitizer always shows weak tissue penetration, and the dependence of oxygen for generating ROS also restricts their therapeutic effect. In addition, the reactivity of ROS is very nonspecific, which causes by-effect and lowers their antibacterial performance [83–87]. Therefore, the supramolecular self-assembly strategy has been employed to improve the therapeutic performance of aPDT in two ways, including enhancing the generation of ROS and regulating the interaction between aPDT systems and bacteria.

Due to a strong tendency of π-π stacking in an aqueous solution, the singlet states of photosensitizer are quenched, leading to low ROS generation efficiency. Host–guest complexation of photosensitizer by macrocycles is an efficient way to enhance ROS generation by preventing aggregation of the hydrophobic photosensitizer. For instance, Zhang and coworkers exploited the host–guest complexation of cucurbit[7]uril (CB[7]) and porphyrin derivatives (TPOR) to enhance the antibacterial efficiency [88]. As the naphthalene-methylpyridinium moiety on TPOR can form stable host–guest complexation with CB[7], the supramolecular TPOR/(CB[7])4 assembly was obtained when mixing TORP and CB[7]. The TPOR/(CB[7])4 assembly showed different self-assembling behavior, comparing the self-assembly of TPOR (Figure 7). The TPOR showed more intense fluorescence due to suppression of self-quenching of the excited state, and the $^1O_2$ generation was also enhanced. Thus, the bacterial inhibition ability was efficiently enhanced by promoting ROS generation. Furthermore, metal ions, including Zn(II) and Pd(II), were introduced into the porphyrin core to enhance its ability to produce ROS and improve the efficiency of PDT [89].

Supramolecular assembly strategy is applied to manipulate the absorbance wavelength of conjugated polythiophene to enhance its aPDT effect under red light irradiation by Xing and coworkers. They fabricated hybrid hydrogels by co-assembling of tri(ethylene glycol)-functionalized polyisocyanide (PIC) and conformation-sensitive conjugated polythiophene, poly(3-(3′-N,N,N-triethylammonium-1′-propyloxy)-4-methyl-2,5-thiophene chloride) (PMNT) (Figure 8). The PIC polymer could serve as a scaffold to trap and align the PMNT backbone into a highly ordered conformation, leading to the generation of a red absorption band. Under red light irradiation, the PMNT/PIC co-assembly showed much higher ROS production than PMNT in its random conformation, leading to efficient aPDT towards various pathogens [90].

Designing aggregation-enhanced ROS production systems is also a novel thread for improving the therapeutic performance of aPDT. Yoon and coworkers developed an aggregation-enhanced photodynamic therapy system for combating antibiotic-resistant bacteria [91]. The self-assembly of the 3-{N-(4-boronobenzyl)-N,N-dimethylammonium} phenoxy-substituted zinc(II) phthalocyanine (PcN4-BA) generated the novel photosensitizer, and the boronic acid functionalized on the PCN4-BA endow the self-assembly with bacteria target capacity (Figure 9). The aggregation of PcN4-BA induces large intermolecular interaction, which further increases the density of electronic states and reduces the energy gap between the excited singlet and triplet states. Furthermore, the enhanced energy overlap between excited singlet and triplet states promotes ISC, resulting in the improvement of ROS generation.

Light harvesting from the antenna dye to photosensitizer was also employed to enhance the therapeutic performance of aPDT by Wang and coworkers [92]. The anionic polythiophene (PTP) and cationic porphyrin (TPPN) formed PTP/TPPN complex through electrostatic interactions, and efficient energy transfer from PTP to TPPN occurred upon white light irradiation, which further promoted ROS generation. After the positive charged PTP/TPPN complex was absorbed on negatively charged bacteria membranes through electrostatic interactions, the singlet oxygen effectively killed the bacteria.

Regulation of the antibacterial activity of biotics was also accomplished by a supramolecular antibiotic switch. Wang and coworkers fabricated the supramolecular antibiotic switch by the host–guest complexation between cucurbit[7]uril (CB[7]) and the cationic pedants of poly(phenylene vinylene) derivative (PPV) [93]. When the host–guest complexation was retained, the supramolecular antibiotic switch showed low ROS generation efficiency, as the encapsulated PPV was prevented from contacting the surrounding oxygen. Further, the encapsulated PPV also could not bind the bacterial membrane, as the cationic pedants were complexed by CB[7]. Upon competitive replacement by amantadine, the antibacterial activity of the PPV was turned on and showed effective aPDT.

Tang and coworkers also regulated the antibacterial photodynamic effect of oligo(phenylene ethynylene) (OPE) through the supramolecular self-assembly of OPE and DNA [94]. DNA-OPE hydrogel was formed through the electrostatic interactions between positively charged OPE and negatively charged DNA. In the gel state, the OPE showed little biocidal activity towards E.coli, as the OPEs are "adsorbed" by DNA via electrostatic interactions in the gel. When the DNA was hydrolyzed by DNase, the released OPE fully exerted its photodynamic effect on the cell.

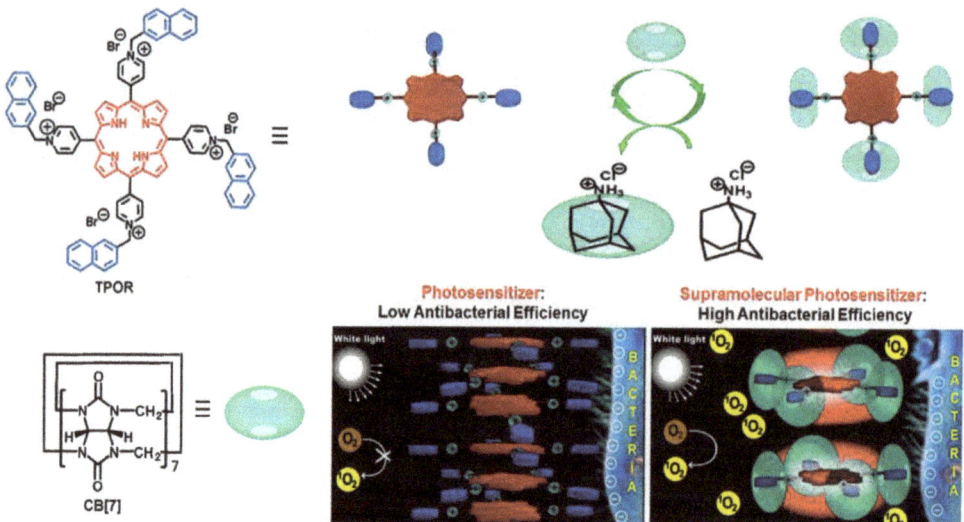

Figure 7. The construction of TPOR/(CB[7])$_4$ supramolecular photosensitizers and the mechanism for the enhanced antibacterial efficiency of TPOR/(CB[7])$_4$ compared with that of TPOR. Reprinted with permission from Ref. [88]. Copyright 2019 Wiley-VCH.

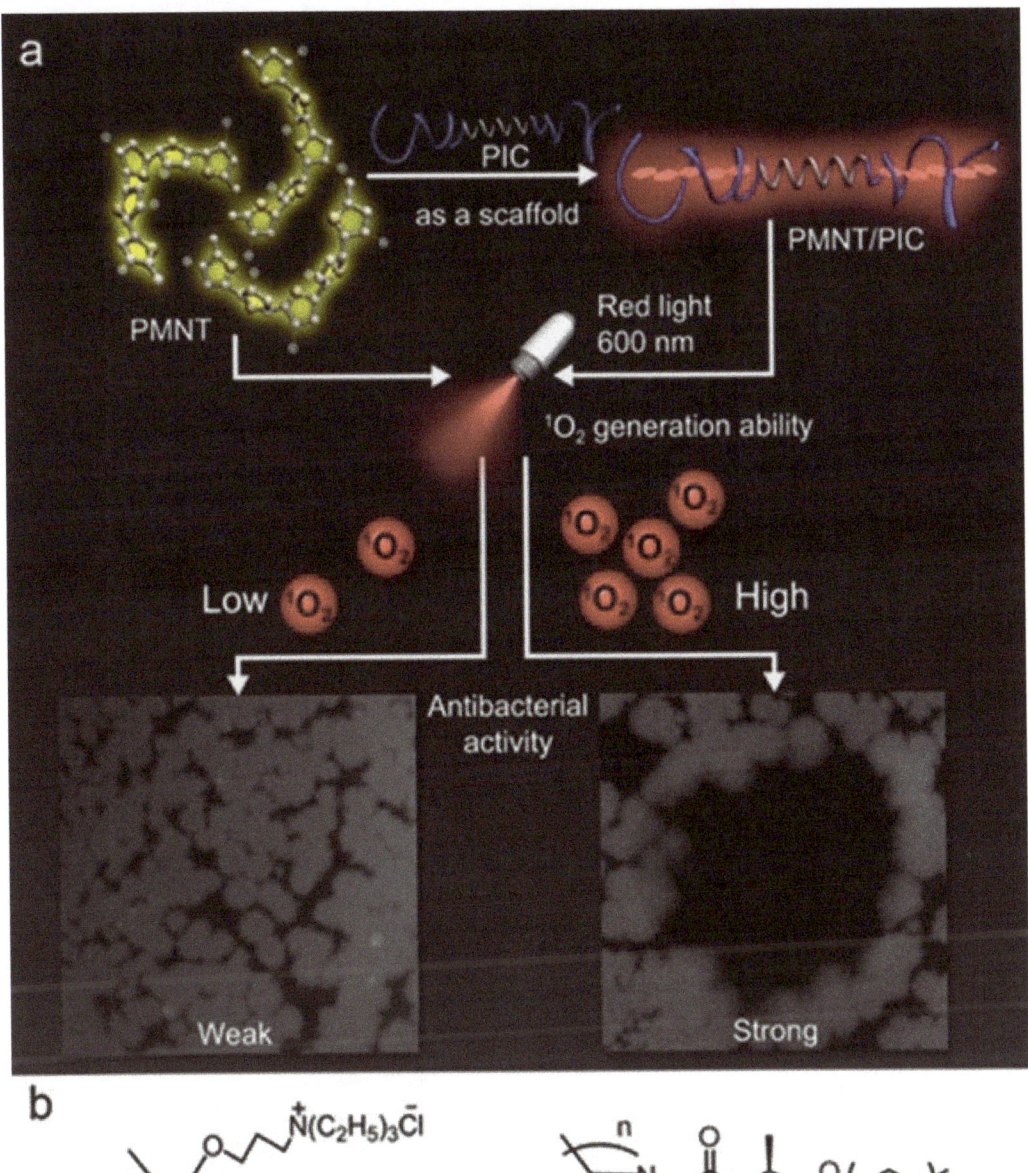

Figure 8. (a) Schematic of the assembly of PMNT/PIC, and its antibacterial activity under red light. (b) Chemical structure of PMNT and PIC. Reprinted with permission from Ref. [90]. Copyright 2021 Wiley-VCH.

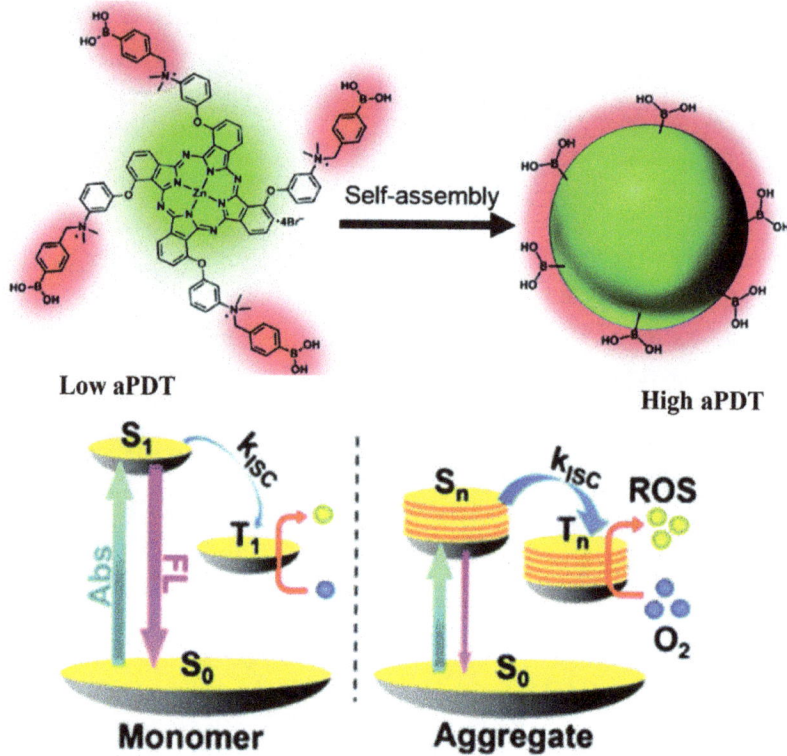

Figure 9. Aggregation-enhanced ROS production of PcN4-BA for combating antibiotic-resistant bacteria. Reprinted with permission from Ref. [91]. Copyright 2020 Royal Society of Chemistry.

A supramolecular self-assembly strategy was also applied to improve the biocompatibility and endow the photosensitizer with acid responsiveness. Recently, Zhang and coworkers fabricated an acid-triggered supramolecular porphyrin photosensitizer to combat bacteria and biofilms. The host–guest complexation of carboxylatopillar[5]arene (CP5) and quaternary ammonium-functionalized tetrafluorophenyl porphyrin (TFPP-QA) formed nanoparticles in aqueous solutions, resulting in the supramolecular photosensitizer [95]. The quaternary ammonium compound showed indiscriminate cytotoxicity to both bacterial cells and normal tissue cells, which greatly limited their clinical applications. After host–guest interactions, the biocompatibility of TFPP-QA was significantly improved. As the CP5 possess excellent pH-responsiveness, the TFPP-QA could release from the nanoparticles when they reach the acidic inflammatory tissue.

Zhang and coworkers also constructed a pH-responsive supramolecular antibacterial photosensitizer AgTPyP@P[5] by the host–guest complexation between a water-soluble photosensitizer silver tetra(N-methyl-4-pyridyl) porphyrin (AgTPyP) and carboxylatopillar[5]arene (P[5]) [96]. The host–guest complexation could further self-assemble to form nanoparticles. The introduction of P[5] could reduce the toxicity of AgTPyP to normal tissues. In the acidic bacteria-infected inflammation tissue microenvironment, the host–guest complexation was dissociated and released the AgTPyP. Upon light irradiation, the ROS produced by AgTPyP could effectively kill bacteria.

Regulating the interaction between aPDT systems and bacteria through supramolecular self-assembly is also a promising way to enhance the therapeutic performance of aPDT and reduce side effects. Xu and coworkers exploited the self-assembly of amphiphilic polymer containing bacteria-target moieties and photosensitizers to combat multi-drug-resistant *Pseudomonas aeruginosa* (MDR-*P. aeruginosa*, Figure 10) [97]. The α-D-galactose on the surface of the nano-assembly targeted the *P. aeruginosa* through carbohydrate–protein interaction, and the bacteria were efficiently killed by the generated singlet oxygen from the photosensitizer Rose Bengal.

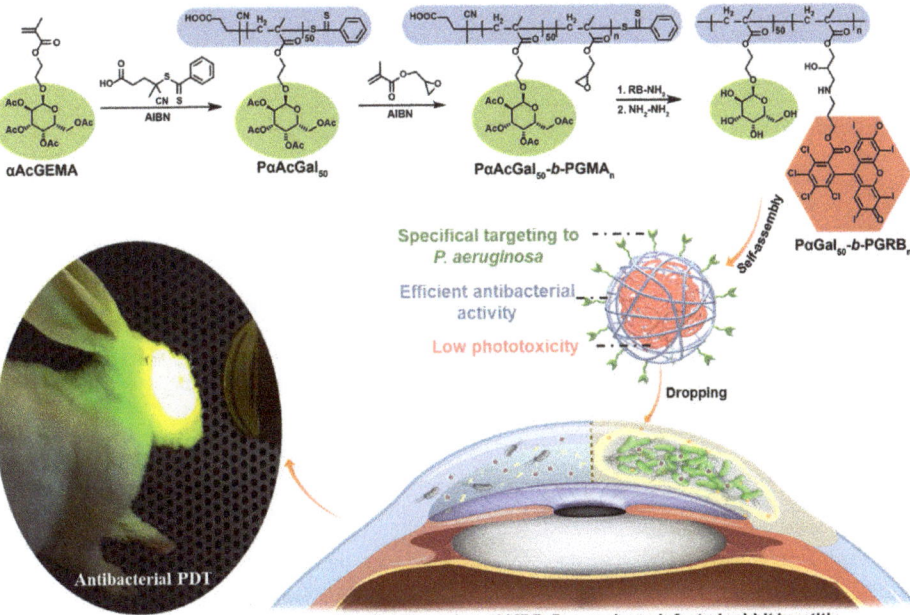

Figure 10. Schematic illustration of MDR bacterial-targeted nanoassembly and its bioapplication in MDR-*P. aeruginosa* biofilm infected rabbit keratitis model. Reprinted with permission from Ref. [97]. Copyright 2021 Wiley-VCH.

Multicharged supramolecular aPDT agent was fabricated by the host–guest complexation between hexa-adamantane-appended ruthenium polypyridyl (Ru2) and polycationic cyclodextrin (CD-QAS) to enhance the specific intercalation between negatively charged bacteria membrane and the aPDT agent [98].

6. Supramolecular Self-Assembly for Improvement of PTT

Photothermal therapy (PTT) is an antibacterial method dependent on the light-induced heat from the photothermal agents. The locally increased temperature (≥ 41 °C) inactivates the microbial cell by inducing protein denaturation and aggregation, disruption of cell membrane integrity, and DNA cross-linking [99–101]. Thus, PTT exhibits no drug resistance and holds great potential to treat drug resistance. However, the shortcomings of PTT still need to be overcome. For instance, the light employed for photothermal agents always shows weak tissue penetration. It also always possesses side effects on normal tissues and low efficiencies [102,103]. In order to overcome these shortcomings, a supramolecular self-assembly strategy has been employed to improve the therapeutic performance of PTT in two ways, including improving the photothermal conversion efficiency and regulating the interaction between PTT systems and bacteria.

Supramolecular radical anions were fabricated to serve as an in situ generated antibacterial near-infrared photothermal agents. The host–guest complexation of PPDI and CB[7] was firstly constructed (Figure 11), which could eliminate nonspecific antibacterial behavior of PPDI thorough preventing the benzyl groups of PPDI from inserting into the membrane of bacteria [104]. Moreover, the host–guest complexation also could promote the stability of PPDI radical anions by preventing them from dimerizing and quenching. When the host–guest complexation was incubated with bacteria with enough reductive ability (E. coli), the supramolecular radical anions were in situ generated, which further generated hyperthermia to induce bacterial cell death under near-infrared light irradiation.

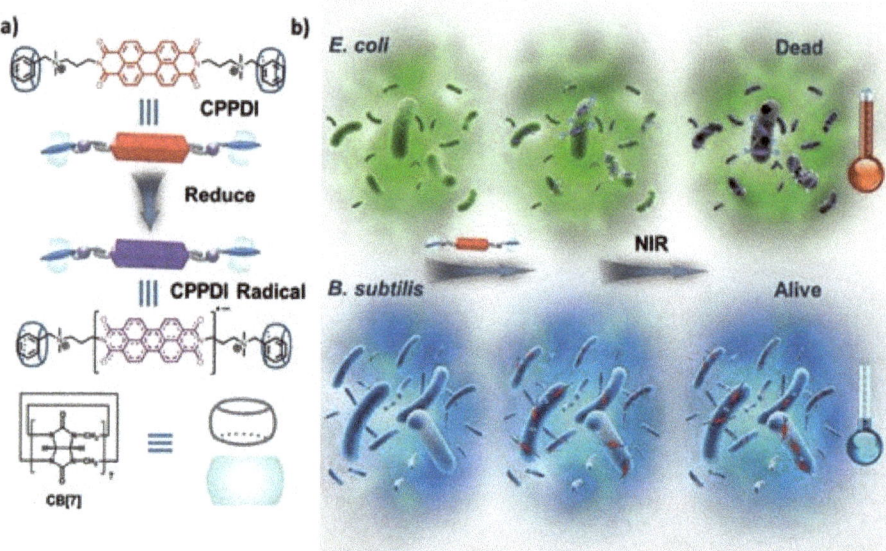

Figure 11. (a) Chemical structures of the supramolecular complex (CPPDI) and CPPDI radical anions. (b) Diagram of photothermal therapy for supramolecular complex with high selectivity towards E. coli over B. subtilis. Reprinted with permission from Ref. [104]. Copyright 2017 Wiley-VCH.

In order to improve the therapeutic performance and selectivity of PPT, E. coli induced in situ supramolecular polymer was constructed by Xu and coworkers [105]. A bifunctional monomer containing two viologen moieties as end groups and a rigid and positively charged 1,4-diazabicyclo[2.2.2]octane unit as the linker (VDV) was designed to fabricate supramolecular polymer with CB[8]. When incubated with E. coli, the CB[8] and VDV in situ formed a supramolecular polymer integrated with supramolecular dimers of viologen cation radicals on the surface of bacteria, as the viologen moieties were reduced to viologen cation radicals by E. coli. The photothermal therapeutic performance of supramolecular dimer of viologen cation radicals was improved by the local enrichment effect of supramolecular polymers and their adsorption onto the bacteria surface. As the viologen moieties were only able to be reduced by E. coli, the supramolecular polymer showed high selective photothermal therapeutic performance toward E. coli.

A NIR-II photothermal antibacterial agent with much deeper tissue penetration and higher maximum permissible exposure was further fabricated by Lee and coworkers using a charge–transfer complex (Figure 12) [106]. The charge–transfer complex was made up of the donor perylene (PER) and acceptor tetracyanoquinodimethane (TCNQ). The PER with high π-electron donating ability may delocalize electron of TCNA, which would narrow the energy gap of charge–transfer complex. The small energy gap would extend the light-absorbing ability in the NIR-II region and promote a high rate of non-radiative transition. The charge–transfer complex showed an excellent PTT effect on E. coli and S. aureus under NIR-II laser irradiation.

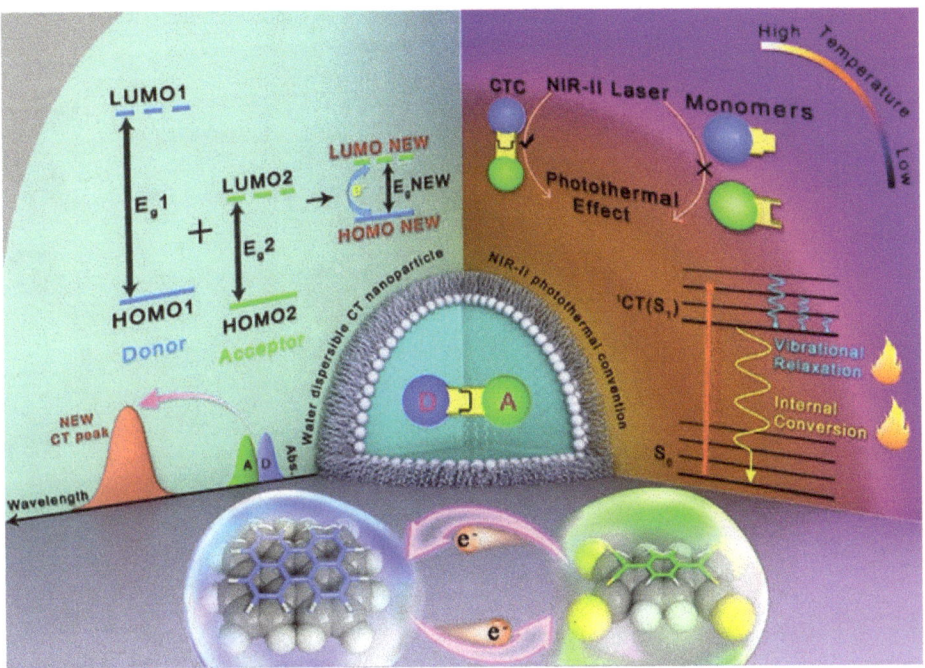

Figure 12. Diagram showing photothermal effect in charge transfer cocrystal nanoparticle. Reprinted with permission from Ref. [106]. Copyright 2021 Wiley-VCH.

Regulating the interaction between PTT agents and bacteria through supramolecular self-assembly is an effective way to enhance the therapeutic performance of PTT. A supramolecular carbohydrate-functionalized two-dimensional (2D) surface was fabricated by Haag and coworkers to selectively capture E. coli and enhance the PTT of the graphene. The supramolecular carbohydrate-functionalized two-dimensional (2D) surface was constructed by host–guest complexation between hepta mannosylated β-CD (ManCD) and the adamant group on thermally reduced graphene oxide. The host–guest complexation on the surface of thermally reduced graphene oxide not only increased the intrinsic water solubility but also endowed the surface with effective E. coli capture capacity through multivalent interactions. Upon IR laser irradiation, the captured E. coli was killed effectively by the photo-induced heat from the graphene oxide [107].

7. Supramolecular Self-Assembly for the Improvement Phagocytosis of Macrophage

The phagocytosis of bacteria by macrophages acts as the front line to protect the human body from bacterial pathogens. The pathogens' clearance efficiency is always low due to the relatively slow encounter and insufficient capture of bacteria by macrophages. Wang and coworkers developed a supramolecular artificial receptor-modified macrophage (SAR-Macrophage) to enhance the recognition and capture of bacteria in the systemic circulation. SAR-Macrophage was fabricated by decorating CB[7] on the macrophage surface via inserting DSPE-PEG-CB[7] (1,2-distearoyl-sn-glycero-3-phosphoethanolamine-poly(ethylene glycol)-CB[7]) into the surface membrane of the cells (Figure 13). E. coli was decorating with adamantyl (ADA) groups by the mannose-ADA via specific mannose–FimH interactions. Based on the strong and multipoint host–guest interactions between CB[7] on SAR-Macrophage and ADA on the E. coli, the SAR-Macrophage could significantly recognize E. coli and catch and internalize more pathogens. The promoted recognition and capture of E. coli induced the M1 polarization of macrophages to generate ROS and effectively killed the intracellular bacteria. This work provides a supramolecular cell-engineering approach for potential antibacterial applications [108].

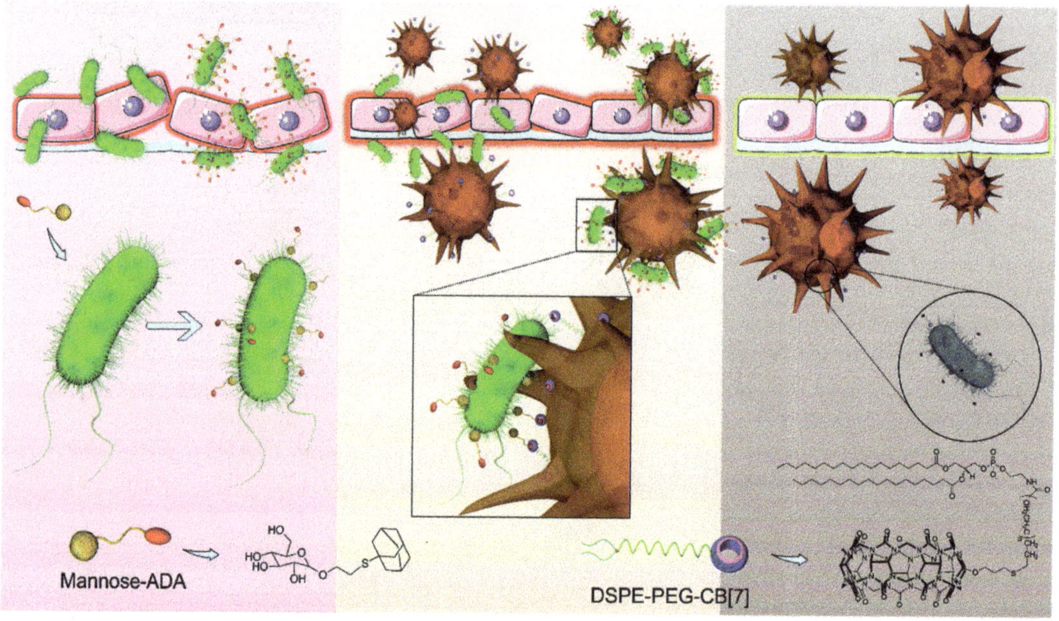

Figure 13. Enhanced antibacterial function of a supramolecular artificial receptor-modified macrophage. Reprinted with permission from Ref. [108]. Copyright 2022 Royal Society of Chemistry.

8. Summary

Antibiotic resistance has caught the attention of researchers from different fields, and plentiful multidisciplinary studies for combating antibiotic resistance have been reported in recent years. Supramolecular assembly is tunable, modular, and responsive, as the formation of assembly is based on noncovalent interactions. Supramolecular self-assembly provides a promising and versatile strategy to construct new antibacterial agents to combat antibiotic-resistant infections. In this review, we hope we have shown the recent development of antibacterial agents based on the supramolecular self-assembly strategies, which include the supramolecular self-assembly of antibiotics, antibacterial peptides, cationic surfactants and polymers, antibacterial photodynamic therapy agents, antibacterial photothermal therapy agents, as well as supramolecular assembly strategy engineered macrophage-bacteria interaction for promoting phagocytosis. These new antibacterial agents showed enhanced antibacterial activity, or the intrinsic shortcomings of the motifs were improved after supramolecular self-assembly.

The current state of the art for antibacterial agents based on supramolecular self-assembly strategy exhibits strong potential to treat drug-resistant bacterial infections in the near future. However, there are still some issues that require to be added to promote the clinical translation of the novel antibacterial agents. First, the new antibacterial agents need to possess excellent biocompatibility. Biocompatibility is a basic requirement for constructing the new antibacterial agents, and the supramolecular self-assembly strategy also could be considered for the cytotoxicity of the potential reported antibacterial agents. Second, the stability of the new antibacterial agents needs to be systematically investigated. The new antibacterial agents exert their enhanced activity through the positive effects induced by supramolecular self-assembly, but the supramolecular self-assembly may show instability in the complex biological media. The revelation of the mechanism of the new antibacterial in the complex biological media will help in developing effective strategies for future applications. Third, the generation of resistance needs to be carefully evaluated during the research stage. As the resistance could be developed rapidly, the evaluation of resistance generation promotes feasibility for clinical tests. Last but not least, more new antibacterial agents based on supramolecular self-assembly need to be fabricated.

Author Contributions: S.G. initiated the review concept and designed the manuscript contents; Y.H. and Y.Z. helped in the revision of the manuscript; Y.T. and B.Y. supervised the entire project. All the authors conjointly reviewed and edited the manuscript. All authors have read and agreed to the published version of the manuscript.

Funding: This research was funded by the National Natural Science Foundation of China (Grants 22101310, 21974084), and Innovation Capability Support Program of Shaanxi (Program no. 2021TD-42). And The APC was funded by the National Natural Science Foundation of China (Grants 22101310).

Institutional Review Board Statement: Not applicable.

Informed Consent Statement: Not applicable.

Data Availability Statement: Data sharing not applicable.

Acknowledgments: The authors are grateful for the support of the funding from the National Natural Science Foundation of China (Grants 22101310, 21974084), and Innovation Capability Support Program of Shaanxi (Program no. 2021TD-42).

Conflicts of Interest: The author does not declare a conflict of interest.

References

1. Willyard, C. The drug-resistant bacteria that pose the greatest health threats. *Nature* **2017**, *543*, 15. [CrossRef] [PubMed]
2. Gupta, A.; Mumtaz, S.; Li, C.-H.; Hussain, I.; Rotello, V.M. Combatting antibiotic-resistant bacteria using nanomaterials. *Chem. Soc. Rev.* **2019**, *48*, 415–427. [CrossRef] [PubMed]
3. Hobson, C.; Chan, A.N.; Wright, G.D. The Antibiotic Resistome: A Guide for the Discovery of Natural Products as Antimicrobial Agents. *Chem. Rev.* **2021**, *121*, 3464–3494. [CrossRef] [PubMed]
4. Miethke, M.; Pieroni, M.; Weber, T.; Brönstrup, M.; Hammann, P.; Halby, L.; Arimondo, P.B.; Glaser, P.; Aigle, B.; Bode, H.B.; et al. Towards the sustainable discovery and development of new antibiotics. *Nat. Rev. Chem.* **2021**, *5*, 726–749. [CrossRef]
5. Namivandi-Zangeneh, R.; Wong, E.H.H.; Boyer, C. Synthetic Antimicrobial Polymers in Combination Therapy: Tackling Antibiotic Resistance. *ACS Infect. Dis.* **2021**, *7*, 215–253. [CrossRef]
6. Wang, Z.; Koirala, B.; Hernandez, Y.; Zimmerman, M.; Park, S.; Perlin, D.S.; Brady, S.F. A naturally inspired antibiotic to target multidrug-resistant pathogens. *Nature* **2022**, *601*, 606–611. [CrossRef]
7. Allen, H.K.; Donato, J.; Wang, H.H.; Cloud-Hansen, K.A.; Davies, J.; Handelsman, J. Call of the wild: Antibiotic resistance genes in natural environments. *Nat. Rev. Microbiol.* **2010**, *8*, 251–259. [CrossRef]
8. Berendonk, T.U.; Manaia, C.M.; Merlin, C.; Fatta-Kassinos, D.; Cytryn, E.; Walsh, F.; Bürgmann, H.; Sørum, H.; Norström, M.; Pons, M.-N.; et al. Tackling antibiotic resistance: The environmental framework. *Nat. Rev. Microbiol.* **2015**, *13*, 310–317. [CrossRef]
9. Dar-Odeh, N.; Fadel, H.T.; Abu-Hammad, S.; Abdeljawad, R.A.; Abu-Hammad, O.A. Antibiotic Prescribing for Oro-Facial Infections in the Paediatric Outpatient: A Review. *Antibitics* **2018**, *7*, 38. [CrossRef]
10. Mendelson, M.; Sharland, M.; Mpundu, M. Antibiotic resistance: Calling time on the 'silent pandemic'. *JAC Antimicrob. Resist.* **2022**, *4*, dlac016. [CrossRef]
11. Prestinaci, F.; Pezzotti, P.; Pantosti, A. Antimicrobial resistance: A global multifaceted phenomenon. *Pathog. Glob. Health* **2015**, *109*, 309–318. [CrossRef] [PubMed]
12. Taubes, G. The Bacteria Fight Back. *Science* **2008**, *321*, 356–361. [CrossRef] [PubMed]
13. Engler, A.C.; Wiradharma, N.; Ong, Z.Y.; Coady, D.J.; Hedrick, J.L.; Yang, Y.-Y. Emerging trends in macromolecular antimicrobials to fight multi-drug-resistant infections. *Nano Today* **2012**, *7*, 201–222. [CrossRef]
14. Hirakawa, H.; Kurushima, J.; Hashimoto, Y.; Tomita, H. Progress Overview of Bacterial Two-Component Regulatory Systems as Potential Targets for Antimicrobial Chemotherapy. *Antibiotics* **2020**, *9*, 635. [CrossRef]
15. Rossiter, S.E.; Fletcher, M.H.; Wuest, W.M. Natural Products as Platforms to Overcome Antibiotic Resistance. *Chem. Rev.* **2017**, *117*, 12415–12474. [CrossRef]
16. Walsh, C.T.; Wencewicz, T.A. Prospects for new antibiotics: A molecule-centered perspective. *J. Antibiot.* **2014**, *67*, 7–22. [CrossRef]
17. Gao, L.; Wang, H.; Zheng, B.; Huang, F. Combating antibiotic resistance: Current strategies for the discovery of novel antibacterial materials based on macrocycle supramolecular chemistry. *Giant* **2021**, *7*, 100066. [CrossRef]
18. Jeong, D.; Joo, S.-W.; Shinde, V.V.; Cho, E.; Jung, S. Carbohydrate-Based Host-Guest Complexation of Hydrophobic Antibiotics for the Enhancement of Antibacterial Activity. *Molecules* **2017**, *22*, 1311. [CrossRef]

19. Seiple, I.B.; Zhang, Z.; Jakubec, P.; Langlois-Mercier, A.; Wright, P.M.; Hog, D.T.; Yabu, K.; Allu, S.R.; Fukuzaki, T.; Carlsen, P.N.; et al. A platform for the discovery of new macrolide antibiotics. *Nature* **2016**, *533*, 338–345. [CrossRef]
20. Stanton, T.B. A call for antibiotic alternatives research. *Trends Microbiol.* **2013**, *21*, 111–113. [CrossRef]
21. Garland, M.; Loscher, S.; Bogyo, M. Chemical Strategies to Target Bacterial Virulence. *Chem. Rev.* **2017**, *117*, 4422–4461. [CrossRef] [PubMed]
22. Walsh, C. Where will new antibiotics come from? *Nat. Rev. Microbiol.* **2003**, *1*, 65–70. [CrossRef] [PubMed]
23. Cohen, M.L. Changing patterns of infectious disease. *Nature* **2000**, *406*, 762–767. [CrossRef] [PubMed]
24. O'Daniel, P.I.; Peng, Z.; Pi, H.; Testero, S.A.; Ding, D.; Spink, E.; Leemans, E.; Boudreau, M.A.; Yamaguchi, T.; Schroeder, V.A.; et al. Discovery of a New Class of Non-β-lactam Inhibitors of Penicillin-Binding Proteins with Gram-Positive Antibacterial Activity. *J. Am. Chem. Soc.* **2014**, *136*, 3664–3672. [CrossRef] [PubMed]
25. Kim, W.; Zhu, W.; Hendricks, G.L.; Van Tyne, D.; Steele, A.D.; Keohane, C.E.; Fricke, N.; Conery, A.L.; Shen, S.; Pan, W.; et al. A new class of synthetic retinoid antibiotics effective against bacterial persisters. *Nature* **2018**, *556*, 103–107. [CrossRef] [PubMed]
26. Andersson, D.I.; Hughes, D. Antibiotic resistance and its cost: Is it possible to reverse resistance? *Nat. Rev. Microbiol.* **2010**, *8*, 260–271. [CrossRef] [PubMed]
27. Liu, M.; Zhang, L.; Wang, T. Supramolecular Chirality in Self-Assembled Systems. *Chem. Rev.* **2015**, *115*, 7304–7397. [CrossRef] [PubMed]
28. Biedermann, F.; Schneider, H.-J. Experimental Binding Energies in Supramolecular Complexes. *Chem. Rev.* **2016**, *116*, 5216–5300. [CrossRef]
29. Huang, F.; Anslyn, E.V. Introduction: Supramolecular Chemistry. *Chem. Rev.* **2015**, *115*, 6999–7000. [CrossRef]
30. Tu, Y.; Peng, F.; Adawy, A.; Men, Y.; Abdelmohsen, L.K.E.A.; Wilson, D.A. Mimicking the Cell: Bio-Inspired Functions of Supramolecular Assemblies. *Chem. Rev.* **2016**, *116*, 2023–2078. [CrossRef]
31. Williams, G.T.; Haynes, C.J.E.; Fares, M.; Caltagirone, C.; Hiscock, J.R.; Gale, P.A. Advances in applied supramolecular technologies. *Chem. Soc. Rev.* **2021**, *50*, 2737–2763. [CrossRef] [PubMed]
32. Zhou, J.; Rao, L.; Yu, G.; Cook, T.R.; Chen, X.; Huang, F. Supramolecular cancer nanotheranostics. *Chem. Soc. Rev.* **2021**, *50*, 2839–2891. [CrossRef] [PubMed]
33. Fischbach, M.A.; Walsh, C.T. Antibiotics for Emerging Pathogens. *Science* **2009**, *325*, 1089–1093. [CrossRef] [PubMed]
34. Bassetti, S.; Tschudin-Sutter, S.; Egli, A.; Osthoff, M. Optimizing antibiotic therapies to reduce the risk of bacterial resistance. *Eur. J. Intern. Med.* **2022**, *99*, 7–12. [CrossRef]
35. Cox, G.; Ejim, L.; Stogios, P.J.; Koteva, K.; Bordeleau, E.; Evdokimova, E.; Sieron, A.O.; Savchenko, A.; Serio, A.W.; Krause, K.M.; et al. Plazomicin Retains Antibiotic Activity against Most Aminoglycoside Modifying Enzymes. *ACS Infect. Dis.* **2018**, *4*, 980–987. [CrossRef]
36. Kirsch, S.H.; Haeckl, F.P.J.; Müller, R. Beyond the approved: Target sites and inhibitors of bacterial RNA polymerase from bacteria and fungi. *Nat. Prod. Rep.* **2022**, *39*, 1226–1263. [CrossRef]
37. Lakemeyer, M.; Zhao, W.; Mandl, F.A.; Hammann, P.; Sieber, S.A. Thinking Outside the Box—Novel Antibacterials to Tackle the Resistance Crisis. *Angew. Chem. Int. Ed.* **2018**, *57*, 14440–14475. [CrossRef]
38. Jadhav, R.W.; Kobaisi, M.A.; Jones, L.A.; Vinu, A.; Bhosale, S.V. The Supramolecular Self-Assembly of Aminoglycoside Antibiotics and their Applications. *Angew. Chem. Int. Ed.* **2019**, *8*, 1154–1166. [CrossRef]
39. Suárez, D.F.; Consuegra, J.; Trajano, V.C.; Gontijo, S.M.L.; Guimarães, P.P.G.; Cortés, M.E.; Denadai, Â.L.; Sinisterra, R.D. Structural and thermodynamic characterization of doxycycline/β-cyclodextrin supramolecular complex and its bacterial membrane interactions. *Colloids Surf. B* **2014**, *118*, 194–201. [CrossRef]
40. Deng, J.-Z. Methicillin/per-6-(4-methoxylbenzyl)-amino-6-deoxy-β-cyclodextrin 1:1 complex and its potentiation in vitro against methicillin-resistant Staphylococcus aureus. *J. Antibiot.* **2013**, *66*, 517–521. [CrossRef]
41. Entenza, J.M.; Hohl, P.; Heinze-Krauss, I.; Glauser, M.P.; Moreillon, P. BAL9141, a Novel Extended-Spectrum Cephalosporin Active against Methicillin-Resistant *Staphylococcus aureus* in Treatment of Experimental Endocarditis. *Antimicrob. Agents Chemther.* **2002**, *46*, 171–177. [CrossRef] [PubMed]
42. Fung-Tomc, J.C.; Clark, J.; Minassian, B.; Pucci, M.; Tsai, Y.-H.; Gradelski, E.; Lamb, L.; Medina, I.; Huczko, E.; Kolek, B.; et al. In Vitro and In Vivo Activities of a Novel Cephalosporin, BMS-247243, against Methicillin-Resistant and -Susceptible Staphylococci. *Antimicrob. Agents Chemther.* **2002**, *46*, 971–976. [CrossRef] [PubMed]
43. Choi, J.M.; Park, K.; Lee, B.; Jeong, D.; Dindulkar, S.D.; Choi, Y.; Cho, E.; Park, S.; Yu, J.-h.; Jung, S. Solubility and bioavailability enhancement of ciprofloxacin by induced oval-shaped mono-6-deoxy-6-aminoethylamino-β-cyclodextrin. *Carbohydr. Polym.* **2017**, *163*, 118–128. [CrossRef] [PubMed]
44. Li, M.; Neoh, K.G.; Xu, L.; Yuan, L.; Leong, D.T.; Kang, E.-T.; Chua, K.L.; Hsu, L.Y. Sugar-Grafted Cyclodextrin Nanocarrier as a "Trojan Horse" for Potentiating Antibiotic Activity. *Pharm. Res.* **2016**, *33*, 1161–1174. [CrossRef]
45. Dawn, A.; Chandra, H.; Ade-Browne, C.; Yadav, J.; Kumari, H. Multifaceted Supramolecular Interactions from C-Methylresorcin[4]arene Lead to an Enhancement in In Vitro Antibacterial Activity of Gatifloxacin. *Chem. Eur. J.* **2017**, *23*, 18171–18179. [CrossRef]
46. Wolfmeier, H.; Pletzer, D.; Mansour, S.C.; Hancock, R.E.W. New Perspectives in Biofilm Eradication. *ACS Infect. Dis.* **2018**, *4*, 93–106. [CrossRef]

47. Blackman, L.D.; Qu, Y.; Cass, P.; Locock, K.E.S. Approaches for the inhibition and elimination of microbial biofilms using macromolecular agents. *Chem. Soc. Rev.* **2021**, *50*, 1587–1616. [CrossRef]
48. Guo, S.; Huang, Q.; Chen, Y.; Wei, J.; Zheng, J.; Wang, L.; Wang, Y.; Wang, R. Synthesis and Bioactivity of Guanidinium-Functionalized Pillar[5]arene as a Biofilm Disruptor. *Angew. Chem. Int. Ed.* **2021**, *60*, 618–623. [CrossRef]
49. Forier, K.; Raemdonck, K.; De Smedt, S.C.; Demeester, J.; Coenye, T.; Braeckmans, K. Lipid and polymer nanoparticles for drug delivery to bacterial biofilms. *J. Control. Release* **2014**, *190*, 607–623. [CrossRef]
50. Li, Y.; Liu, Y.; Ren, Y.; Su, L.; Li, A.; An, Y.; Rotello, V.; Zhang, Z.; Wang, Y.; Liu, Y.; et al. Coating of a Novel Antimicrobial Nanoparticle with a Macrophage Membrane for the Selective Entry into Infected Macrophages and Killing of Intracellular Staphylococci. *Adv. Funct. Mater.* **2020**, *30*, 2004942. [CrossRef]
51. Ho, D.-K.; Murgia, X.; De Rossi, C.; Christmann, R.; Hüfner de Mello Martins, A.G.; Koch, M.; Andreas, A.; Herrmann, J.; Müller, R.; Empting, M.; et al. Squalenyl Hydrogen Sulfate Nanoparticles for Simultaneous Delivery of Tobramycin and an Alkylquinolone Quorum Sensing Inhibitor Enable the Eradication of P. aeruginosa Biofilm Infections. *Angew. Chem. Int. Ed.* **2020**, *59*, 10292–10296. [CrossRef] [PubMed]
52. Feng, W.; Li, G.; Kang, X.; Wang, R.; Liu, F.; Zhao, D.; Li, H.; Bu, F.; Yu, Y.; Moriarty, T.F.; et al. Cascade-Targeting Poly(amino acid) Nanoparticles Eliminate Intracellular Bacteria via On-Site Antibiotic Delivery. *Adv. Mater.* **2022**, *34*, 2109789. [CrossRef] [PubMed]
53. Gan, B.H.; Gaynord, J.; Rowe, S.M.; Deingruber, T.; Spring, D.R. The multifaceted nature of antimicrobial peptides: Current synthetic chemistry approaches and future directions. *Chem. Soc. Rev.* **2021**, *50*, 7820–7880. [CrossRef] [PubMed]
54. Li, W.; Separovic, F.; O'Brien-Simpson, N.M.; Wade, J.D. Chemically modified and conjugated antimicrobial peptides against superbugs. *Chem. Soc. Rev.* **2021**, *50*, 4932–4973. [CrossRef]
55. Mookherjee, N.; Anderson, M.A.; Haagsman, H.P.; Davidson, D.J. Antimicrobial host defence peptides: Functions and clinical potential. *Nat. Rev. Drug Discov.* **2020**, *19*, 311–332. [CrossRef]
56. Abbas, M.; Ovais, M.; Atiq, A.; Ansari, T.M.; Xing, R.; Spruijt, E.; Yan, X. Tailoring supramolecular short peptide nanomaterials for antibacterial applications. *Coord. Chem. Rev.* **2022**, *460*, 214481. [CrossRef]
57. Yan, Y.; Li, Y.; Zhang, Z.; Wang, X.; Niu, Y.; Zhang, S.; Xu, W.; Ren, C. Advances of peptides for antibacterial applications. *Colloids Surf. B* **2021**, *202*, 111682. [CrossRef]
58. Zou, P.; Chen, W.-T.; Sun, T.; Gao, Y.; Li, L.-L.; Wang, H. Recent advances: Peptides and self-assembled peptide-nanosystems for antimicrobial therapy and diagnosis. *Biomater. Sci.* **2020**, *8*, 4975–4996. [CrossRef]
59. Chen, J.; Meng, Q.; Zhang, Y.; Dong, M.; Zhao, L.; Zhang, Y.; Chen, L.; Chai, Y.; Meng, Z.; Wang, C.; et al. Complexation of an Antimicrobial Peptide by Large-Sized Macrocycles for Decreasing Hemolysis and Improving Stability. *Angew. Chem. Int. Ed.* **2021**, *60*, 11288–11293. [CrossRef]
60. Liao, F.-H.; Wu, T.-H.; Yao, C.-N.; Kuo, S.-C.; Su, C.-J.; Jeng, U.S.; Lin, S.-Y. A Supramolecular Trap to Increase the Antibacterial Activity of Colistin. *Angew. Chem. Int. Ed.* **2020**, *59*, 1430–1434. [CrossRef]
61. Zhao, Q.; Zhao, Y.; Lu, Z.; Tang, Y. Amino Acid-Modified Conjugated Oligomer Self-Assembly Hydrogel for Efficient Capture and Specific Killing of Antibiotic-Resistant Bacteria. *ACS Appl. Mater. Interfaces* **2019**, *11*, 16320–16327. [CrossRef] [PubMed]
62. Jadhav, M.; Kalhapure, R.S.; Rambharose, S.; Mocktar, C.; Govender, T. Synthesis, characterization and antibacterial activity of novel heterocyclic quaternary ammonium surfactants. *Ind. Eng. Chem.* **2017**, *47*, 405–414. [CrossRef]
63. Zakharova, L.Y.; Pashirova, T.N.; Doktorovova, S.; Fernandes, A.R.; Sanchez-Lopez, E.; Silva, A.M.; Souto, S.B.; Souto, E.B. Cationic Surfactants: Self-Assembly, Structure-Activity Correlation and Their Biological Applications. *Int. J. Mol. Sci.* **2019**, *20*, 5534. [CrossRef]
64. Zhou, C.; Wang, Y. Structure–Activity relationship of cationic surfactants as antimicrobial agents. *Curr. Opin. Colloid Interface Sci.* **2020**, *45*, 28–43. [CrossRef]
65. Ahmady, A.R.; Hosseinzadeh, P.; Solouk, A.; Akbari, S.; Szulc, A.M.; Brycki, B.E. Cationic gemini surfactant properties, its potential as a promising bioapplication candidate, and strategies for improving its biocompatibility: A review. *Adv. Colloid Interface Sci.* **2022**, *299*, 102581. [CrossRef]
66. Jennings, M.C.; Buttaro, B.A.; Minbiole, K.P.C.; Wuest, W.M. Bioorganic Investigation of Multicationic Antimicrobials to Combat QAC-Resistant Staphylococcus aureus. *ACS Infect. Dis.* **2015**, *1*, 304–309. [CrossRef]
67. Buffet-Bataillon, S.; Tattevin, P.; Bonnaure-Mallet, M.; Jolivet-Gougeon, A. Emergence of resistance to antibacterial agents: The role of quaternary ammonium compounds—A critical review. *Int. J. Antimicrob. Agents* **2012**, *39*, 381–389. [CrossRef]
68. Jennings, M.C.; Minbiole, K.P.C.; Wuest, W.M. Quaternary Ammonium Compounds: An Antimicrobial Mainstay and Platform for Innovation to Address Bacterial Resistance. *ACS Infect. Dis.* **2015**, *1*, 288–303. [CrossRef]
69. Obłąk, E.; Futoma-Kołoch, B.; Wieczyńska, A. Biological activity of quaternary ammonium salts and resistance of microorganisms to these compounds. *World J. Microbiol. Biotechnol.* **2021**, *37*, 22. [CrossRef]
70. Bai, H.; Fu, X.; Huang, Z.; Lv, F.; Liu, L.; Zhang, X.; Wang, S. Supramolecular Germicide Switches through Host-Guest Interactions for Decelerating Emergence of Drug-Resistant Pathogens. *ChemistrySelect* **2017**, *2*, 7940–7945. [CrossRef]
71. Zhou, C.; Wang, D.; Cao, M.; Chen, Y.; Liu, Z.; Wu, C.; Xu, H.; Wang, S.; Wang, Y. Self-Aggregation, Antibacterial Activity, and Mildness of Cyclodextrin/Cationic Trimeric Surfactant Complexes. *ACS Appl. Mater. Interfaces* **2016**, *8*, 30811–30823. [CrossRef] [PubMed]

72. Wang, L.; Zhao, Q.; Zhang, Z.; Lu, Z.; Zhao, Y.; Tang, Y. Fluorescent Conjugated Polymer/Quarternary Ammonium Salt Co-assembly Nanoparticles: Applications in Highly Effective Antibacteria and Bioimaging. *ACS Appl. Bio Mater.* **2018**, *1*, 1478–1486. [CrossRef] [PubMed]
73. Shen, Y.; Li, S.; Qi, R.; Wu, C.; Yang, M.; Wang, J.; Cai, Z.; Liu, K.; Yue, J.; Guan, B.; et al. Assembly of Hexagonal Column Interpenetrated Spheres from Plant Polyphenol/Cationic Surfactants and Their Application as Antimicrobial Molecular Banks. *Angew. Chem. Int. Ed.* **2021**, *61*, e202110938. [CrossRef]
74. Li, T.; Wang, P.; Guo, W.; Huang, X.; Tian, X.; Wu, G.; Xu, B.; Li, F.; Yan, C.; Liang, X.-J.; et al. Natural Berberine-Based Chinese Herb Medicine Assembled Nanostructures with Modified Antibacterial Application. *ACS Nano* **2019**, *13*, 6770–6781. [CrossRef] [PubMed]
75. Huang, X.; Wang, P.; Li, T.; Tian, X.; Guo, W.; Xu, B.; Huang, G.; Cai, D.; Zhou, F.; Zhang, H.; et al. Self-Assemblies Based on Traditional Medicine Berberine and Cinnamic Acid for Adhesion-Induced Inhibition Multidrug-Resistant Staphylococcus aureus. *ACS Appl. Mater. Interfaces* **2020**, *12*, 227–237. [CrossRef]
76. Zhang, D.X.; Du, J.; Wang, R.; Luo, J.; Jing, T.F.; Li, B.X.; Mu, W.; Liu, F.; Hou, Y. Core/Shell Dual-Responsive Nanocarriers via Iron-Mineralized Electrostatic Self-Assembly for Precise Pesticide Delivery. *Adv. Funct. Mater.* **2021**, *31*, 2102027. [CrossRef]
77. Shen, S.; Huang, Y.; Yuan, A.; Lv, F.; Liu, L.; Wang, S. Electrochemical Regulation of Antibacterial Activity Using Ferrocene-Containing Antibiotics. *CCS Chem.* **2021**, *3*, 129–135. [CrossRef]
78. Yan, S.; Chen, S.; Gou, X.; Yang, J.; An, J.; Jin, X.; Yang, Y.-W.; Chen, L.; Gao, H. Biodegradable Supramolecular Materials Based on Cationic Polyaspartamides and Pillar[5]arene for Targeting Gram-Positive Bacteria and Mitigating Antimicrobial Resistance. *Adv. Funct. Mater.* **2019**, *29*, 1904683. [CrossRef]
79. Li, J.; Zhang, K.; Ruan, L.; Chin, S.F.; Wickramasinghe, N.; Liu, H.; Ravikumar, V.; Ren, J.; Duan, H.; Yang, L.; et al. Block Copolymer Nanoparticles Remove Biofilms of Drug-Resistant Gram-Positive Bacteria by Nanoscale Bacterial Debridement. *Nano Lett.* **2018**, *18*, 4180–4187. [CrossRef]
80. Amos-Tautua, B.M.; Songca, S.P.; Oluwafemi, O.S. Application of Porphyrins in Antibacterial Photodynamic Therapy. *Molecules* **2019**, *24*, 2456. [CrossRef]
81. Meisel, P.; Kocher, T. Photodynamic therapy for periodontal diseases: State of the art. *J. Photochem. Photobiol. B Biol.* **2005**, *79*, 159–170. [CrossRef] [PubMed]
82. Peron, D.; Bergamo, A.; Prates, R.; Vieira, S.S.; de Tarso Camillo de Carvalho, P.; Serra, A.J. Photodynamic antimicrobial chemotherapy has an overt killing effect on periodontal pathogens? A systematic review of experimental studies. *Lasers Med. Sci.* **2019**, *34*, 1527–1534. [CrossRef] [PubMed]
83. Sobotta, L.; Skupin-Mrugalska, P.; Piskorz, J.; Mielcarek, J. Porphyrinoid photosensitizers mediated photodynamic inactivation against bacteria. *Eur. J. Med. Chem.* **2019**, *175*, 72–106. [CrossRef] [PubMed]
84. Eleraky, N.E.; Allam, A.; Hassan, S.B.; Omar, M.M. Nanomedicine Fight against Antibacterial Resistance: An Overview of the Recent Pharmaceutical Innovations. *Pharmaceutics* **2020**, *12*, 142. [CrossRef] [PubMed]
85. Cieplik, F.; Deng, D.; Crielaard, W.; Buchalla, W.; Hellwig, E.; Al-Ahmad, A.; Maisch, T. Antimicrobial photodynamic therapy–What we know and what we don't. *Crit. Rev. Microbiol.* **2018**, *44*, 571–589. [CrossRef] [PubMed]
86. Hopper, C. Photodynamic therapy: A clinical reality in the treatment of cancer. *Lancet Oncol.* **2000**, *1*, 212–219. [CrossRef]
87. Jiang, L.; Gan, C.R.R.; Gao, J.; Loh, X.J. A Perspective on the Trends and Challenges Facing Porphyrin-Based Anti-Microbial Materials. *Small* **2016**, *12*, 3609–3644. [CrossRef]
88. Liu, K.; Liu, Y.; Yao, Y.; Yuan, H.; Wang, S.; Wang, Z.; Zhang, X. Supramolecular Photosensitizers with Enhanced Antibacterial Efficiency. *Angew. Chem. Int. Ed.* **2013**, *52*, 8285–8289. [CrossRef]
89. Chen, L.; Bai, H.; Xu, J.-F.; Wang, S.; Zhang, X. Supramolecular Porphyrin Photosensitizers: Controllable Disguise and Photoinduced Activation of Antibacterial Behavior. *ACS Appl. Mater. Interfaces* **2017**, *9*, 13950–13957. [CrossRef]
90. Yuan, H.; Zhan, Y.; Rowan, A.E.; Xing, C.; Kouwer, P.H.J. Biomimetic Networks with Enhanced Photodynamic Antimicrobial Activity from Conjugated Polythiophene/Polyisocyanide Hybrid Hydrogels. *Angew. Chem. Int. Ed.* **2020**, *59*, 2720–2724. [CrossRef]
91. Lee, E.; Li, X.; Oh, J.; Kwon, N.; Kim, G.; Kim, D.; Yoon, J. A boronic acid-functionalized phthalocyanine with an aggregation-enhanced photodynamic effect for combating antibiotic-resistant bacteria. *Chem. Sci.* **2020**, *11*, 5735–5739. [CrossRef] [PubMed]
92. Xing, C.; Xu, Q.; Tang, H.; Liu, L.; Wang, S. Conjugated Polymer/Porphyrin Complexes for Efficient Energy Transfer and Improving Light-Activated Antibacterial Activity. *J. Am. Chem. Soc.* **2009**, *131*, 13117–13124. [CrossRef] [PubMed]
93. Bai, H.; Yuan, H.; Nie, C.; Wang, B.; Lv, F.; Liu, L.; Wang, S. A Supramolecular Antibiotic Switch for Antibacterial Regulation. *Angew. Chem. Int. Ed.* **2015**, *54*, 13208–13213. [CrossRef] [PubMed]
94. Cao, A.; Tang, Y.; Liu, Y.; Yuan, H.; Liu, L. A strategy for antimicrobial regulation based on fluorescent conjugated oligomer–DNA hybrid hydrogels. *Chem. Commun.* **2013**, *49*, 5574–5576. [CrossRef]
95. Xia, L.; Tian, J.; Yue, T.; Cao, H.; Chu, J.; Cai, H.; Zhang, W. Pillar[5]arene-Based Acid-Triggered Supramolecular Porphyrin Photosensitizer for Combating Bacterial Infections and Biofilm Dispersion. *Adv. Healthc. Mater.* **2021**, *11*, 2102015. [CrossRef]
96. Yue, T.; Xia, L.; Tian, J.; Huang, B.; Chen, C.; Cao, H.; Zhang, W. A carboxylatopillar[5]arene-based pH-triggering supramolecular photosensitizer for enhanced photodynamic antibacterial efficacy. *Chem. Comm.* **2022**, *58*, 2991–2994. [CrossRef]

97. Zhu, Y.; Wu, S.; Sun, Y.; Zou, X.; Zheng, L.; Duan, S.; Wang, J.; Yu, B.; Sui, R.; Xu, F.-J. Bacteria-Targeting Photodynamic Nanoassemblies for Efficient Treatment of Multidrug-Resistant Biofilm Infected Keratitis. *Adv. Funct. Mater.* **2021**, *32*, 2111066. [CrossRef]
98. Dai, X.; Zhang, B.; Yu, Q.; Liu, Y. Multicharged Supramolecular Assembly Mediated by Polycationic Cyclodextrin for Efficiently Photodynamic Antibacteria. *ACS Appl. Bio Mater.* **2021**, *4*, 8536–8542. [CrossRef]
99. Ray, P.C.; Khan, S.A.; Singh, A.K.; Senapati, D.; Fan, Z. Nanomaterials for targeted detection and photothermal killing of bacteria. *Chem. Soc. Rev.* **2012**, *41*, 3193–3209. [CrossRef]
100. Dumas, A.; Couvreur, P. Palladium: A future key player in the nanomedical field? *Chem. Sci.* **2015**, *6*, 2153–2157. [CrossRef]
101. Wang, C.; Xu, H.; Liang, C.; Liu, Y.; Li, Z.; Yang, G.; Cheng, L.; Li, Y.; Liu, Z. Iron Oxide @ Polypyrrole Nanoparticles as a Multifunctional Drug Carrier for Remotely Controlled Cancer Therapy with Synergistic Antitumor Effect. *ACS Nano* **2013**, *7*, 6782–6795. [CrossRef] [PubMed]
102. Wang, D.; Kuzma, M.L.; Tan, X.; He, T.-C.; Dong, C.; Liu, Z.; Yang, J. Phototherapy and optical waveguides for the treatment of infection. *Adv. Drug Deliv. Rev.* **2021**, *179*, 114036. [CrossRef] [PubMed]
103. Tian, Q.; Jiang, F.; Zou, R.; Liu, Q.; Chen, Z.; Zhu, M.; Yang, S.; Wang, J.; Wang, J.; Hu, J. Hydrophilic Cu9S5 Nanocrystals: A Photothermal Agent with a 25.7% Heat Conversion Efficiency for Photothermal Ablation of Cancer Cells In Vivo. *ACS Nano* **2011**, *5*, 9761–9771. [CrossRef] [PubMed]
104. Yang, Y.; He, P.; Wang, Y.; Bai, H.; Wang, S.; Xu, J.-F.; Zhang, X. Supramolecular Radical Anions Triggered by Bacteria In Situ for Selective Photothermal Therapy. *Angew. Chem. Int. Ed.* **2017**, *56*, 16239–16242. [CrossRef]
105. Yin, Z.; Yang, Y.; Yang, J.; Song, G.; Hu, H.; Zheng, P.; Xu, J.-F. Supramolecular Polymerization Powered by *Escherichia coli*: Fabricating a Near-Infrared Photothermal Antibacterial Agent In Situ. *CCS Chem.* **2021**, *3*, 3615–3625. [CrossRef]
106. Tian, S.; Bai, H.; Li, S.; Xiao, Y.; Cui, X.; Li, X.; Tan, J.; Huang, Z.; Shen, D.; Liu, W.; et al. Water-Soluble Organic Nanoparticles with Programable Intermolecular Charge Transfer for NIR-II Photothermal Anti-Bacterial Therapy. *Angew. Chem. Int. Ed.* **2021**, *60*, 11758–11762. [CrossRef]
107. Qi, Z.; Bharate, P.; Lai, C.-H.; Ziem, B.; Böttcher, C.; Schulz, A.; Beckert, F.; Hatting, B.; Mülhaupt, R.; Seeberger, P.H.; et al. Multivalency at Interfaces: Supramolecular Carbohydrate-Functionalized Graphene Derivatives for Bacterial Capture, Release, and Disinfection. *Nano Lett.* **2015**, *15*, 6051–6057. [CrossRef]
108. Cheng, Q.; Xu, M.; Sun, C.; Yang, K.; Yang, Z.; Li, J.; Zheng, J.; Zheng, Y.; Wang, R. Enhanced antibacterial function of a supramolecular artificial receptor-modified macrophage (SAR-Macrophage). *Mater. Horiz.* **2022**, *9*, 934–941. [CrossRef]

Article

Effect of Plasmonic Gold Nanoprisms on Biofilm Formation and Heat Shock Proteins Expression in Human Pathogenic Bacteria

Rihab Lagha [1,2], Fethi Ben Abdallah [1,2,*], Amine Mezni [3] and Othman M. Alzahrani [1]

1. Department of Biology, College of Sciences, Taif University, P.O. Box 11099, Taif 21944, Saudi Arabia; rihab.k@tu.edu.sa (R.L.); o.alzahrani@tu.edu.sa (O.M.A.)
2. Research Unit UR17ES30: Virology and Antiviral Strategies, Higher Institute of Biotechnology, University of Monastir, Monastir 5000, Tunisia
3. Department of Chemistry, College of Sciences, Taif University, P.O. Box 11099, Taif 21944, Saudi Arabia; aminemezni@tu.edu.sa
* Correspondence: fetyben@tu.edu.sa

Abstract: Gold nanoparticles have gained interest in biomedical sciences in the areas of nano-diagnostics, bio-labeling, drug delivery, and bacterial infection. In this study, we examined, for the first time, the antibacterial and antibiofilm properties of plasmonic gold nanoprisms against human pathogenic bacteria using MIC and crystal violet. In addition, the expression level of GroEL/GroES heat shock proteins was also investigated by western blot. Gold nanoparticles were characterized by TEM and EDX, which showed equilateral triangular prisms with an average edge length of 150 nm. Antibacterial activity testing showed a great effect of AuNPs against pathogenic bacteria with MICs values ranging from 50 µg/mL to 100 µg/mL. Nanoparticles demonstrated strong biofilm inhibition action with a percentage of inhibition ranging from 40.44 to 82.43%. Western blot analysis revealed that GroEL was an AuNPs-inducible protein with an increase of up to 66.04%, but GroES was down-regulated with a reduction of up to 46.81%. Accordingly, plasmonic gold nanoprisms, could be a good candidate for antibiotics substitution in order to treat bacterial infections.

Keywords: plasmonic gold nanoprisms; antibacterial; antibiofilm; GroEL/GroES expression; pathogenic bacteria

1. Introduction

Pathogenic bacteria are microorganisms that cause infectious diseases. They can transmit from person to person directly or indirectly through animal and insect vectors, as well as through polluted water and food [1]. Antibiotic therapy continues to lose its effectiveness due to the spread of drug resistance in bacterial pathogens. The infections caused by drug-resistant bacteria result in additional very expensive medical costs [2,3]. Faced with this critical situation, the development of new antibacterial agents and therapeutic strategies is urgently necessary.

Most pathogenic bacteria are resistant to antibacterial compounds due to their capacity to produce biofilm [4]. In addition, the biofilm produced by sessile bacteria is responsible for several chronic diseases [5]. Biofilms formed by pathogenic bacteria are one of their major virulence factors that contribute to >80% of human infections [6]. Biofilm is a complex community composed of nucleic acids, lipids, proteins, and polysaccharides [7]. In biofilms, pathogenic bacteria become more persistent in the host environment and more resistant to antibiotics [8,9] that is a major medical problem. Thereby, the discovering of new compounds to combat biofilm is of great importance.

Heat shock proteins (HSPs) are major proteins that may be involved in bacterial biofilm [10]. These HSPs are implicated in the bacterial response to environmental stresses and protect pathogens against phagocytic cells [11]. In bacteria, one of the major molecular chaperones is the GroE machine (GroES and GroEL) [12]. GroEL assists in correct folding and assembly of proteins and is involved in diverse cellular processes, including DNA replication, UV mutagenesis, bacterial growth, RNA transcription, and flagella synthesis [13]. GroEL, together with GroES, translocate protein across membrane barriers. These HSPs act by preventing protein denaturation and to reactivate partially denatured proteins [14–17].

Nanoparticles have been used successfully in the delivery of therapeutic agents [18]. Due to their physicochemical properties, nanoparticles become more effective against multidrug-resistant pathogens [19]. AuNPs have gained interest in biomedical sciences due to their low toxicity to human cells, chemical, and biological stability, shape, surface properties, catalytic and antibacterial activities, especially in the areas of nano-diagnostics, bio-labeling, and drug delivery [18]. Safe nanomaterial could be a novel approach to inhibit and/or eradicate biofilm-related bacterial infections. In addition, the penetrating power of nanoparticles plays a major role in their action within the biofilm. In addition, the chance of developing resistance to nanoparticles is lower compared to conventional antibiotics [20].

This work aimed to study the antibacterial and biofilm inhibition properties of plasmonic gold nanoprisms against human pathogenic bacteria. In addition, the expression level of heat shock proteins GroEL and GroES was also investigated.

2. Results

2.1. AuNPs Characterization

AuNPs were characterized using TEM (Figure 1a) that shows equilateral triangular prisms with an average edge length of 150 nm (Figure 1b). At a molar ratio of PVP/Au = 0.05, the PVP concentration is sufficient to stabilize the gold nanoparticles. Therefore, the gold nanoparticles nucleate more quickly and form monodisperse triangular gold nanoprisms. According to the energy dispersive spectrum (EDX) analysis, le Tr-Au NPs consist of only gold, and the copper element came from a copper grid.

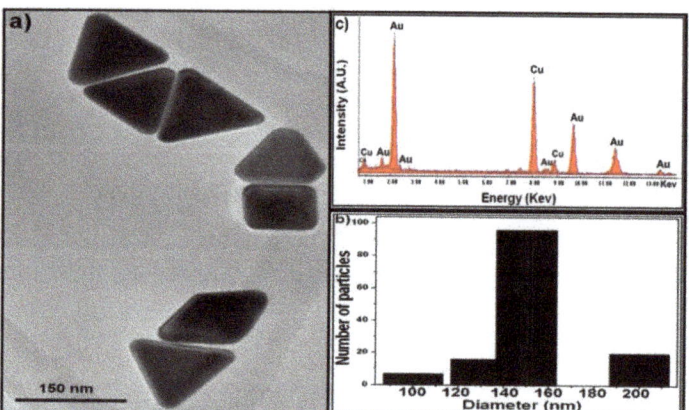

Figure 1. Characterization of AuNPs. (a) TEM image, (b) particle size distribution, and (c) EDX spectrum of triangular gold nanoprisms. Adapted with permission from [21], Copyright 2014 American Chemical Society.

2.2. Antibacterial Property of AuNPs

The minimal inhibition concentrations (MICs) and the minimal bactericidal concentrations (MBCs) were used to evaluate the antibacterial activity of AuNPs against the investigated bacteria.

The value of MIC was 50 µg/mL for S. Typhimurium and V. cholerae, and was 100 µg/mL for E. coli, E. faecalis, P. aeruginosa, S. aureus, B. cereus, S. sonnei, and N. gonorrhoeae. However, the MBC values were 0.4 mg/mL for all the tested bacteria excepted from S. aureus and S. sonnei, which were 0.8 mg/mL.

2.3. Biofilm Formation

Bacteria were evaluated for their ability to produce biofilm on polystyrene surfaces (Table 1). Results showed that almost all of the strains were low-grade positive producers with OD570 values varied from 0.119 to 0.599. In addition, only *S. aureus* was a highly positive producer, while *B. cereus* and *V. cholerae* were not able to form biofilms.

Table 1. Antibiofilm activity of AuNPs on polystyrene.

Strains	Control OD570 ± SD	Biofilm Phenotype	AuNPs OD570 ± SD	Biofilm Phenotype	Inhibition (%)
S. Typhimurium	0.37 ± 0.026	low-grade positive	0.065 ± 0.073	negative	82.43
B. cereus	0.046 ± 0.003	negative	-	-	-
E. coli	0.119 ± 0.042	low-grade positive	0.12 ± 0.062	low-grade positive	0
E. faecalis	0.599 ± 0.067	low-grade positive	0.206 ± 0.032	low-grade positive	65.60
S. sonnei	0.142 ± 0.022	low-grade positive	0.131 ± 0.046	low-grade positive	0
S. aureus	3.22 ± 0.088	highly positive	0.6 ± 0.058	low-grade positive	81.36
N. gonorrhoeae	0.361 ± 0.014	low-grade positive	0.215 ± 0.023	low-grade positive	40.44
P. aeruginosa	0.409 ± 0.076	low-grade positive	0.084 ± 0.016	negative	79.46
V. cholerae	0.032 ± 0.002	negative	-	-	-

2.4. Biofilm Inhibition

Biofilm inhibitory activity of AuNPs was assessed on the strains that showed a biofilm formation potential on polystyrene surface.

According to Table 1, AuNPs showed great antibiofilm activity. On polystyrene, the percentage of biofilm inhibition ranged from 40.44 to 82.43%. The greatest antibiofilm activity was detected from S. Typhimurium, which passed from low-grade positive to biofilm negative. In addition to *S. aureus* that changed from highly positive to and low-grade positive and *P. aeruginosa* that changed from highly positive to biofilm negative. Moreover, the two low-grade positive *E. faecalis* and *N. gonorrhoeae* have conserved their initial biofilm phenotype despite the observed inhibition. However, AuNPs did not show any activity on *E. coli* and *S. sonnei*.

2.5. Effect of AuNPs on GroES and GroEL Expression

The expression level of GroES before and after treatment with AuNPs was analyzed using western blot (Figure 2a). Control cells showed that GroES was expressed in all tested bacteria with different degrees except for *E. coli* and *V. cholerae*. The very high and low expression levels of this protein were observed in *S. typhimurium* and *S. sonnei*, respectively.

Under the AuNPs effect, the analysis by Image-J (Figure 2b) demonstrated that the quantity of GroES protein was reduced in S. Typhimurium, B. cereus, S. aureus, and N. gonorrhoeae with percentages of 46.81%, 30.17%, 38.35%, and 28.43%, respectively. However, the amount of GroES protein was increased in S. sonnei (86.33%) and P. aeruginosa (11.40%). In addition, this protein remained stable in E. faecalis but was expressed in E. coli and V. cholerae.

Concerning GroEL (Figure 3a), the results showed that this protein was expressed with different degrees only in the non-treated S. Typhimurium, B. cereus, E. faecalis, S. aureus, and N. gonorrhoeae.

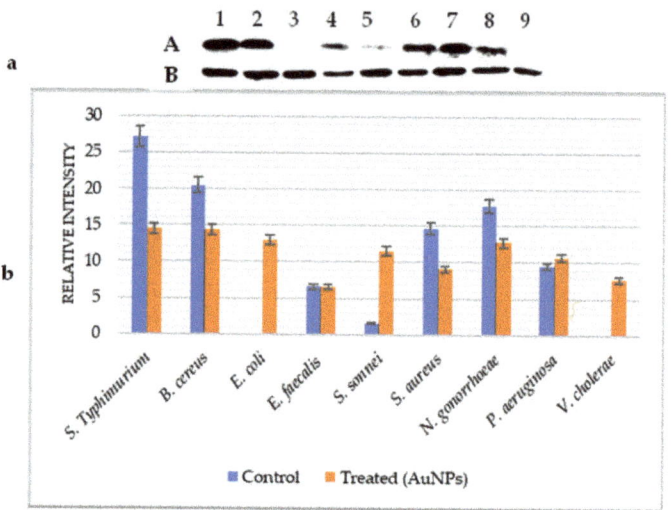

Figure 2. GroES heat shock protein expression under AuNPs effect. (**a**): Western blot analysis; (**b**): Image-J bands quantification; (A): Control; (B): Treated (AuNPs). S. Typhimurium (1), B. cereus (2), E. coli (3), E. faecalis (4), S. sonnei (5), S. aureus (6), N. gonorrhoeae (7), P. aeruginosa (8), and V. cholerae (9).

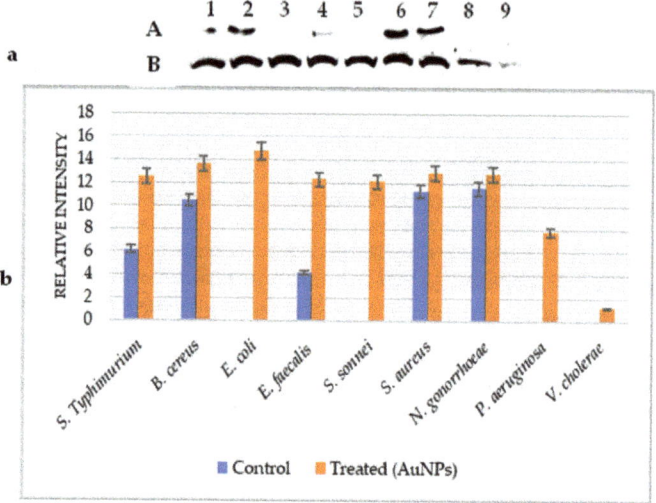

Figure 3. GroEl heat shock protein expression under AuNPs effect. (**a**): Western blot analysis; (**b**): Image-J bands quantification; (A): Control; (B): Treated (AuNPs) S. Typhimurium (1), B. cereus (2), E. coli (3), E. faecalis (4), S. sonnei (5), S. aureus (6), N. gonorrhoeae (7), P. aeruginosa (8), and V. cholerae (9).

After treatment with AuNPs, GroEL was induced in all tested bacteria. Its level of expression was considerably increased in *E. faecalis* (66.04%), S. Typhimurium (50.83%), and *B. cereus* (23.35%). This Hsp was slightly induced in *S. aureus* (12.5%) and *N. gonorrhoeae* (9.64%). The very high and low expression levels of GroEL were detected in *E. coli* and *V. cholerae*, respectively (Figure 3b).

3. Discussion

The increased resistance of pathogenic bacteria to antibiotics and the lack of new antibacterial drugs have prompted researchers to focus their efforts to explore nanotechnological measures against microbial infections for therapeutic applications. AuNPs have attracted considerable interest because of their promising applications in the development of novel antibacterial molecules [1]. This study investigates for the first time the effect of plasmonic gold nanoprisms on biofilm formation and GroEL/GroES proteins expression in some pathogenic bacteria as well as their antibacterial properties.

The antibacterial activity showed that AuNPs are effective against tested pathogenic bacteria. The MICs values varied from 50 µg/mL, for *S. typhimurium* and *V. cholerae*, to 100 µg/mL for *E. coli*, *E. faecalis*, *P. aeruginosa*, *S. aureus*, *B. cereus*, *S. sonnei*, and *N. gonorrhoeae*. Accordingly, Gram-negative and Gram-positive bacteria react in the same way face to AuNPs. However, the study of Zawrah et al. [22] indicated that AuNPs are less effective against Gram-positive bacteria due to the nature of the cell wall. Nanoparticles, with a positive charge, attach with the negatively charged microorganisms by the electrostatic attraction in the cell wall membrane [23].

AuNPs characterized using TEM appear as equilateral triangular prisms with an average edge length of 150 nm. Several reports have shown that smaller NPs have higher antibacterial activity [24–26]. Indeed, smaller particles affect the larger surface area of the bacteria, enter easily into the bacterial cell and affect the intracellular processes. Thereby they have more bactericidal activity than larger [27]. However, in this work, the triangular prisms of gold showed great antimicrobial property, despite their high length, against Gram-positive and Gram-negative bacteria. This finding corroborates the data of Sohm et al. [28] showing that larger NPs are more effective and indicating that size alone is not the most important factor of their toxicity. The effectiveness of gold triangular nanoprisms may be due to their shape. According to Smitha and Gopchandran [29], the antibacterial properties of triangular AuNPs have shown better activity against Gram-positive and Gram-negative bacteria compared to spherical AuNPs, which indicates that the shape could play a significant role in the potential antibacterial activity of AuNPs.

Biofilm, as a major virulence factor, is responsible for about 80% of human infections, and the Gram-negative bacterium *P. aeruginosa*, *E. coli*, and the Gram-positive *S. aureus* are the most frequent [30]. Due to the resistance to antibiotics, the immune system, and the spread of infection, the treatment of biofilm becomes more complex [31]. Thereby, the discovering of novel therapeutic molecules, like plasmonic gold nanoprisms, to inhibit biofilm is of great importance. In this work, almost all of the tested bacteria were low-grade positive producers on polystyrene surfaces with OD570 values ranging from 0.119 to 0.599. In addition, only *S. aureus* was a highly positive producer, while *B. cereus* and *V. cholerae* were not able to form biofilm. AuNPs showed great antibiofilm activity with the percentage of inhibition ranging from 40.44% for *N. gonorrhoeae* to 82.43% for *S. typhimurium*. Thereby, this nanoparticle could be a good alternative for the treatment of the formation of biofilm by pathogenic bacteria. Indeed, the penetration ability of AuNPs to get inside the biofilm and to disperse is of great importance as the biofilm layer provides an impermeable barrier to many antibiotics. Our results are in agreement with those of Singh et al. [32], who showed that AuNPs possess a remarkable reduction in the biofilm formation by *P. aeruginosa* and *E. coli*. The highest antibiofilm activity of AuNPs was observed in *S. typhimurium* and *S. aureus* that formed the strongest biofilm among the tested bacteria. However, this result is in discordance with the report of Ahmed et al. [33], who observed lower disruption of biofilms in the strong biofilm producer *K. pneumoniae* compared to the other isolates suggesting that the thickness and composition of biofilm play a key role in the penetration of AuNPs. This may be attributed to the shape of AuNPs as equilateral triangular prisms.

Hsps are key elements in the bacterial response to stress and environmental changes in order to maintain cell homeostasis in addition to their important role in pathogenesis [34]. GroEL/GroES play a major role in protein folding even during non-stressed growth conditions, although their action becomes more important during stress. In this work, we investigated the differential expression of GroEL and GroES under the effect of AuNPs. Firstly, before treatment, GroES was expressed with different levels in all tested bacteria except for *E. coli* and *V. cholerae*, but GroEL was expressed only in *S. typhimurium*, *B. cereus*, *E. faecalis*, *S. aureus*, and *N. gonorrhoeae* which indicates that GroES is more implicated in normal growth bacteria than GroEL. Under AuNPs effect, the amount of GroEL was considerably increased in all strains indicating that this protein plays an important role in maintaining the cell in such condition, which is in agreement with the report of Makumire et al. [35]. Thereby it can be considered as an AuNPs-inducible protein [36]. Concerning GroES, its production was increased only in *S. sonnei* and *P. aeruginosa* but was decreased or remained stable in other strains suggesting that this protein was altered or downregulated by AuNPs. According to Kustos et al. [37], under stress conditions such as nanoparticles, protein synthesis is inhibited, and cell division is interrupted. In parallel, the expression of various proteins increases; these are the so-called stress proteins as GroEL in this study. Knowing that the GroE machine is involved in protein folding and other mechanisms, the overexpression of GroEL may have served as compensation for the lack of GroES expression to maintain cell homeostasis. However, alteration observed in GroES expression may indicate that the GroE machine was altered, and thereby the protein folding mechanism of bacteria may be altered, which can explain the high activity of AuNPs.

4. Materials and Methods

4.1. Bacterial Strains

Nine (9) human pathogenic bacteria were tested in this work included: *S. typhimurium* (ATCC 1408), *E. coli* (ATCC 35218), *S. sonnei* (ATCC 25931), *E. faecalis*, *P. aeruginosa* PAO1, *S. aureus* (ATCC 25923), *B. cereus* (ATCC 11778), *V. cholerae* (ATCC 9459), and *N. gonorrhoeae* (ATCC 49226).

4.2. Gold Nanoparticles

4.2.1. Synthesis

To synthesize gold nanoparticles, a triethylene glycol reagent (ACROS Organics, 98%) was used as a solvent. The precursor of gold was hydrogen tetrachloroaurate (III) trihydrate (HAuCl$_4$·3H$_2$O; Sigma-Aldrich, St. Louis, MO, USA). In addition, polyvinyl-pyrrolidone (PVP) (K30, Sigma–Aldrich) was used as a surfactant [21]. For the experiment, 25 mL of triethylene glycol suspension containing 0.038 mmol of HAuCl$_4$·3H$_2$O and a given quantity of PVP were heated at 150 °C for 30 min under shaking. The molar ratio of PVP to HAuCl$_4$ (R(PVP/Au)) was 0.05. The formed gold particles and the final colloidal solution had a blue color. The product was centrifuged, washed many times with ethanol/acetone (2:1) solution, and scattered in ethanol.

4.2.2. Characterization

The morphological analysis of the gold particles was determined by transmission electron microscopy (TEM) (JEOL-JFC 1600). An elemental analysis was conducted using energy-dispersive X-ray spectrograph (EDX) attached to the TEM. The selected area electron diffraction (SAED) was also performed on the microscope. The Perkin-Elmer Lambda 11 UV/VIS spectrophotometer was used to determine the optical absorption spectra of diluted Au NPs solution.

4.3. Antibacterial Activity of Au NPs: Minimum Inhibitory and Minimum Bactericidal Concentration

MICs and MBCs were determined three times on 96-well microtiter plates (Nunc, Roskilde, Denmark) [38]. The bacterial inoculums (0.5 McFarland standards turbidity) were prepared from 12 h broth cultures. Then, a serial two-fold dilution of the AuNPs was prepared in 5 mL of nutrient broth with a concentration ranging from 0.012 to 7.2 mg/mL.

The plates were prepared by adding 100 μL from the AuNPs serial dilutions, 95 μL of nutrient broth, and 5 μL of the bacterial inoculum in each well. The negative control comprising 5 µL of the inoculum and 195 µL of nutrient broth without AuNPs was placed in the last well. The plates at a final volume of 200 µL were then incubated at 37 °C for 18–24 h.

The lowest concentration of the AuNPs at which the growth of the cells was inhibited was interpreted as MIC. However, the lowest concentration of AuNPs, required to kill ≥99.9% of the initial bacterial cells was interpreted as MBC and was determined by subculturing 20 µL from clear wells of the MIC test on MHA [39].

4.4. Biofilm Formation on Polystyrene

The capacity of bacteria to form biofilm was evaluated by crystal violet assay using U-bottomed 96-well microtiter plates [40]. Each bacterium was investigated three times, and sterile TSB was served as control. An automated Multiskan reader (GIO. DE VITA E C, Rome, Italy) was used to measure the wells' optical density at 570 nm (OD570). Biofilm formation was defined as highly positive (OD570 ≥ 1), low-grade positive (0.1 ≤ OD570 < 1), or negative (OD570 < 0.1).

4.5. Biofilm Inhibition

The capacity of AuNPs to prevent bacterial strains from forming biofilms was investigated. Experimentally, 100 µL of AuNPs mixed in TSB (2% glucose) were placed in each well of a U-bottomed 96-well microtiter plate, along with 100 µL of bacterial suspensions (10^8 CFU/mL). The final AuNPs concentrations were equivalent to the MIC, and the final volume per well was 200 µL. The experiments were carried out three times. Crystal violet was used to measure the formed biofilm after incubation of microplates at 37 °C for 24 h [41]. The inoculum volume and AuNPs were replaced with TSB and sterile water in the control wells, respectively. The percentage of biofilm inhibition was determined according to the formula described by Jadhav et al. [42].

$$\% \text{ Inhibition} = 100 - \left(\frac{\text{OD570 sample}}{\text{OD570 control}} \times 100 \right) \quad (1)$$

4.6. Western Blot for GroEL and GroES Analysis

Expression of GroEL and GroES under AuNPs conditions were analyzed using Western blot. For experiments, 11µg of whole-cell proteins (control and AuNPs treated bacteria) were analyzed by SDS-PAGE [43] and blotted onto nitrocellulose membranes (Millipore, Bedford, MA, USA). The membranes were blocked overnight at 4 °C in PBS-5% skimmed milk, 0.1% Tween 20. The membranes were incubated for 1 h with anti-GroEL (Abcam, ab90522) and anti-GroES (Abcam, ab69823) polyclonal rabbit-antibodies were diluted to 1/1000 and 1/5000, respectively. Then, the membranes were incubated with a goat anti-rabbit IgG peroxidase-conjugated monoclonal antibody (Sigma, St Louis, MO, USA) diluted at 1/10,000. The bound antibodies were visualized by ECL (GE Healthcare, Uppsala, Sweden). Western blots were analyzed with imaging (Image-J 1.50) to determine the intensity of each band [44].

5. Conclusions

The results developed in this work support the medical application of AuNPs as a therapeutic molecule for the treatment of infections caused by pathogenic bacteria. Plasmonic gold nanoprisms showed, for the first time, great antibacterial and antibiofilm activities against human pathogenic bacteria. In addition to the alteration observed in GroES expression as a part of GroE machine implicated in the bacterial protein folding mechanism. Despite their stability and low toxicity, the application of AuNPs needs more in vitro testing and in vivo clinical trials to establish their safety, efficacity, and possible adverse effects.

Author Contributions: F.B.A. conceived and designed the experiments; F.B.A. and R.L. performed the experiments; A.M. the synthesis of AuNPs; O.M.A. funding acquisition; F.B.A. and R.L. writing—review and editing. All authors have read and agreed to the published version of the manuscript.

Funding: The Authors appreciated Taif University Researchers Supporting Project number (TURSP-2020/262), Taif University, Taif, Saudi Arabia.

Institutional Review Board Statement: Not applicable.

Informed Consent Statement: Not applicable.

Data Availability Statement: Data is contained within the article.

Acknowledgments: The authors are thankful to Taif University for supplying essential facilities and acknowledge the support of Taif University Researchers Supporting Project number (TURSP-2020/262), Taif University, Taif, Saudi Arabia.

Conflicts of Interest: The authors declare no conflict of interest.

References

1. Katas, H.; Lim, C.; Nor Azlan, A.; Buang, F.; Mh Busra, M. Antibacterial activity of biosynthesized gold nanoparticles using biomolecules from lignosus rhinocerotis and chitosan. *Saudi. Pharm. J.* **2019**, *27*, 283–292. [CrossRef] [PubMed]
2. Rossolini, G.; Arena, F.; Pecile, P.; Pollini, S. Update on the antibiotic resistance crisis. *Curr. Opin. Pharmacol.* **2014**, *18*, 56–60. [CrossRef]
3. Li, X.; Robinson, S.; Gupta, A.; Saha, K.; Jiang, Z.; Moyano, D.; Sahar, A.; Riley, M.; Rotello, V. Functional gold nanoparticles as potent antimicrobial agents against multi-drug-resistant bacteria. *ACS. Nano.* **2014**, *8*, 10682–10686. [CrossRef]
4. Lenchenko, E.; Blumenkrants, D.; Sachivkina, N.; Shadrova, N.; Ibragimova, A. Morphological and adhesive properties of Klebsiella pneumoniae biofilms. *Vet. World.* **2020**, *13*, 197–200. [CrossRef] [PubMed]
5. Brady, R.; Leid, J.; Calhoun, J.; Costerton, J.; Shirtliff, M. Osteomyelitis and the role of biofilms in chronic infection. *FEMS. Immunol. Med. Microbiol.* **2008**, *52*, 13–22. [CrossRef] [PubMed]
6. Römling, U.; Balsalobre, C. Biofilm infections, their resilience to therapy and innovative treatment strategies. *J. Intern. Med.* **2012**, *272*, 541–561. [CrossRef]
7. Hall-Stoodley, L.; Costerton, J.W.; Stoodley, P. Bacterial biofilms: From the natural environment to infectious diseases. *Nat. Rev. Microbiol.* **2004**, *2*, 95–108. [CrossRef]
8. Simões, M.; Bennett, R.N.; Rosa, E.A.S. Understanding antimicrobial activities of phytochemicals against multidrug resistant bacteria and biofilms. *Nat. Prod. Rep.* **2009**, *26*, 746–757. [CrossRef] [PubMed]
9. Stewart, P.S. Mechanisms of antibiotic resistance in bacterial biofilms. *Int. J. Med. Microbiol.* **2002**, *292*, 107–113. [CrossRef]
10. Vinod Kumar, K.; Lall, C.; Vimal Raj, R.; Vedhagiri, K.; Kartick, C.; Surya, P.; Natarajaseenivasan, K.; Vijayachari, P. Overexpression of heat shock groel stress protein in Leptospiral biofilm. *Microb. Pathog.* **2017**, *102*, 8–11. [CrossRef]
11. Stamm, L.; Gherardini, F.; Parrish, E.; Moomaw, C. Heat shock response of spirochetes. *Infect. Immun.* **1991**, *59*, 1572–1575. [CrossRef]
12. Susin, M.; Baldini, R.; Gueiros-Filho, F.; Gomes, S. GroES/GroEL and Dnak/Dnaj have distinct roles in stress responses and during cell cycle progression in *Caulobacter crescentus*. *J. Bacteriol.* **2006**, *188*, 8044–8053. [CrossRef]
13. Maleki, F.; Khosravi, A.; Nasser, A.; Taghinejad, H.; Azizian, M. Bacterial heat shock protein activity. *J. Clin. Diagn. Res.* **2016**, *10*, BE01–BE03. [CrossRef]
14. Gomes, S.L.; Simaõ, R.C.G. Stress response: Heat. In *Encyclopedia of Microbiology*; Schaechter, M., Ed.; Elsevier Academic Press: Oxford, UK, 2009; pp. 464–474.
15. Yafout, M.; Ousaid, A.; Khayati, Y.; El Otmani, I. Gold nanoparticles as a drug delivery system for standard chemotherapeutics: A new lead for targeted pharmacological cancer treatments. *Sci. Afr.* **2021**, *11*, e00685. [CrossRef]
16. Thambiraj, S.; Hema, S.; Ravi Shankaran, D. Functionalized gold nanoparticles for drug delivery applications. *Mater. Today Proc.* **2018**, *5*, 16763–16773. [CrossRef]
17. Kong, F.; Zhang, J.; Li, R.; Wang, Z.; Wang, W.; Wang, W. Unique roles of gold nanoparticles in drug delivery, targeting and imaging applications. *Molecules* **2017**, *22*, 1445. [CrossRef] [PubMed]
18. Zhang, L.; Gu, F.; Chan, J.; Wang, A.; Langer, R.; Farokhzad, O. Nanoparticles in medicine: Therapeutic applications and developments. *Clin. Pharmacol. Ther.* **2008**, *83*, 761–769. [CrossRef] [PubMed]
19. Kandi, V.; Kandi, S. Antimicrobial properties of nanomolecules: Potential candidates as antibiotics in the era of multi-drug resistance. *Epidemiol. Health* **2015**, *37*, e2015020. [CrossRef] [PubMed]
20. Chhibber, S.; Nag, D.; Bansal, S. Inhibiting biofilm formation by Klebsiella pneumoniae B5055 using an iron antagonizing molecule and a bacteriophage. *BMC Microbiol.* **2013**, *13*, 174. [CrossRef]
21. Mezni, A.; Dammak, T.; Fkiri, A.; Mlayah, A.; Abid, Y.; Smiri, L.S. Photochemistry at the surface of gold nanoprisms from surface-enhanced raman scattering blinking. *J. Phys. Chem. C* **2014**, *118*, 17956–17967. [CrossRef]

22. Zawrah, M.F.; AbdEl-Moez, S.I. Antimicrobial activities of gold nanoparticles against major foodborne pathogens. *Life Sci.* **2011**, *8*, 37–44.
23. Dibrov, P.; Dzioba, J.; Gosink, K.; Häse, C. Chemiosmotic mechanism of antimicrobial activity of ag + in Vibrio cholerae. *Antimicrob. Agents Chemother.* **2002**, *46*, 2668–2670. [CrossRef] [PubMed]
24. Penders, J.; Stolzoff, M.; Hickey, D.; Andersson, M.; Webster, T. Shape-dependent antibacterial effects of non-cytotoxic gold nanoparticles. *Int. J. Nanomed.* **2017**, *12*, 2457–2468. [CrossRef] [PubMed]
25. Cheon, J.Y.; Kim, S.J.; Rhee, Y.H.; Kwon, O.H.; Park, W.H. Shape-dependent antimicrobial activities of silver nanoparticles. *Int. J. Nanomed.* **2019**, *14*, 2773–2780. [CrossRef] [PubMed]
26. Cui, L.; Chen, P.; Chen, S.; Yuan, Z.; Yu, C.; Ren, B.; Zhang, K. In situ study of the antibacterial activity and mechanism of action of silver nanoparticles by surface-enhanced Raman spectroscopy. *Anal. Chem.* **2013**, *85*, 5436–5443. [CrossRef]
27. Shrivastava, S.; Bera, T.; Roy, A.; Singh, G.; Ramachandrarao, P.; Dash, D. Characterization of enhanced antibacterial effects of novel silver nanoparticles. *Nanotechnology.* **2007**, *18*, 225103. [CrossRef]
28. Sohm, B.; Immel, F.; Bauda, P.; Pagnout, C. Insight into the primary mode of action of tio2 nanoparticles on *Escherichia coli* in the dark. *Proteomics.* **2014**, *15*, 98–113. [CrossRef]
29. Smitha, S.L.; Gopchandran, K.G. Surface Enhanced Raman scattering, antibacterial and antifungal active triangular gold nanoparticles. *Spectrochim. Acta A Mol. Biomol. Spectrosc.* **2013**, *102*, 114–119. [CrossRef]
30. Joo, H.-S.; Otto, M. Molecular basis of in vivo biofilm formation by bacterial pathogens. *Chem. Biol.* **2012**, *19*, 1503–1513. [CrossRef]
31. Paharik, A.E.; Horswill, A.R. The Staphylococcal Biofilm: Adhesins, Regulation, and Host Response. *Microbiol. Spectr.* **2016**, *4*, 10. [CrossRef]
32. Singh, P.; Pandit, S.; Beshay, M.; Mokkapati, V.R.S.S.; Garnaes, J.; Olsson, M.E.; Sultan, A.; Mackevica, A.; Mateiu, R.V.; Lütken, H.; et al. Anti-biofilm effects of gold and silver nanoparticles synthesized by the rhodiola rosea rhizome extracts. *Artif. Cells. Nanomed. Biotechnol.* **2018**, *46*, S886–S899. [CrossRef]
33. Ahmed, A.; Khan, A.K.; Anwar, A.; Ali, S.A.; Shah, M.R. Biofilm inhibitory effect of chlorhexidine conjugated gold nanoparticles against Klebsiella pneumoniae. *Microb. Pathog.* **2016**, *98*, 50–56. [CrossRef]
34. Bohne, J.; Sokolovic, Z.; Goebel, W. Transcriptional regulation of PRFA and PRFA-regulated virulence genes in listeria monocytogenes. *Mol. Microbiol.* **1994**, *11*, 1141–1150. [CrossRef] [PubMed]
35. Makumire, S.; Revaprasadu, N.; Shonhai, A. Dnak protein alleviates toxicity induced by citrate-coated gold nanoparticles in Escherichia coli. *PLoS ONE* **2015**, *10*, e0121243.
36. Ben Abdallah, F.; Ellafi, A.; Lagha, R.; Bakhrouf, A.; Namane, A.; Rousselle, J.-C.; Lenormand, P.; Kallel, H. Identification of outer membrane proteins of Vibrio parahaemolyticus and Vibrio alginolyticus altered in response to γ-irradiation or long-term starvation. *Res. Microbiol.* **2010**, *161*, 869–875. [CrossRef] [PubMed]
37. Kustos, I.; Kocsis, B.; Kilár, F. Bacterial outer membrane protein analysis by electrophoresis and microchip technology. *Expert Rev. Proteomics* **2007**, *4*, 91–106. [CrossRef]
38. Bagamboula, C.; Uyttendaele, M.; Debevere, J. Inhibitory effect of thyme and basil essential oils, carvacrol, thymol, estragol, linalool and p-cymene towards Shigella sonnei and S. flexneri. *J. Food Microbiol.* **2004**, *21*, 33–42. [CrossRef]
39. El-Deeb, B.; Elhariry, H.; Mostafa, N.Y. Antimicrobial Activity of Silver and Gold Nanoparticles Biosynthesized Using Ginger Extract. *Res. J. Pharm. Biol. Chem. Sci.* **2016**, *7*, 1085.
40. Gulluce, M.; Sahin, F.; Sokmen, M. Antimicrobial and antioxidant properties of the essential oils and methanol extract from Mentha longifolia L. *Food. Chem.* **2007**, *103*, 1449–1456. [CrossRef]
41. Oulkheir, S.; Aghrouch, M.; El Mourabit, F.; Dalha, F.; Graich, H.; Amouch, F.; Ouzaid, K.; Moukale, A.; Chadli, S. Antibacterial Activity of Essential Oils Extracts from Cinnamon, Thyme, Clove and Geranium Against a Gram Negative and Gram-Positive Pathogenic Bacteria. *J. Dis. Med. Plants* **2017**, *3*, 1–5. [CrossRef]
42. Jadhav, S.; Shah, R.; Bhave, M.; Palombo, E.A. Inhibitory activity of yarrow essential oil on *Listeria* planktonic cells and biofilms. *J. Food Control* **2013**, *29*, 125–130. [CrossRef]
43. Laemmli, U.K. Cleavage of structural proteins during the assembly of the head of bacteriophage t4. *Nature* **1970**, *227*, 680–685. [CrossRef] [PubMed]
44. Gasmi, N.; Ayed, A.; Ammar, B.; Zrigui, R.; Nicaud, J.-M.; Kallel, H. Development of a cultivation process for the enhancement of human interferon alpha 2b production in the oleaginous yeast, Yarrowia Lipolytica. *Microb. Cell. Factories* **2011**, *10*, 90. [CrossRef] [PubMed]

pharmaceuticals

Article

Novel In Vivo Assessment of Antimicrobial Efficacy of Ciprofloxacin Loaded Mesoporous Silica Nanoparticles against *Salmonella typhimurium* Infection

Maher N. Alandiyjany [1,2], Ahmed S. Abdelaziz [3], Ahmed Abdelfattah-Hassan [4,5], Wael A. H. Hegazy [6], Arwa A. Hassan [7], Sara T. Elazab [8], Eman A. A. Mohamed [9], Eman S. El-Shetry [10], Ayman A. Saleh [11], Naser A. ElSawy [12] and Doaa Ibrahim [13,*]

1. Laboratory Medicine Department, Faculty of Applied Medical Sciences, Umm Al-Qura University, Makkah 21955, Saudi Arabia; mnandiyjany@uqu.edu.sa
2. Quality and Development Affair, Batterjee Medical College, Jeddah 21442, Saudi Arabia
3. Department of Pharmacology, Faculty of Veterinary Medicine, Zagazig University, Zagazig 44519, Egypt; asabdelaziz@vet.zu.edu.eg
4. Department of Anatomy and Embryology, Faculty of Veterinary Medicine, Zagazig University, Zagazig 44511, Egypt; aabdelfattah@vet.zu.edu.eg
5. Biomedical Sciences Program, University of Science and Technology, Zewail City of Science and Technology, October Gardens, 6th of October, Giza 12578, Egypt
6. Department of Microbiology and Immunology, Faculty of Pharmacy, Zagazig University, Zagazig 44511, Egypt; waelmhegazy@daad-alumni.de
7. Department of Pharmacology & Toxicology, Faculty of Pharmacy & Pharmaceutical Industries, Sinai University, El-Arish 45511, Egypt; arwa.ahmed@su.edu.eg
8. Department of Pharmacology, Faculty of Veterinary Medicine, Mansoura University, Mansoura 35516, Egypt; sarataha1@mans.edu.eg
9. Department of Microbiology, Faculty of Veterinary Medicine, Zagazig University, Zagazig 44519, Egypt; eman.zewail@hotmail.com
10. Department of Human Anatomy and Embryology, Faculty of Medicine, Zagazig University, Zagazig 44511, Egypt; emanelshetry@zu.edu.eg
11. Department of Animal Wealth Development, Veterinary Genetics & Genetic Engineering, Faculty of Veterinary Medicine, Zagazig University, Zagazig 44519, Egypt; lateefsaleh@yahoo.com
12. Department of Anatomy & Embryology, Faculty of Medicine, Zagazig University, Zagazig 44511, Egypt; naser_elsawy@ymail.com
13. Department of Nutrition and Clinical Nutrition, Faculty of Veterinary Medicine, Zagazig University, Zagazig 44511, Egypt
* Correspondence: doibrahim@vet.zu.edu.eg

Abstract: *Salmonella enterica* serovar Typhimurium (*S. typhimurium*) is known for its intracellular survival, evading the robust inflammation and adaptive immune response of the host. The emergence of decreased ciprofloxacin (CIP) susceptibility (DCS) requires a prolonged antibiotic course with increased dosage, leading to threatening, adverse effects. Moreover, antibiotic-resistant bacteria can persist in biofilms, causing serious diseases. Hence, we validated the in vitro and in vivo efficacy of ciprofloxacin-loaded mesoporous silica nanoparticles (CIP–MSN) using a rat model of salmonella infection to compare the oral efficacy of 5 mg/kg body weight CIP–MSN and a traditional treatment regimen with 10 mg/kg CIP postinfection. Our results revealed that mesoporous silica particles can regulate the release rate of CIP with an MIC of 0.03125 mg/L against DCS *S. typhimurium* with a greater than 50% reduction of biofilm formation without significantly affecting the viable cells residing within the biofilm, and a sub-inhibitory concentration of CIP–MSN significantly reduced *invA* and *FimA* gene expressions. Furthermore, oral supplementation of CIP–MSN had an insignificant effect on all blood parameter values as well as on liver and kidney function parameters. MPO and NO activities that are key mediators of oxidative stress were abolished by CIP–MSN supplementation. Additionally, CIP–MSN supplementation has a promising role in attenuating the elevated secretion of pro-inflammatory cytokines and chemokines in serum from *S. typhimurium*-infected rats with a reduction in pro-apoptotic gene expression, resulting in reduced *S. typhimurium*-induced hepatic apoptosis. This counteracted the negative effects of the *S. typhimurium* challenge, as seen in a corrected histopathological picture of both the intestine and liver, along with increased bacterial clearance. We concluded that, compared with a normal ciprofloxacin treatment regime, MSN particles loaded with a half-dose of ciprofloxacin exhibited controlled release of the antibiotic, which can prolong the antibacterial effect.

Keywords: Salmonella typhimurium; ciprofloxacin; drug-loaded nanoparticles; qRT-PCR; histopathological examination

1. Introduction

Salmonella enterica serovar Typhimurium is a Gram-negative member of the family Enterobacteriaceae and is considered the second-most common cause of food poisoning associated with the consumption of contaminated food or water. It causes gastroenteritis, typhoid, and paratyphoid diseases in humans, and it can infect a wide range of hosts, including reptiles, birds, and mammals [1]. It is known that this bacterial pathogen is capable of intracellular survival by replicating inside the host cell in specialized vacuoles called *Salmonella*-containing vacuoles (SCVs). This clever mechanism causes persistent infection by allowing evasion of the robust inflammation and adaptive immune response of the host [2]. In addition, *Salmonella* can invade and translocate across the gut–epithelial barrier and infect the phagocytes, gaining access to the lymphatics and bloodstream, which allows the bacteria to spread to the liver and spleen [3]. Conventional antibacterial regimens have difficulty treating salmonella infection because of the intracellular survival and defenses of the bacteria [4]. Moreover, the increased prevalence of antimicrobial resistance limits the use of traditional antimicrobial agents, such as ampicillin, chloramphenicol, and trimethoprim-sulfonamide combinations, leading to the emergence of antimicrobial resistance among bacterial strains isolated from humans [5]. Fluoroquinolones, especially ciprofloxacin, have become the alternative option for treating salmonella infections [6]. However, the recent emergence of decreased ciprofloxacin susceptibility (DCS) or even resistance has led to treatment failures [7]. Delivering free antibiotics intracellularly has several limitations as a prolonged antibiotic course of 7 d, and the increased dosage required for treating salmonella infection leads to threatening, adverse effects [8]. Additionally, the WHO described a growing concern about *Salmonella*'s ability to form biofilms of fimbriae components [9], along with alarming antimicrobial resistance that presents a major threat to human and veterinary medicine [10].

These limitations have highlighted the importance of developing a biocompatible drug-delivery system with a high-loading capacity that controls drug release, allowing reduced dosage without compromising efficacy in treating intracellular pathogens [11]. Improvement of the overall pharmacokinetics, reduction of antimicrobial resistance, enhancement of the solubility of some antibiotics, and a wider therapeutic index are some benefits of developing drug-delivery systems [12]. Mesoporous silica nanoparticles show great promise as a biomedical application [13]. Their fabrication method at low temperatures enables them to carry biologically active agents as a drug-delivery system [14]. In addition, their unique properties include good biocompatibility, low hemolytic effect, and low toxicity. The high surface area and highly permeable, porous shell enable high drug-loading capacity and delayed release of antibacterial agents, which reduce the frequency of the dosages [15]. As shown in many previous reports, mesoporous silica nanoparticles (MSNs) as antibiotic delivery systems have been proven to boost the antimicrobial efficacy and safety profile of ciprofloxacin [16,17]. However, these reported studies were either conducted to evaluate in vitro antibacterial and cytotoxicity activity [17], or were in vivo survival assays prior to a murine oral *Salmonella typhimurium* infection model [16].

The above-mentioned studies provided a new sense of hope to further validate the in vitro efficacy of ciprofloxacin-loaded mesoporous silica nanoparticles against biofilm formation and the fold change in the mRNA expression of *iva*A and *Fim*A genes. Furthermore, the in vivo efficacy of a reduced dose of ciprofloxacin-loaded mesoporous silica nanoparticles compared with a normal ciprofloxacin treatment regime showed good biocompatibility and lower cytotoxicity after measuring the hematological profile and the liver and kidney function parameters. Interaction between the immune system and the ciprofloxacin-loaded mesoporous silica nanoparticles was evaluated through the antioxidant profile, pro-inflammatory cytokines, chemokines, and apoptosis regulator genes. Additionally, pathological changes and the bacterial clearance effect were evaluated.

2. Results

2.1. Characterization of Mesoporous Silica Nanoparticles (MSNs), Ciprofloxacin Loading, and Release

Characterization of mesoporous silica nanoparticles with ciprofloxacin (CIP–MSN) was carried out through transmission electron microscopy (TEM) (Figure 1A,B) at the National Center for Radiation Research and Technology (NCRRT), Atomic Energy Authority, Egypt. The loaded ciprofloxacin amount was increased by increasing the initial 5 mg/mL concentration of ciprofloxacin until it reached the maximum loading capacity of 1 mg for every 5 mg of MSN (Figure 1C). The in vitro release of ciprofloxacin from the CIP–MSN was gradual over 12 h, beginning with a 25% release in the first 2 h, followed by a cumulative and ongoing release until nearly 90% of the drug was released at 12 h after incubation (Figure 1D).

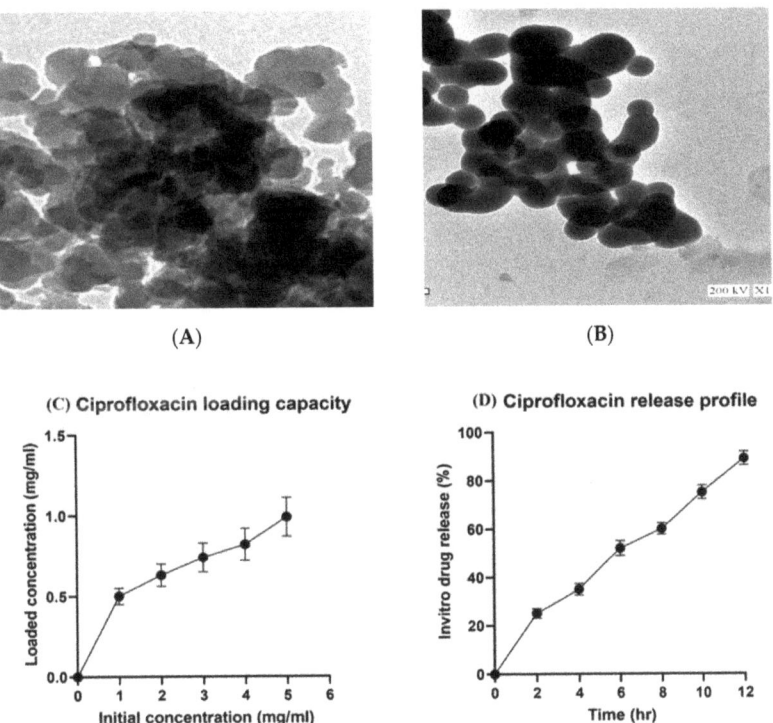

Figure 1. (**A**,**B**) Characterization of CIP–MSNs NPs by transmission electron microscopy (TEM); (**C**) Ciprofloxacin loading amount in relation to initial drug concentration (mg/mL); (**D**) percentage of ciprofloxacin in vitro release from CIP–MSNs over 12 h.

2.2. In Vitro Antibacterial Activity of CIP–MSN against Decreased CIP Susceptible Salmonella typhimurium

The antibacterial activity of prepared CIP–MSN was tested against the *Salmonella typhimurium* strain with decreased CIP susceptibility and compared with that of MSN and CIP. The results showed a significant variation in the inhibition zone of CIP–MSNs, along with both ciprofloxacin and unloaded mesoporous silica nanoparticles. The maximum antibacterial activity was observed in CIP–MSN with 40.5 ± 0.4 mm diameter of inhibition zone compared with 3.4 ± 0.37 and 8.7 ± 0.2 mm diameter inhibition zones of MSN and CIP, respectively (Table 1). The *Salmonella typhimurium* strain with decreased CIP susceptibility had an MIC for ciprofloxacin of 1.0 mg/L, compared with CIP–MSN. The antibacterial efficacy was confirmed with an MIC of CIP–MSN of 0.03125 mg/L, while MSN exhibited no inhibitory effect on visible bacterial growth (Table 1). Furthermore, the MBC value was two-fold higher than MIC values, indicating their bactericidal effect.

Table 1. Zone of inhibition diameter, MIC, and MBC of ciprofloxacin and MSN particles loaded with ciprofloxacin evaluated by agar well diffusion, and broth microdilution methods for *Salmonella typhimurium*.

	Zone of Inhibition (mm)	MIC (mg/L)	MBC (mg/L)
CIP	$8.7^b \pm 0.2$	1.0	2.0
MSN	$3.4^c \pm 0.37$	ND	ND
CIP–MSN	$40.5^a \pm 0.4$	0.03125	0.0625

MIC and MBC are presented in mg/L, and zones of inhibition are presented in mm. CIP–MSN: MSN particles loaded with ciprofloxacin; CIP: ciprofloxacin; MSN: Mesoporous silica particles; MIC: minimum inhibitory concentration; MBC: minimum bactericidal concentration; ND: not determined. Breakpoints for *Salmonella typhimurium* susceptibility profile according to EUCAST guidelines [18]. [a–c] Means within the same column carrying different superscripts are significantly different at $p < 0.05$.

2.3. Biofilm Inhibition and Transcriptional Modulatory Effect of CIP–MSN

Our results revealed that a sub-MIC concentration of CIP–MSN significantly reduced biofilm formation to 45% through a CV assay where 58% of cells remained viable upon performing an antibiofilm assay with the XTT method. This confirmed the effectiveness of a sub-MIC (1/2 MIC) concentration at inhibiting the extracellular polymeric substances (EPS) that form the biofilm matrix, with remaining viable cells residing within the biofilm. However, upon comparison to CIP and MSN, similar patterns were obtained with the CV and XTT assay, as it showed a nonsignificant reduction in biofilm (89 and 92%, respectively) ($p > 0.05$) (Figure 2A).

The modulatory effect of CIP–MSN on the invasion and major fimbrial subunit associated genes were investigated by reverse transcriptase expression qPCR of *inv*A and *Fim*A in biofilm *S. typhimurium* culture (Figure 2B). Both genes were downregulated as the expression levels of *inv*A and *Fim*A in CIP–MSN-treated *S. typhimurium* were 0.21-fold and 0.13-fold, respectively, which were more effective than CIP alone (0.6-fold change for *inv*A and *Fim*A genes). As expected, the MSN-treated *S. typhimurium* did not produce a detectable change in *inv*A and *Fim*A expression compared with untreated *S. typhimurium* ($p > 0.05$).

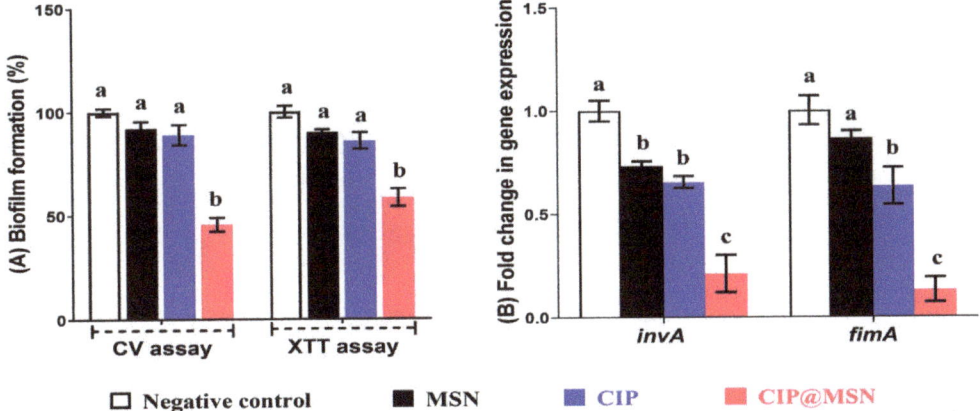

Figure 2. CIP–MSN reduces the biofilm formation of *S. typhimurium* compared with CIP alone, MSN, and NC(negative control; untreated bacteria). (**A**) Biofilm formation of tested *S. typhimurium* in the presence of CIP–MSN was detected by crystal violet staining and quantified by measuring the OD590, and cell viability within biofilm was detected by XTT assay and quantified by measuring the OD490; (**B**) Relative gene expression of *S. typhimurium* virulence genes *FimA* and *invA* upon treatment with CIP–MSN were calculated using the ΔΔCT method and expressed as fold change. 16S rRNA was used as the endogenous control. Each column shows the mean ± SD of three independent experiments. [a–c] Means within the same column carrying different superscripts are significantly different at $p < 0.05$.

2.4. Hematological, Biochemical, Antioxidant, and Immunological Effect of CIP–MSN on Blood, and Serum Constituents

Red blood cells (RBCs), hemoglobin concentration (Hb), and packed cell volume (PCV) were significantly elevated in CIP–MSN compared with the untreated challenged group, with red blood cells (RBCs) and hemoglobin concentration (Hb) levels nonsignificantly different from NC (Table 2). Moreover, based on the biochemical analysis of the blood serum constituents, the challenging with *S. typhimurium* adversely affected the liver and kidney function parameters, as represented by significantly increased alanine aminotransferase (ALT), aspartate aminotransferase (AST), urea, and creatinine levels at both 7 and 14 d postchallenge, while serum AST, ALT, uric acid, and creatinine levels were significantly reduced in the CIP–MSN-treated group regardless of challenge at both 7 and 14 d postchallenge, and their levels were nonsignificantly different from NC (Table 2). Concerning oxidative stress mediators, a significant decrease ($p < 0.05$) in NO and MPO levels at both time intervals in the group supplemented with CIP–MSN, when compared with the treated and control groups' MPO levels, was nonsignificantly different from the negative group at 14 d postchallenge (Table 2). The results of pro-inflammatory cytokine and chemokine analysis in the group challenged with *S. typhimurium* showed a significant increase in blood serum levels of CXCL10, CXCL11, IFN-γ, IL-6, TNF-α, and CRP at both time points (Table 2). Meanwhile, supplementation with CIP–MSN caused a significant reduction in blood serum levels of CXCL10, CXCL11, IFN-γ, IL-6, TNF-α, and CRP at 7 and 14 d when compared with the treated and control groups. No significant difference was found in CXCL11, IFN-γ, IL-6, and TNF-α serum levels in the CIP–MSN-treated groups relative to the unchallenged NC group (Table 2).

Table 2. Effects of MSN particles loaded with ciprofloxacin compared with ciprofloxacin on hematological, biochemical, and oxidative stress mediators and the immunological parameters of male rats (7 and 14 days postchallenge with *Salmonella typhimurium*).

Groups	At 7 Days Postinfection							At 14 Days Postinfection						
	NC	PC	CIP	MSN	CIP@MSN	p Value	SEM	NC	PC	CIP	MSN	CIP-MSN	p Value	SEM
RBCs	12.5 [a]	7.96 [b]	9.17 [ab]	8.23 [b]	11.19 [a]	<0.001	0.09	12.63 [a]	9.66 [c]	10.30 [b]	10.20 [b]	11.60 [ab]	0.09	12.63 [a]
Hb	12.9 [a]	6.3 [b]	11.2 [a]	7.10 [b]	11.56 [a]	<0.001	0.07	12.80 [a]	8.80 [c]	11.98 [a]	10.26 [b]	12.00 [a]	0.12	12.80 [a]
PCV	41.4 [a]	24.5 [c]	37.43 [b]	26.60 [c]	39.5 [ab]	0.03	0.14	42.30 [a]	29.69 [c]	38.6 [ab]	37.75 [b]	39.5 [a]	0.13	42.30 [a]
Biochemical biomarkers for tissue injury analysis														
ALT (U/L)	48.3 [c]	74.7 [a]	58.7 [b]	65.63 [ab]	50.86 [c]	<0.001	0.21	46.7 [d]	85.1 [a]	52.36 [c]	57.99 [b]	48.36 [cd]	0.25	48.3 [c]
AST (U/L)	25.2 [b]	42.3 [a]	27.6 [b]	26.30 [b]	25.53 [b]	<0.001	0.19	24.7 [c]	48.4 [a]	25.8 [c]	34.96 [b]	23.46 [c]	0.16	25.2 [b]
Urea (µmol/L)	32.1 [c]	52.3 [a]	38.76 [b]	50.90 [a]	35.53 [bc]	<0.001	0.12	31.6 [c]	59.8 [a]	32.6 [c]	41.90 [b]	32.1 [c]	0.14	32.1 [c]
Creatinine (mg/dL)	1.33 [b]	2.73 [a]	1.45 [b]	1.95 [ab]	1.41 [b]	<0.001	0.09	1.34 [c]	3.07 [a]	1.38 [c]	2.35 [b]	1.33 [c]	0.08	1.33 [b]
Oxidative stress mediators analysis														
NO	153.8 [d]	532.60 [a]	246.40 [b]	480.60	185.36 [c]	0.027	0.25	150.5 [e]	559.1 [a]	202.16 [d]	140.22 [b]	177.66 [c]	0.31	153.8 [d]
MPO	2.21 [e]	10.80 [a]	7.62 [c]	8.65 [b]	6.11 [d]	0.03	0.09	2.16 [c]	10.08 [a]	6.76 [b]	3.57 [c]	2.21 [c]	0.10	2.21 [c]
CRP	1.12 [e]	54.30 [a]	24.4 [c]	50.36 [b]	16.26 [d]	0.14	0.14	1.16 [e]	53.3 [a]	17.4 [c]	44.30 [b]	9.2 [d]	<0.001	0.08
Chemokines and pro-inflammatory cytokines analysis														
CXCL10	156.00 [e]	393.70 [a]	224.80 [c]	385.64 [b]	181.6 [d]	0.03	0.24	152.6 [d]	416.3 [a]	197.9 [b]	400.02 [a]	161.83 [c]	0.02	0.24
CXCL11	108.86 [e]	238.70 [a]	161.30 [c]	220.30 [b]	130.53 [d]	0.02	0.29	108.5 [d]	243.4 [a]	140.6 [b]	239.23 [a]	113.16 [c]	0.01	0.31
IFN-γ	38.36 [e]	99.50 [a]	53.26 [c]	82.20 [b]	43.00 [d]	<0.001	0.11	38.7 [c]	104.4 [a]	46.6 [b]	100.23 [a]	38.36 [c]	0.03	0.23
IL-6	16.28 [c]	47.36 [a]	47.93 [a]	45.636 [a]	34.86 [b]	<0.001	0.13	15.86 [e]	47.56 [a]	34.56 [b]	49.36 [a]	16.59 [c]	<0.001	0.14
TNF-α	11.5 [c]	23.61 [a]	16.50 [b]	20.36 [a]	16.4 [b]	<0.001	0.10	11.7 [c]	21.96 [a]	15.1 [b]	22.90 [a]	13.2 [c]	<0.001	0.06

NC (negative control): rats received a control diet without any addition; PC (positive control): rats received a control diet without any addition and were orally challenged with *S. typhimurium*; CIP(ciprofloxacin): rats received a control diet, were orally challenged with *S. typhimurium*, and received 10 mg/kg CIP orally twice a day for 3 days; and CIP–MSN (MSN particles loaded with ciprofloxacin): rats received a control diet, were orally challenged with *S. typhimurium*, and received 5 mg/kg CIP orally twice a day for 3 days. RBCs: red blood cells; Hb: hemoglobin concentration; PCV: packed cell volume. ALT: alanine aminotransferase; AST: aspartate aminotransferase; NO: nitric oxide; MPO: myeloperoxidase; CXCL10: C-X-C motif chemokine ligand 10; CCL11:CC motif chemokine ligand 11; IFN-γ: interferon gamma; IL-6: interleukin-6; TNF-α: Tumor Necrosis Factor-alpha; CRP:C-reactive protein. [a–e] Means in the same row with different letters are significantly different at ($p < 0.05$).

2.5. Inhibitory Effect of CIP–MSN on Hepatic Salmonella typhimurium Load

Quantitative, real-time PCR counting of *S. typhimurium* in the liver and spleen samples at 7 and 14 d postchallenge revealed significant lowering in the CIP–MSN-treated groups compared with the PC group ($p < 0.05$), and the bacterial load decreased steadily over time, whereas the reduction in $\log_{10}$ copies of *S. typhimurium* populations was evidenced by 3.2 and 2.9 $\log_{10}$ CFU/g at 7 and 14 d postinfection, respectively, in the liver sample, and 5.3 and 3.7 $\log_{10}$ CFU/g at 7 and 14 d postinfection, respectively, in the spleen sample (Figure 3). Changes in the $\log_{10}$ values of *S. typhimurium* CFU/g in MSN- and CIP-supplemented groups showed no statistically significant differences at 7 d postchallenge, and the bacterial load was significantly reduced in the CIP-treated group at 14 d postchallenge when compared to the PC group in both hepatic and splenic samples (Figure 3).

2.6. Pro-Inflammatory Cytokines Transcriptional Modulatory Effect of CIP–MSN

The mRNA expressions of pro-inflammatory cytokines interleukin-6 (IL-6), interleukin-1β (IL-1β), and tumor necrosis factor alpha (TNF-α) were examined at 7 and 14 d postinfection (Figure 4). The challenge with *S. typhimurium* adversely increased the mRNA expression of pro-inflammatory cytokines at both 7 and 14 d postchallenge (assigned a value of 1 arbitrary unit). The level of IL-6 and IL-1β expression in the splenic tissue was significantly lower ($p < 0.05$) in the group that had received CIP–MSN, as compared with the other treatment and control groups; the most pronounced downregulation in the level of IL-6 (about a 0.3-fold reduction) was noticed at 14 d postinfection. Moreover, dietary supplementation of CIP–MSN decreased the transcriptional levels of TNF-α in a time-dependent manner compared to the PC group, as it reached about a 0.4-fold reduction by the end of the experiment (14 d postinfection).

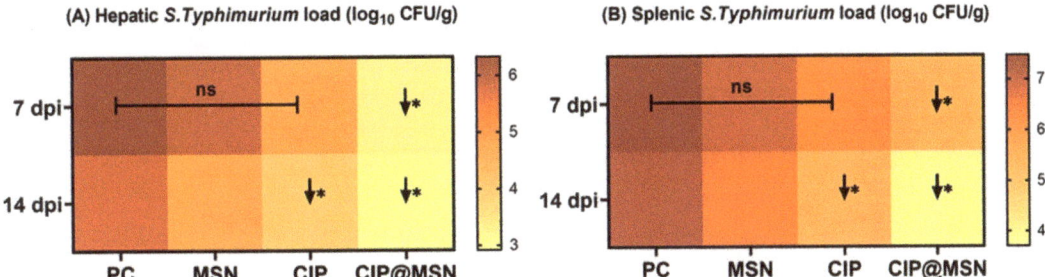

Figure 3. In vivo evaluation of CIP–MSN treatment on *S. typhimurium* bacterial load in (**A**) hepatic tissue and (**B**) splenic tissue at 7 and 14 days post-infection (7 and 14 dpi) by real-time PCR quantification of DNA copies, and represented as $\log_{10}$ of the CFU per gram of tissue. PC (positive control): rats received a control diet without any addition and were orally challenged with *S. typhimurium*; MSN(mesoporous silica particles): rats received a control diet, were orally challenged with *S. typhimurium*, and received 10 mg/kg MSN orally twice a day for 3 days; CIP(ciprofloxacin): rats received a control diet, were orally challenged with *S. typhimurium*, and received 10 mg/kg CIP orally twice a day for 3 days; and CIP–MSN (MSN particles loaded with ciprofloxacin): rats received a control diet, were orally challenged with *S. typhimurium*, and received 5 mg/kg CIP orally twice a day for 3 days. Data are expressed as means ± SE (error bars). Arrows correspond to significant decrease (↓*) relative to the PC group ($p < 0.05$), and NS represents nonsignificant differences relative to the PC group (p-value > 0.05).

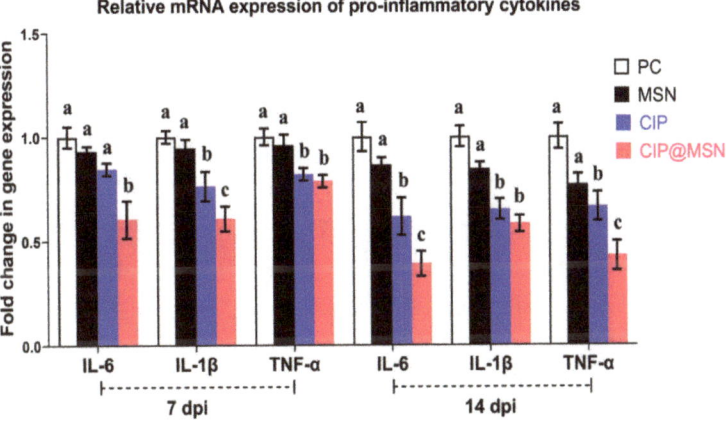

Figure 4. Relative mRNA expression levels of pro-inflammatory cytokines; interleukin-6 (IL-6), interleukin-β (IL-β) and tumor necrosis factor alpha (TNF-α) in the splenic tissue of rats treated with CIP–MSN compared with CIP alone at 7 and 14 days postinfection with *S. typhimurium* (7 and 14 dpi). The expression levels were calculated using the $2^{-\Delta\Delta Ct}$ method and expressed as fold change, and glyceraldehyde-3-phosphate dehydrogenase (GAPDH) was used as the endogenous control. PC (positive control): rats received a control diet without any addition and were orally challenged with *S. typhimurium*; MSN(mesoporous silica particles): rats received a control diet, were orally challenged with *S. typhimurium*, and received 10 mg/kg MSN orally twice a day for 3 days; CIP(ciprofloxacin): rats received a control diet, were orally challenged with *S. typhimurium*, and received 10 mg/kg CIP orally twice a day for 3 days; and CIP–MSN (MSN particles loaded with ciprofloxacin): rats received a control diet, were orally challenged with *S. typhimurium*, and received 5 mg/kg CIP orally twice a day for 3 days. Each column shows the mean ± SD of three independent experiments. [a–c] Means within the same column carrying different superscripts are significantly different at $p < 0.05$.

2.7. Modulation of Pro-Apoptoticgenes Expression

The expression levels of six pro-apoptotic genes (COX-2, caspase-3, P450, iNOS, Bcl-2, and BAX) of the CIP–MSN-treated and control groups at different time points (7 and 14 dpi) were investigated by qRT-PCR (Figure 5A,B).

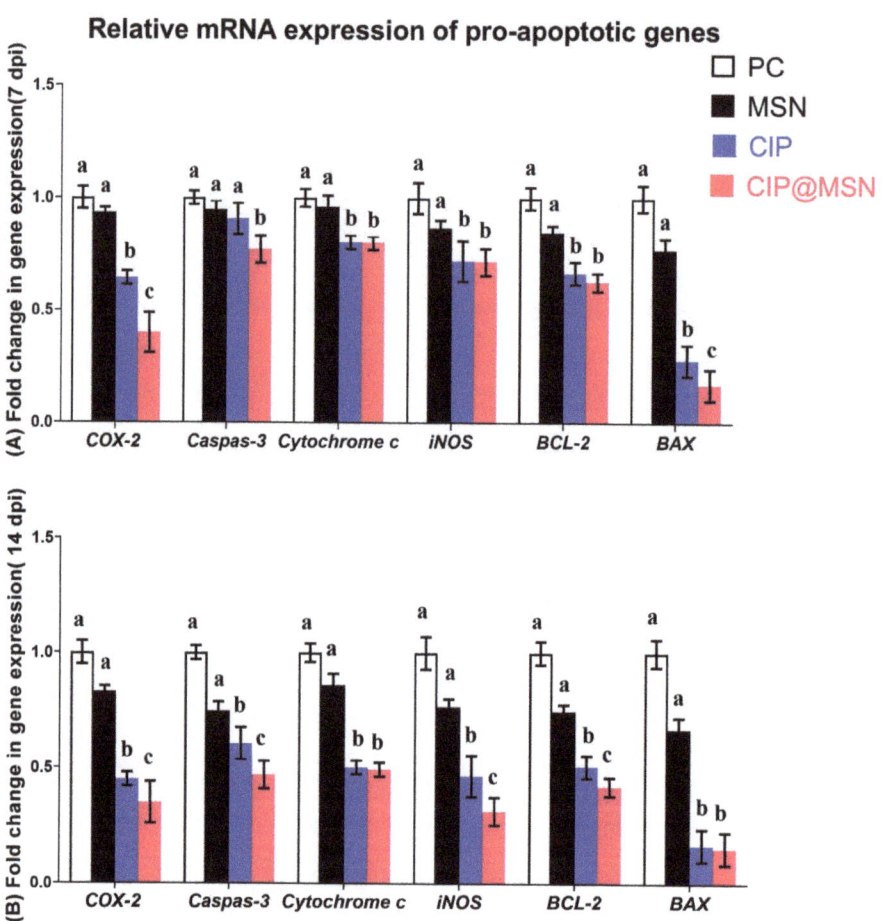

Figure 5. Relative mRNA expression levels of pro-apoptotic genes; Cyclooxygenase-2(*COX-2*), caspase-3, Cytochrome P450, inducible nitric oxide synthase (*iNOS*), B-cell lymphoma-2 (*Bcl-2*), and Bcl-2-associated X (*BAX*)in the splenic tissue of rats treated with CIP–MSN compared with CIP alone at (**A**) 7 and (**B**) 14 days postinfection with *S. typhimurium* (7 and 14 dpi).The expression levels were calculated using the $-2^{\Delta\Delta CT}$ method and expressed as fold change, and glyceraldehyde-3-phosphate dehydrogenase (GAPDH) was used as the endogenous control. PC (positive control): rats received a control diet without any addition and were orally challenged with *S. typhimurium*; MSN(mesoporous silica particles): rats received a control diet, were orally challenged with *S. typhimurium*, and received 10 mg/kg MSN orally twice a day for 3 days; CIP(ciprofloxacin): rats received a control diet, were orally challenged with *S. typhimurium*, and received 10 mg/kg CIP orally twice a day for 3 days; and CIP–MSN (MSN particles loaded with ciprofloxacin): rats received a control diet, were orally challenged with *S. typhimurium*, and received 5 mg/kg CIP orally twice a day for 3 days. Each column shows the mean ± SD of three independent experiments. [a–c] Means within the same column carrying different superscripts are significantly different at $p < 0.05$.

Gene expression analysis found significant differences between the CIP–MSN-treated and control groups. Interestingly, we found that expression of COX-2 was greatly decreased in the CIP–MSN-treated group at 7 and 14 dpi, with 0.4- and 0.3-fold changes, respectively, compared with CIP treatment alone. CIP–MSN dietary supplementation downregulated the expression of caspase-3 significantly at both time intervals, with a cumulative effect (0.7- and 0.4-fold changes at 7 and 14 dpi, respectively). The CIP–MSN-treated group significantly downregulated expression of P450, and iNOS with no notable difference between the CIP–MSN- and CIP-treated groups at both time points. The lowest transcriptional expression of iNOS level was observed in the CIP–MSN-supplemented group at 14 dpi of about 0.3-fold when compared with the CIP-treated group and the PC group. Bcl-2 expression was significantly downregulated in the CIP–MSN-treated group compared with control groups, reaching a 0.4-fold decrease at 14 dpi. Meanwhile, the most downregulation occurred for the BAX gene, as observed in the CIP–MSN-supplemented group (about a 0.1-fold reduction) compared to the PC group at both time points.

2.8. Histopathological Evaluation

Histopathological findings of intestinal and liver tissues post-*S. typhimurium* infection are presented in Figure 6. In the NC group, intestinal histomorphological structures showed normal mucosa, submucosa, musculosa, and serosa (Figure 6A). Meanwhile, in the PC group, the majority of intestinal sections showed enteritis, which was represented by dilated blood vessels, leukocytic infiltrations, metaplastic changes of the epithelial lining into goblet cells, and desquamated villous epithelium (Figure 6B). Additionally, the group treated with CIP alone displayed apparently normal intestinal layers (Figure 6C). However, goblet cell hyperplasia and denuded epithelium were detected in some examined sections. In the group treated with CIP–MSN, apparently normal mucosa, submucosa, and musculosa in most sections of intestine were detected (Figure 6D). Liver sections revealed normal cytoarchitectures of the hepatic cords, sinusoids, and stromal components in the NC group (Figure 6E). In the PC group, liver sections revealed unexpected multifocal necrotic areas that were mostly replaced by macrophages, and lymphocytes with severely dilated sinusoids accompanied by atrophied hepatic cords were seen within most examined sections (Figure 6F). In the group treated with CIP alone, liver sections showed some apparently normal hepatic parenchyma (Figure 6G), and in the group treated with CIP–MSN, preserved hepatic cords and hepatic vasculatures were more prominent in most liver tissues (Figure 6H).

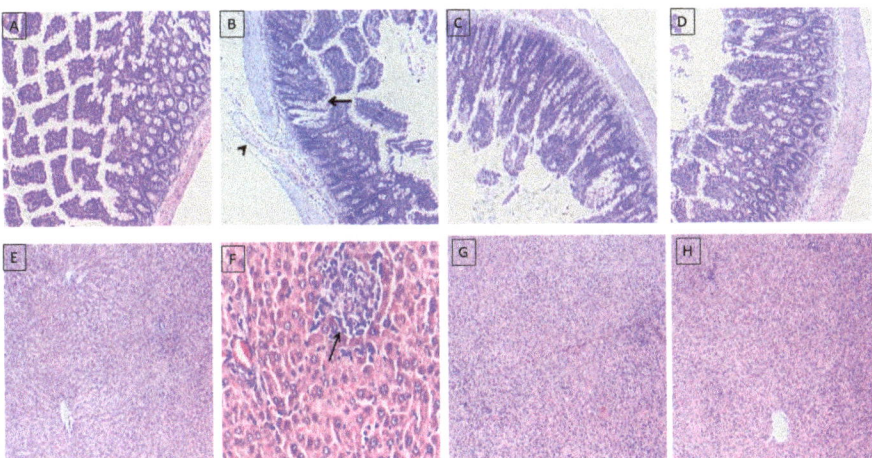

Figure 6. Representative photomicrographs using H&E staining of histological sections for intestines (**A–D**), and livers (**E–H**) of rats treated with CIP–MSN compared with CIP alone at 14 days postinfection

with *S. typhimurium*. PC (positive control): rats received a control diet without any addition and were orally challenged with *S. typhimurium*; MSN(mesoporous silica particles): rats received a control diet, were orally challenged with *S. typhimurium*, and received 10 mg/kg MSN orally twice a day for 3 days; CIP(ciprofloxacin): rats received a control diet, were orally challenged with *S. typhimurium*, and received 10 mg/kg CIP orally twice a day for 3 days; and CIP–MSN (MSN particles loaded with ciprofloxacin): rats received a control diet, were orally challenged with *S. typhimurium*, and received 5 mg/kg CIP orally twice a day for 3 days.

3. Discussion

Salmonella enterica serovar Typhimurium is a common cause of persistent infection owing to its capacity for intracellular survival, biofilm formation, and evasion of the robust inflammatory response of the host, causing decreased ciprofloxacin susceptibility. Thus, an optimal strategy to treat these infections should deliver drugs with prolonged release from a single dose with good biocompatibility and lower toxicity. Previous reports demonstrating MSNs as antibiotic-delivery vehicles have been performed in vitro. There is a previous study reporting the in vivo survival in an assay of a murine oral *Salmonella typhimurium* infection model [16]. To the best of our knowledge, there are no previous reports validating in vitro antivirulence efficacy against biofilm formation; in vivo biosafety in blood parameters; and biochemical biomarkers of tissue injury or inflammation, pro-inflammatory cytokines and chemokines accompanied by pro-apoptotic gene expression, and histopathological effect.

Our results showed that the small particle sizes of the CIP-loaded MSN (CIP–MSN) play an important role in the internalization into cells, as reported previously, where submicron-sized particles can be taken up by M-cells and macrophages present in Peyer's patches [19]. Accordingly, mesoporous silica particles can regulate the release rate of an antimicrobial agent [20]. Our in vitro antibacterial activity confirmed that the combination of mesoporous silica nanoparticles and CIP was responsible for higher antimicrobial activity when compared with the drug alone, and *Salmonella typhimurium* with decreased ciprofloxacin susceptibility (MIC 1.0 mg/L) was converted to 0.03125 mg/L. This indicated a reduction of antimicrobial resistance; the bacteria reverted to being susceptible to ciprofloxacin, according to the MIC break point of <0.125 mg/L. (EUCAST guidelines were applied for category interpretation for the different antibiotics [18] Silica nanoparticles alone showed no detrimental effects on bacteria [21].

The biofilm formation is a bacterial survival strategy leading to increased resistance to antibiotics [22]. Moreover, antibiotic-resistant bacteria can persist in biofilms and result in enhanced tolerance of adverse environmental conditions, causing serious infectious diseases [23]. To support the in vitro antibacterial enhancement effect of CIP–MSN, nanomaterials were proposed as an interventional strategy for the management of biofilm formation because of their high surface area-to-volume ratio and unique chemical and physical properties [24]. Our study reported a greater than 50% reduction in biofilm formation, with the remaining viable cells residing within the biofilm upon using a subinhibitory concentration of CIP–MSN, which was in accordance with another study that applied silver nanoparticle-doped nanoporous silica and reported a 70% reduction in biofilm survival [25].

S. typhimurium has been shown to produce a major fimbrial subunit and invasin A on the bacterial surface. Both of these vital virulence factors, encoded by the *FimA* and *invA* genes, play roles in mediating bacterial adherence to eukaryotic cells and facilitating the entry into intestinal epithelial cells, respectively, which are critical steps in successful colonization and pathogenesis [26]. Herein, the effect of CIP-loaded MSN on the reduction of biofilm formation was supported by significantly reduced *invA* gene expressions and *FimA* coding a major fimbrial subunit. This indicated the possible effect of this drug-delivery cargo on decreasing the *Salmonella* adherence and invasion of host cells and tissues [27,28].

The safety of the oral supplementation with a reduced dose of drug-delivery cargo was estimated through the evaluation of the hematological and biochemical parameters of rats. The pathophysiological status of the body related to infection and therapy was indicated by analysis of the hematological profile [29] and biochemical biomarkers for tissue injury or inflammation [30]. Herein, the administration of this drug-delivery cargo showed an insignificant effect on all blood parameter values and liver and kidney function parameters, indicating its good biocompatibility, low hemolytic activity, and lack of cytotoxicity in the liver and kidney function test. Consistent with a previous study [31], *Salmonella typhimurium* infection adversely affected the liver and kidney function parameters, resulting in a dramatic increase in ALT, AST, uric acid, and creatinine levels in serum, while these parameters were diminished with a reduced dose in CIP–MSN-treated groups, regardless of challenge, compared with the normal ciprofloxacin treatment regime.

The enhancing effects on the protective humoral immune response were indicated by detection of the key mediators of oxidative stress known to eliminate the invading bacterial pathogens, owing to the bactericidal action of MPO and NO produced from activated neutrophils and monocytes in the blood [32]. Our data revealed a significant elevation in oxidative stress mediators in response to *S. typhimurium* challenging, even after treatment with ciprofloxacin only, indicating an increased number of neutrophilic granulocytes that was consistent with a previous study [33]. We also demonstrated that the CIP–MSN had an insignificant effect on the oxidative stress enzyme activity in the challenged group when compared with the control positive group and the normal ciprofloxacin treatment regime, indicating that this drug-delivery model exhibited no harmful effects on the blood serum. Interestingly, we found that MPO and NO production were reduced in the CIP–MSN-treated group with a lower dose compared with the normal ciprofloxacin treatment regime at 7 and 14 d postchallenge, indicating its important role in abrogating the *S. typhimurium*-induced stimulation of phagocytes and the subsequent health benefit. Tissue injury and inflammation in animals are indicated by elevated secretion of pro-inflammatory cytokines and chemokines by macrophages in the blood serum and liver [34]. Our result showed that *Salmonella typhimurium* challenge significantly increased serum level and liver mRNA expression of pro-inflammatory cytokines and chemokines, which was documented in a previous study [33], and these elevations were abolished by CIP–MSN supplementation at a reduced dose compared to the normal ciprofloxacin treatment regime.

Caspases play a very important role in apoptosis, and their activation takes place upon assembly of an intracellular complex known as inflammasome, which is responsible for the processing and maturation of pro-inflammatory cytokines, such as IL-1β and IL-18 [35,36]. Thus, they act in inflammation and innate immune host defense against microbial pathogens [37]. Previously, *Salmonella typhimurium*w as reported to induce the caspase-dependent death of macrophages upon infection, with the release of pro-inflammatory cytokines that colonize the Peyer's patches (PPs) and cross the intestinal barrier [38]. These reports prompted us to investigate whether the protective effects of a reduced dose of CIP–MSN supplementation were associated with reduced *S. typhimurium*-induced apoptosis in hepatic cells by analyzing the mRNA expression of pro-apoptotic genes, such as *COX-2*, caspase-3, cytochrome P450, *iNOS*, *Bcl-2*, and *BAX*. As expected, our results revealed increased mRNA levels of pro-apoptotic proteins induced by *Salmonella typhimurium*, indicating infection and increased apoptosis [39]. Accumulating evidence has shown that oxidative stress-related apoptosis is involved in pathogen-infection-induced tissue injury [40]. CIP–MSN administration at half-dose, as compared with the normal ciprofloxacin treatment regime, significantly decreased apoptosis by inhibiting mRNA levels of pro-apoptotic proteins. Consistent with this result, a previous study showed that apoptosis induced by *Salmonella typhimurium* was abolished by therapy supplementation [33].

S. typhimurium infection in rats resulted in salmonella-infected phagocytes gaining access to the lymphatics and bloodstream, allowing the bacteria to spread to the liver and the spleen [3]. Previous studies confirmed that the combination of mesoporous silica nanoparticles or silica xerogel and drugs was responsible for effective antibacterial activity when compared with the drug alone, resulting in a more effective clearance of *Salmonella enteric* serovar Typhimurium infection from mouse spleen and liver than the same dose of free drug [14]. In accordance, our results indicated that CIP–MSN at half-dose can boost the clearance rate of *Salmonella enteric* serovar Typhimurium infection in liver and spleen tissues with a significant log reduction of bacterial load than the higher dose of free drug. This counteracted the negative effects of the *S. typhimurium* challenge through a corrected histopathological picture of both the intestine and liver and increased bacterial clearance. Post-*Salmonella typhimurium* infection, the histological pathological architecture of intestinal tissues showed a diffuse inflammatory cell infiltration and complete desquamated epithelial tissues. Meanwhile, with administration of CIP, the severity of intestinal inflammation and liver damage was greatly reduced. Moreover, intestinal and liver tissues that were nearly restored to normal condition were more prominent in the group supplemented with CIP–MSN after *Salmonella typhimurium* infection, indicating its better efficacy in treating infection. Accordingly, treatment with ciprofloxacin and thymol oils against *Shigella flexneri* reduced intestinal infiltration of inflammatory cells in male albino rats and thus reduced the severity of gastric ulcer [41]. Additionally, administration of ciprofloxacin-loaded gold nanoparticles decreased the load of *Enterococcus faecalis* in the liver and kidneys of mice and consequently reduced the severity of tissue damage [42].

4. Materials and Methods

4.1. Synthesis and Characterization of Mesoporous Silica Nanoparticles (MSNs) and Ciprofloxacin Loading

The mesoporous nanosilica was prepared as previously described at the National Center for Radiation Research and Technology (NCRRT), Atomic Energy Authority, Egypt [43]. The purchased ciprofloxacin (Cipro, Fluka, 98%) was prepared as solution having a concentration of 5 mg/mL, and loaded by incubating with 5 mg of MSN particles for 12 h as previously described [16]. The drug loading capacity was evaluated by measuring the concentration of the free drug in the mixture before and after loading using standard calibration curves obtained by UV–Vis spectroscopy, and calculated by the following equation: loading capacity = weight of ciprofloxacin in MSN particles/weight of MSN particles. In vitro release of ciprofloxacin from MSN particles was determined as previously described [16], in which the dispersed CIP–MSN solution was placed in a dialysis sac, then placed into 50 mL of PBS solution, and shaken at 37 °C. The amount of ciprofloxacin released at different time intervals was evaluated spectrophotometrically at 275 nm (Nanodrop, ND1000, Thermo Scientific, Waltham, MA, USA).

4.2. Antibacterial Effect of Ciprofloxacin—Loaded Mesoporous Silica Nanoparticles

The *Salmonellaenterica* serovar Typhimurium ATCC 14028 strain used in this experiment was previously found to be a multivirulent and multidrug-resistant bacterium [44] that has decreased CIP susceptibility, with a minimum inhibitory concentration [9] value of 1 mg/L [7]. Stocks were maintained in 20% (v/v) glycerol at -80 °C until needed. Bacterial strains were grown on Tryptone Soya Agar (TSA; Oxoid, UK) overnight at 37 °C.

4.2.1. Agar Well Diffusion Assay

The antibacterial activities of CIP–MSN, MSN, and CIP (10 µg/mL concentration) were evaluated against *Salmonella typhimurium*, in which bacterial suspension in sterile saline was prepared to match the optical density of 0.5 MacFarland (1.5×10^8 CFU/mL), and then were grown in Mueller–Hinton (MH) agar (Oxoid Ltd., Hampshire, UK). Wells (8 mm) were cut into each inoculated agar plate, and a 100-µL aliquot of each compound was pipetted into each well. MSN were replaced with sterile water as a negative control for bacterial growth. The plates were incubated at 37 °C for 24 h. After incubation, zones of growth inhibition were measured, and results were expressed as mean $\pm$ standard deviation (SD) to determine the antimicrobial potency of the screened compound [45].

4.2.2. Minimum Inhibitory Concentration

The minimum inhibitory concentration was determined using micro broth dilution methods [46]. The concentration of CIP–MSN, MSN, and CIP were serially diluted two-fold, from 1 to 512 µg/mL, and were inoculated with the suspension of standardized inoculum, and then incubated at 37 °C for 24 h. MIC were determined as the lowest concentration that showed no visible growth, while the MBC was determined by culturing 10 µL of each clear well on the Mueller–Hinton agar plates, and the lowest concentration showing a 99.9% reduction in the initial inoculum after overnight incubation was determined as the MBC.

4.3. Ciprofloxacin Loaded Mesoporous Silica Nanoparticles Effect on Biofilm Formation

A single colony was inoculated in Mueller–Hinton broth (10^6 CFU/mL) in microtiter plates with sub-inhibitory concentrations (1/2 × the MIC) of either CIP–MSN, MSN, or CIP, and antibiotic-free medium was used as a negative control. Then, the biofilm formation was performed with the protocol developed previously [47]. The crystal violet was added to measure the extracellular polymeric substances (EPSs) in the biofilm, and the biofilm mass optical density was measured by a microplate ELISA reader (Huma Reader HS, Wiesbaden, Germany) at a wavelength of 590 nm [48]. The XTT cell viability assay kit, according to the manufacturer's protocol, was employed to measure the viability of cells residing within the matrix [49]. The change in color due to the viability of cells was measured using a microplate reader (Huma Reader HS, German) at a wavelength of 490nm. Experiments were performed in triplicate, and any inferences of nanoparticles in the measurement was deducted from the absorbance imposed by samples, and then the average value was reported with ±SD. The biofilm formation and viable cells were tested in triplicate in independent experiments and interpreted as the ratio of CIP–MSN, MSN, and CIP to the untreated negative control.

4.4. Expression of Genes Associated Virulence in Biofilm Culture

qRT-PCR was carried out with the biofilm culture grown in the presence of subinhibitory concentrations of either CIP–MSN, MSN, or CIP, with the untreated negative control as described in a previous section. Then, the bacterial suspension was mixed with RNAprotect Bacteria Reagent (Qiagen, Hilden, Germany) and centrifuged at 5000× g for 10 min. RNA extraction was performed using a QIAamp RNeasy Mini kit (Qiagen, Germany, GmbH) according to the manufacturer's instructions. Genomic DNA was removed from the samples by treatment with 1U DNase I, RNase-free (Thermo Scientific) for 60 min at 37 °C. Real-time PCR amplification reaction was prepared in a final volume 25 µL containing 10 µL of the 2x HERA SYBR® Green RT-qPCR Master Mix (Willowfort, UK), 1 µL of RT Enzyme Mix (20X), 0.5 µL of each primer of 20 pmol concentration, 5 µL of RNase- and DNase-free water, and 3 µL of RNA template. The primer sequences used for the *invA* (invasion protein) and, *FimA* (major fimbrial subunit) virulence genes in biofilm culture are shown in Table 1. The 16rRNA gene was used as an internal control for the normalization of the mRNA expression. The PCR products were analyzed using a Step One Real-Time PCR System (Applied Biosystems, California, CA, USA). The comparative Ct method was used to analyze the relative expression of targeted genes, and normalized to the untreated negative control, which was assigned a value of 1 arbitrary unit [50].

4.5. Experimental Design, and Oral Challenging with Salmonella typhimurium

This study was conducted in accordance with the regulations approved by the Institutional Animal Care and Use, Faculty of Veterinary Medicine, Zagazig University, and to confirm the freeing of rats from any *Salmonella* spp, bacteriological examinations in accordance with the International Organization for Standardization [51] were conducted on a total of 75 male rats (housed in a standard housing condition). Rats were randomly divided into five different experimental groups with three replicates per each group after one week of adaption.

The experimental treatments were as follows: rats in the first group(negative control, NC) were fed with basal diets without any addition and administered sterile water; rats in the second group (positive control, PC) received a control diet without any addition and were challenged orally with *Salmonella typhimurium* strain (1×10^6 CFU/mL) [16]; those in third and fourth groups received CIP–MSN or MSN orally 12 hr after *Salmonella typhimurium* challenging with 5 mg/kg body weight twice a day for 3 days [16], while the traditional treatment regimen was used in the last group with 10 mg/kg CIP orally twice a day for 3 days. The rats were observed for 14 days postinfection.

4.6. Hematological, Biochemical, Oxidative Stress Mediators, and Immunological Measurements

At 7 and 14 d postchallenge, three rats from each treatment were randomly chosen for aseptic collection of blood samples from the tail vein. The collected blood was divided into two equal parts: The first part was collected on heparin as an anticoagulant to determine the blood hematology: red blood cells (RBCs), hemoglobin concentration (Hb), and packed cell volume (PCV) [52]. The second part was immediately centrifuged at 3500 rpm for 15 min, and the serum was used for biochemical biomarkers for liver and kidney injury: alanine aminotransferase (ALT), aspartate aminotransferase (AST),urea, and creatinine, using commercial kits (Span Diagnostic Ltd., Sachin, India). Evaluation of mediators of oxidative stress: nitric oxide (NO) and myeloperoxidase (MPO)were analyzed using commercial kits (Jiancheng Biotechnology Institute, Nanjing, China) [53]. Immunological evaluation of serum chemokines: C-X-C motif chemokine ligand 10 (CXCL10), CC motif chemokine ligand 11(CCL11), pro-inflammatory cytokines; interferon gamma (IFN-γ), interleukin-6 (IL-6), and tumor necrosis factor alpha (TNF-α) were analyzed spectrophotometrically by enzyme-linked immunosorbent assay (ELISA) kits (Cusabio Biotech Co. Ltd., Wuhan, China),and latex-enhanced nephelometry was used for detection of C-reactive protein (CRP) [54].

4.7. Quantification of S. typhimurium DNA Copies

Three rats from each experimental group were randomly chosen and slaughtered at 7 and 14 d postchallenge. Liver and spleen tissues were aseptically removed and stored at −80 °C until used; after that, the tissues homogenized, and DNA was extracted according to the manufacturer's instructions using a QIAamp DNA Stool Mini Kit (Qiagen GmbH, Hilden, Germany), and a NanoDrop2000 spectrophotometer (Thermo Fisher Scientific Inc., Waltham, MA, USA) was used for assessing DNA purity and concentrations. Real-time PCR (RT-PCR) assays were conducted in a Stratagene MX3005P real-time PCR machine for the quantification of DNA copies using generated standard calibration curves from a pure *S. typhimurium* strain, and then interpolating the Ct values of DNA from the liver and spleen samples into the standard curves, and detecting the $\log_{10}$ of the CFU numbers

4.8. Pro-Inflammatory Cytokines, and Pro-Apoptotic Gene Expression Analysis by Real-Time PCR

Splenic tissues were also aseptically removed from the same three slaughtered rats, as mentioned in the previous step, and homogenized into RNA later (Sigma, St. Louis, MO, USA) for analyzing the differential gene expressions and immune-related parameters by the RT-qPCR assay. Briefly, total RNA was extracted from the spleen according to the manufacturer's instructions using Qiagen RNA extraction kits (Cat, No. 74104). Total RNA purity was measured using a NanoDrop_ND-1000 Spectrophotometer (Nano-Drop Technologies, Wilmington, DE, USA). The expression levels of pro-inflammatory cytokines genes: interleukin-6 (*IL-6*),interleukin-1β (*IL-1β*),and tumor necrosis factor-alpha (*TNF-α*), and pro-apoptotic genes:Cyclooxygenase-2 (COX-2), caspase-3, cytochrome-C, inducible nitric oxide synthase (*iNOS*), B-cell lymphoma-2 (*Bcl-2*), and Bcl-2-associated X (*BAX*) as listed in Table 3 were conducted in a Stratagene MX3005P real-time PCR machine using a one-step QuantiTect SYBR Green RT-PCR Kit (Qiagen GmbH, Hilden, Germany) according to the manufacturer's procedures, and normalized using glyceraldehyde-3-phosphate dehydrogenase (GAPDH) as an internal housekeeping gene. The relative gene expression data were analyzed using the 2−ΔΔCt method and normalized to the untreated negative control which was assigned a value of 1 arbitrary unit [50].

Table 3. Primer sequences utilized for rRT-PCR analysis of targeted gene expression.

Target Gene	Primer Sequence (5′-3′)	Accession No./Reference
iNOS	F-ACCTTCCGGGCAGCCTGTGA R-CAAGGAGGGTGGTGCGGCTG-3′	NM_012611
COX-2	F-GCTCAGCC ATACAGCAAATCC R-GGGAGTCGGGCAAT CATCAG	NM_017232
Caspase-3	F-GCAGCTAACCTCAGAGAGACATTC R-ACGAGTAAGGTCATTTTTATTCCTGACTT	NM_012922
Bcl-2	F-TGCGCTCAGCCCTGTG R-GGTAGCGACGAGAGAAGTCATC	NM_016993
BAX	F-CAAGAAGCTGAGCGAGTGTCT R-CAATCATCCTCTGCAGCTCCATATT	NM_017059
Cytochrome C	F-TTTGAATTCCTCATTAGTAGCTTTTTTGG R-CCATCCCTACGCATCCTTTAC	NM_012839
IL-1β	F-TGACAGACCCCAAAAGATTAAGG R-CTCATCTGGACAGCCCAAGTC	NM_031512.2
IL-6	F-CCACCAGGAACGAAAGTCAAC R-TTGCGGAGAGAAACTTCATAGCT	NM_012589.2
TNF-α	F-CAGCCGATTTGCCATTTCA R-AGGGCTCTTGATGGCAGAGA	L19123.1
β-actin	F-CGCAGTTGGTTGGAGCAAA R-ACAATCAAAGTCCTCAGCCACAT	V01217.1
GAPDH	F-TGCTGGTGCTGAGTATGTCG-3′ R-TTGAGAGCAATGCCAGCC-3′	NM_017008
invA.	F-ACAGTGCTCGTTTACGACCTGAAT R-AGACGACTGGTACTGATCGATAAT	[55]
FimA	F-TTGCGAGTCTGATGTTTGTCG 62 R-CACGCTCACCGGAGTAGGAT	[55]
16S rRNA.	F-AGGCCTTCGGGTTGTAAAGT R-GTTAGCCGGTGCTTCTTCTG	[55]

iNOS: Inducible nitric oxide synthase; COX-2: Cyclo-oxygenase-2; IL: interleukin, TNF-α: tumor necrosis factor-alpha, TGF-β: transforming growth factor-beta, COX-2: cyclooxygenase-2, Bcl-2: B-cell lymphoma-2, BAX: Bcl-2-associated X protein, invA: Invasion protein A, FimA: Major fimbrial subunit.

4.9. Histopathologic Evaluation

Immediately after the end of the experiment (at 14 dpi), tissue specimens were collected from the liver and intestine tissues, control group, CIP–MSN-, MSN-, and CIP-treated rats (n = 3/group). Collected specimens were fixed for 48 h in 10% formalin solution, followed by the routine processing of the specimens, as previously described [44,56]. Thin sections (5 µm) were microtomed and stained with H&E stain and examined under light microscopy. Tissues were blindly examined and evaluated by an experienced pathologist.

4.10. Statistical Analysis

The data was analyzed by general linear model (GLM) after confirming the homogeneity among experimental groups using Levene's test, and normality using Shapiro–Wilk's test was performed. All data were presented as Mean ± SD. Post-hoc Tukey's tests were performed to determine if there were significant differences among groups ($p < 0.05$). All statistical analysis and graphical outputs were generated by GraphPad Prism software (Version 8, GraphPad Software Inc.).

5. Conclusions

Based on our data, we conclude that, when compared with a normal ciprofloxacin treatment regime, MSN particles loaded with ciprofloxacin at half-dose exhibited a controlled release of the antibiotic that aids in prolonging the antibacterial effect, enhancing in vitro antivirulence efficacy against biofilm formation with reduced levels of adherence and invasion protein expression. Another benefit was its insignificant effect on all blood parameter values, and biochemical biomarkers for tissue injury or inflammation indicated its good biocompatibility and lower cytotoxicity. Additionally, a lower dose of CIP–MSN supplementation has a promising role in reducing inflammatory response and oxidative stress by abolishing the elevated secretion of key mediators of oxidative stress, pro-inflammatory cytokines, and chemokines in the blood serum and liver resulting from *S. typhimurium* infection. Additional evidence supporting the protective effects of a reduced dose of CIP–MSN supplementation established that its association with a reduction in mRNA expression of pro-apoptotic genes resulted in reduced *S. typhimurium*-induced hepatic apoptosis. This counteracted the negative effects of the *S. typhimurium* challenge through a corrected histopathological picture of both the intestine and liver and increased bacterial clearance. Finally, our results indicated that supplementation with CIP–MSN drug-delivery cargo led to a lower antibiotic dose requirement and might be a preventive strategy to alleviate *Salmonella typhimurium*-induced liver injury in humans and animals.

Author Contributions: Conceptualization, M.N.A., A.S.A., A.A.-H., A.A.H., W.A.H.H., S.T.E., E.A.A.M., E.S.E.-S., A.A.S., N.A.E. and D.I.; methodology, M.N.A., A.S.A., A.A.-H., A.A.H., W.A.H.H., S.T.E., E.A.A.M., A.A.S., N.A.E. and D.I.; validation, M.N.A., A.S.A., A.A.-H., A.A.H., W.A.H.H., S.T.E., E.A.A.M., E.S.E.-S., A.A.S., N.A.E. and D.I.; formal analysis, M.N.A., A.S.A., A.A.-H., A.A.H., S.T.E., E.A.A.M., A.A.S., N.A.E. and D.I.; investigation, M.N.A., A.S.A., A.A.-H., A.A.H., S.T.E., E.A.A.M., A.A.S., N.A.E. and D.I.; resources, M.N.A., A.S.A., A.A.-H., A.A.H., W.A.H.H., S.T.E., E.A.A.M., A.A.S., E.S.E.-S., N.A.E. and D.I.; data curation, M.N.A., A.S.A., A.A.-H., A.A.H., S.T.E., E.A.A.M., E.S.E.-S., A.A.S., N.A.E. and D.I.; writing—original draft preparation, M.N.A., A.S.A., A.A.-H., A.A.H., W.A.H.H., S.T.E., E.A.A.M., A.A.S., N.A.E. and D.I.; writing—review and editing, M.N.A., A.S.A., A.A.-H., A.A.H., S.T.E., E.A.A.M., A.A.S., N.A.E. and D.I.; visualization, M.N.A., A.S.A., A.A.-H., A.A.H., W.A.H.H., S.T.E., E.A.A.M., A.A.S., N.A.E., E.S.E.-S. and D.I.; supervision, M.N.A., A.S.A., A.A.-H., A.A.H., W.A.H.H., S.T.E., E.A.A.M., A.A.S., N.A.E. and D.I.; project administration, M.N.A., A.S.A., A.A.-H., A.A.H., W.A.H.H., S.T.E., E.A.A.M., A.A.S., N.A.E. and D.I.; funding acquisition, M.N.A., A.S.A., A.A.-H., A.A.H., S.T.E., W.A.H.H., E.A.A.M., E.S.E.-S., A.A.S., N.A.E. and D.I. All authors have read and agreed to the published version of the manuscript.

Funding: This research was funded by Umm Al-Qura University grant number 19-MED-1-01-0023.

Institutional Review Board Statement: The animal study protocol was approved by the Institutional Animal Care and Use Committee of the Faculty of Veterinary Medicine at Zagazig University (protocol code ZU-IACUC2/f/71/2021).

Informed Consent Statement: Not applicable.

Data Availability Statement: Data is contained within the article.

Acknowledgments: The authors would like to thank the Deanship of Scientific Research at Umm Al-Qura University, Saudi Arabia, for support by project number (19-MED-1-01-0023).

Conflicts of Interest: The authors declare no conflict of interest.

References

1. Scallan, E.; Hoekstra, R.M.; Angulo, F.J.; Tauxe, R.V.; Widdowson, M.-A.; Roy, S.L.; Jones, J.L.; Griffin, P.M. Foodborne illness acquired in the United States—Major pathogens. *Emerg. Infect. Dis.* **2011**, *17*, 7. [CrossRef] [PubMed]
2. Monack, D.M.; Mueller, A.; Falkow, S. Persistent bacterial infections: The interface of the pathogen and the host immune system. *Nat. Rev. Microbiol.* **2004**, *2*, 747–765. [CrossRef] [PubMed]
3. Vazquez-Torres, A.; Jones-Carson, J.; Bäumler, A.J.; Falkow, S.; Valdivia, R.; Brown, W.; Le, M.; Berggren, R.; Parks, W.T.; Fang, F.C. Extraintestinal dissemination of Salmonella by CD18-expressing phagocytes. *Nature* **1999**, *401*, 804–808. [CrossRef] [PubMed]
4. Melzak, K.A.; Melzak, S.A.; Gizeli, E.; Toca-Herrera, J.L. Cholesterol organization in phosphatidylcholine liposomes: A surface plasmon resonance study. *Materials* **2012**, *5*, 2306–2325. [CrossRef]
5. Ahmed, A.M.; Younis, E.E.; Ishida, Y.; Shimamoto, T. Genetic basis of multidrug resistance in *Salmonella enterica* serovars Enteritidis and Typhimurium isolated from diarrheic calves in Egypt. *Acta Trop.* **2009**, *111*, 144–149. [CrossRef]
6. Foley, S.; Lynne, A.; Nayak, R. *Salmonella* challenges: Prevalence in swine and poultry and potential pathogenicity of such isolates. *J. Anim. Sci.* **2008**, *86*, E149–E162. [CrossRef]

7. Hassing, R.-J.; Goessens, W.; Mevius, D.J.; van Pelt, W.; Mouton, J.W.; Verbon, A.; Van Genderen, P. Decreased ciprofloxacin susceptibility in *Salmonella* Typhi and Paratyphi infections in ill-returned travellers: The impact on clinical outcome and future treatment options. *Eur. J. Clin. Microbiol. Infect. Dis.* **2013**, *32*, 1295–1301. [CrossRef]
8. Ahmed, A.I.; van der Heijden, F.M.; van den Berkmortel, H.; Kramers, K. A man who wanted to commit suicide by hanging himself: An adverse effect of ciprofloxacin. *Gen. Hosp. Psychiatry* **2011**, *33*, 82.e85–82.e87. [CrossRef]
9. Barnhart, M.M.; Chapman, M.R. Curli biogenesis and function. *Annu. Rev. Microbiol.* **2006**, *60*, 131–147. [CrossRef]
10. World Health Organization. *Antimicrobial Resistance: Global Report on Surveillance*; World Health Organization: Geneva, Switzerland, 2014.
11. Slowing, I.I.; Vivero-Escoto, J.L.; Wu, C.-W.; Lin, V.S.-Y. Mesoporous silica nanoparticles as controlled release drug delivery and gene transfection carriers. *Adv. Drug Deliv. Rev.* **2008**, *60*, 1278–1288. [CrossRef]
12. Egger, S.; Lehmann, R.P.; Height, M.J.; Loessner, M.J.; Schuppler, M. Antimicrobial properties of a novel silver-silica nanocomposite material. *Appl. Environ. Microbiol.* **2009**, *75*, 2973–2976. [CrossRef] [PubMed]
13. Sulaiman, G.M.; Mohammed, W.H.; Marzoog, T.R.; Al-Amiery, A.A.A.; Kadhum, A.A.H.; Mohamad, A.B. Green synthesis, antimicrobial and cytotoxic effects of silver nanoparticles using *Eucalyptus chapmaniana* leaves extract. *Asian Pac. J. Trop. Biomed.* **2013**, *3*, 58–63. [CrossRef]
14. Seleem, M.N.; Munusamy, P.; Ranjan, A.; Alqublan, H.; Pickrell, G.; Sriranganathan, N. Silica-antibiotic hybrid nanoparticles for targeting intracellular pathogens. *Antimicrob. Agents Chemother.* **2009**, *53*, 4270–4274. [CrossRef] [PubMed]
15. Jiao, Y.; Sun, Y.; Chang, B.; Lu, D.; Yang, W. Redox-and temperature-controlled drug release from hollow mesoporous silica nanoparticles. *Chem.—Eur. J.* **2013**, *19*, 15410–15420. [CrossRef] [PubMed]
16. Mudakavi, R.J.; Raichur, A.M.; Chakravortty, D. Lipid coated mesoporous silica nanoparticles as an oral delivery system for targeting and treatment of intravacuolar *Salmonella* infections. *RSC Adv.* **2014**, *4*, 61160–61166. [CrossRef]
17. de Juan Mora, B.; Filipe, L.; Forte, A.; Santos, M.M.; Alves, C.; Teodoro, F.; Pedrosa, R.; Ribeiro Carrott, M.; Branco, L.C.; Gago, S. Boosting Antimicrobial Activity of Ciprofloxacin by Functionalization of Mesoporous Silica Nanoparticles. *Pharmaceutics* **2021**, *13*, 218. [CrossRef]
18. E.C.o.A.S. Testing., Breakpoint Tables for Interpretation of Mics and Zone Diameters. Version 7.1, Valid from 2017-03-10. Available online: http://www.eucast.org/clinical_breakpoints/ (accessed on 20 August 2019).
19. Borges, O.; Cordeiro-da-Silva, A.; Romeijn, S.G.; Amidi, M.; de Sousa, A.; Borchard, G.; Junginger, H.E. Uptake studies in rat Peyer's patches, cytotoxicity and release studies of alginate coated chitosan nanoparticles for mucosal vaccination. *J. Control. Release* **2006**, *114*, 348–358. [CrossRef]
20. Rivero, P.J.; Urrutia, A.; Goicoechea, J.; Zamarreño, C.R.; Arregui, F.J.; Matías, I.R. An antibacterial coating based on a polymer/sol-gel hybrid matrix loaded with silver nanoparticles. *Nanoscale Res. Lett.* **2011**, *6*, 305. [CrossRef]
21. Camporotondi, D.; Foglia, M.; Alvarez, G.; Mebert, A.; Diaz, L.; Coradin, T.; Desimone, M. Antimicrobial properties of silica modified nanoparticles. *Microb. Pathog. Strateg. Combat. Sci. Technol. Educ.* **2013**, *2*, 283–290.
22. Van Houdt, R.; Michiels, C. Biofilm formation and the food industry, a focus on the bacterial outer surface. *J. Appl. Microbiol.* **2010**, *109*, 1117–1131. [CrossRef] [PubMed]
23. Kim, S.-H.; Wei, C.-i. Biofilm formation by multidrug-resistant *Salmonella enterica* serotype Typhimurium phage type DT104 and other pathogens. *J. Food Prot.* **2007**, *70*, 22–29. [CrossRef] [PubMed]
24. Morones, J.R.; Elechiguerra, J.L.; Camacho, A.; Holt, K.; Kouri, J.B.; Ramírez, J.T.; Yacaman, M.J. The bactericidal effect of silver nanoparticles. *Nanotechnology* **2005**, *16*, 2346. [CrossRef] [PubMed]
25. Massa, M.A.; Covarrubias, C.; Bittner, M.; Fuentevilla, I.A.; Capetillo, P.; Von Marttens, A.; Carvajal, J.C. Synthesis of new antibacterial composite coating for titanium based on highly ordered nanoporous silica and silver nanoparticles. *Mater. Sci. Eng. C* **2014**, *45*, 146–153. [CrossRef] [PubMed]
26. Naughton, P.J.; Grant, G.; Bardocz, S.; Allen-Vercoe, E.; Woodward, M.J.; Pusztai, A. Expression of type 1 fimbriae (SEF 21) of *Salmonella enterica* serotype enteritidis in the early colonisation of the rat intestine. *J. Med. Microbiol.* **2001**, *50*, 191–197. [CrossRef] [PubMed]
27. Boddicker, J.D.; Ledeboer, N.A.; Jagnow, J.; Jones, B.D.; Clegg, S. Differential binding to and biofilm formation on, HEp-2 cells by *Salmonella enterica* serovar Typhimurium is dependent upon allelic variation in the *fimH* gene of the *fim* gene cluster. *Mol. Microbiol.* **2002**, *45*, 1255–1265. [CrossRef]
28. Ghosh, S.; Mittal, A.; Vohra, H.; Ganguly, N.K. Interaction of a rat intestinal brush border membrane glycoprotein with type-1 fimbriae of *Salmonella typhimurium*. *Mol. Cell. Biochem.* **1996**, *158*, 125–131. [CrossRef]
29. Khan, T.A.; Zafar, F. Haematological study in response to varying doses of estrogen in broiler chicken. *Int. J. Poult. Sci.* **2005**, *4*, 748–751.
30. McGill, M.R. The past and present of serum aminotransferases and the future of liver injury biomarkers. *EXCLI J.* **2016**, *15*, 817–828. [CrossRef]
31. Kengni, F.; Fodouop, S.P.; Tala, D.S.; Djimeli, M.N.; Fokunang, C.; Gatsing, D. Antityphoid properties and toxicity evaluation of *Harungana madagascariensis* Lam (Hypericaceae) aqueous leaf extract. *J. Ethnopharmacol.* **2016**, *179*, 137–145. [CrossRef]
32. Osman, K.M.; El-Enbaawy, M.I.; Ezzeldin, N.A.; Hussein, H.M. Nitric oxide and lysozyme production as an impact to *Clostridium perfringens* mastitis. *Comp. Immunol. Microbiol. Infect. Dis.* **2010**, *33*, 505–511. [CrossRef] [PubMed]

33. Wang, R.; Li, S.; Jia, H.; Si, X.; Lei, Y.; Lyu, J.; Dai, Z.; Wu, Z. Protective Effects of Cinnamaldehyde on the Inflammatory Response, Oxidative Stress, and Apoptosis in Liver of *Salmonella typhimurium*-Challenged Mice. *Molecules* **2021**, *26*, 2309. [CrossRef] [PubMed]
34. Dai, C.; Xiao, X.; Li, D.; Tun, S.; Wang, Y.; Velkov, T.; Tang, S. Chloroquine ameliorates carbon tetrachloride-induced acute liver injury in mice via the concomitant inhibition of inflammation and induction of apoptosis. *Cell Death Dis.* **2018**, *9*, 1164. [CrossRef] [PubMed]
35. Kostura, M.J.; Tocci, M.J.; Limjuco, G.; Chin, J.; Cameron, P.; Hillman, A.G.; Chartrain, N.A.; Schmidt, J.A. Identification of a monocyte specific pre-interleukin 1 beta convertase activity. *Proc. Natl. Acad. Sci. USA* **1989**, *86*, 5227–5231. [CrossRef] [PubMed]
36. Lara-Tejero, M.; Sutterwala, F.S.; Ogura, Y.; Grant, E.P.; Bertin, J.; Coyle, A.J.; Flavell, R.A.; Galán, J.E. Role of the caspase-1 inflammasome in *Salmonella typhimurium* pathogenesis. *J. Exp. Med.* **2006**, *203*, 1407–1412. [CrossRef]
37. Martinon, F.; Tschopp, J. Inflammatory caspases: Linking an intracellular innate immune system to autoinflammatory diseases. *Cell* **2004**, *117*, 561–574. [CrossRef]
38. Brennan, M.A.; Cookson, B.T. *Salmonella* induces macrophage death by caspase-1-dependent necrosis. *Mol. Microbiol.* **2000**, *38*, 31–40. [CrossRef]
39. Sairanen, T.; Szepesi, R.; Karjalainen-Lindsberg, M.L.; Saksi, J.; Paetau, A.; Lindsberg, P.J. Neuronal caspase-3 and PARP-1 correlate differentially with apoptosis and necrosis in ischemic human stroke. *Acta Neuropathol.* **2009**, *118*, 541–552. [CrossRef]
40. Tian, T.; Wang, Z. Pathomechanisms of Oxidative Stress in Inflammatory Bowel Disease and Potential Antioxidant Therapies. *Oxidative Med. Cell. Longev.* **2017**, *2017*, 4535194. [CrossRef]
41. Allam, N.G.; Eldrieny, E.A.E.-A.; Mohamed, A.Z. Effect of combination therapy between thyme oil and ciprofloxacin on ulcer-forming *Shigella flexneri*. *J. Infect. Dev. Ctries.* **2015**, *9*, 486–495. [CrossRef]
42. Nawaz, A.; Ali, S.M.; Rana, N.F.; Tanweer, T.; Batool, A.; Webster, T.J.; Menaa, F.; Riaz, S.; Rehman, Z.; Batool, F. Ciprofloxacin-Loaded Gold Nanoparticles against Antimicrobial Resistance: An In Vivo Assessment. *Nanomaterials* **2021**, *11*, 3152. [CrossRef] [PubMed]
43. Bouchoucha, M.; Cote, M.-F.; C.-Gaudreault, R.; Fortin, M.-A.; Kleitz, F. Size-controlled functionalized mesoporous silica nanoparticles for tunable drug release and enhanced anti-tumoral activity. *Chem. Mater.* **2016**, *28*, 4243–4258. [CrossRef]
44. Ibrahim, D.; Abdelfattah-Hassan, A.; Badawi, M.; Ismail, T.A.; Bendary, M.M.; Abdelaziz, A.M.; Mosbah, R.A.; Mohamed, D.I.; Arisha, A.H.; Abd El-Hamid, M.I. Thymol nanoemulsion promoted broiler chicken's growth, gastrointestinal barrier and bacterial community and conferred protection against *Salmonella* Typhimurium. *Sci. Rep.* **2021**, *11*, 7742. [CrossRef] [PubMed]
45. Selim, M.S.; Hamouda, H.; Hao, Z.; Shabana, S.; Chen, X. Design of γ-AlOOH, γ-MnOOH, and α-Mn$_2$O$_3$ nanorods as advanced antibacterial active agents. *Dalton Trans.* **2020**, *49*, 8601–8613. [CrossRef]
46. Parvekar, P.; Palaskar, J.; Metgud, S.; Maria, R.; Dutta, S. The minimum inhibitory concentration (MIC) and minimum bactericidal concentration (MBC) of silver nanoparticles against *Staphylococcus aureus*. *Biomater. Investig. Dent.* **2020**, *7*, 105–109. [CrossRef]
47. Lamas, A.; Fernandez-No, I.; Miranda, J.; Vázquez, B.; Cepeda, A.; Franco, C. Biofilm formation and morphotypes of *Salmonella enterica* subsp. arizonae differs from those of other *Salmonella enterica* subspecies in isolates from poultry houses. *J. Food Prot.* **2016**, *79*, 1127–1134. [CrossRef]
48. Stepanović, S.; Ćirković, I.; Ranin, L.; S√vabić-Vlahović, M. Biofilm formation by *Salmonella* spp. and Listeria monocytogenes on plastic surface. *Lett. Appl. Microbiol.* **2004**, *38*, 428–432. [CrossRef]
49. Roehm, N.W.; Rodgers, G.H.; Hatfield, S.M.; Glasebrook, A.L. An improved colorimetric assay for cell proliferation and viability utilizing the tetrazolium salt XTT. *J. Immunol. Methods* **1991**, *142*, 257–265. [CrossRef]
50. Xu, H.; Lee, H.-Y.; Ahn, J. Growth and virulence properties of biofilm-forming *Salmonella enterica* serovar Typhimurium under different acidic conditions. *Appl. Environ. Microbiol.* **2010**, *76*, 7910–7917. [CrossRef]
51. EN ISO 6579:2002/A1:2007; Microbiology of Food and Animal Feeding Stuffs: Horizontal Method for the Detection of *Salmonella* spp.—Amendment 1: Annex D: Detection of *Salmonella* spp. in Animal Faeces and in Environmental Samples from the Primary Production Stage. ISO: Geneva, Switzerland, 2007.
52. Hawk, P.B. *Practical Physiological Chemistry*; P. Blakiston's Son & Company: Philadelphia, PA, USA, 1916.
53. Motor, S.; Ozturk, S.; Ozcan, O.; Gurpinar, A.B.; Can, Y.; Yuksel, R.; Yenin, J.Z.; Seraslan, G.; Ozturk, O.H. Evaluation of total antioxidant status, total oxidant status and oxidative stress index in patients with alopecia areata. *Int. J. Clin. Exp. Med.* **2014**, *7*, 1089.
54. Hutchinson, M.K. C-Reactive Protein in Serum by Nephelometry. University of Washington Medical Center, Department of Laboratory Medicine, Immunology Division. 2000. Available online: http://www.cdc.gov/nchs/data/nhanes/nhanes_01_02/l1 1_b_met_c_reactive_protein.pdf (accessed on 11 August 2014).
55. Huyghebaert, G.; Ducatelle, R.; Van Immerseel, F. An update on alternatives to antimicrobial growth promoters for broilers. *Vet. J.* **2011**, *187*, 182–188. [CrossRef]
56. Alandiyjany, M.N.; Kishawy, A.T.; Abdelfattah-Hassan, A.; Eldoumani, H.; Elazab, S.T.; El-Mandrawy, S.A.; Saleh, A.A.; ElSawy, N.A.; Attia, Y.A.; Arisha, A.H. Nano-silica and magnetized-silica mitigated lead toxicity: Their efficacy on bioaccumulation risk, performance, and apoptotic targeted genes in Nile tilapia (*Oreochromis niloticus*). *Aquat. Toxicol.* **2022**, *242*, 106054. [CrossRef] [PubMed]

 *pharmaceuticals*

Article

In Vivo and In Vitro Antimicrobial Activity of Biogenic Silver Nanoparticles against *Staphylococcus aureus* Clinical Isolates

Nashwah G. M. Attallah [1,†], Engy Elekhnawy [2,*,†], Walaa A. Negm [3,*], Ismail A. Hussein [4], Fatma Alzahraa Mokhtar [5] and Omnia Momtaz Al-Fakhrany [2]

1. Department of Pharmaceutical Science, College of Pharmacy, Princess Nourah Bint Abdulrahman University, P.O. Box 84428, Riyadh 11671, Saudi Arabia; ngmohamed@pnu.edu.sa
2. Department of Pharmaceutical Microbiology, Faculty of Pharmacy, Tanta University, Tanta 31527, Egypt; omina.elfakharany@pharm.tanta.edu.eg
3. Department of Pharmacognosy, Faculty of Pharmacy, Tanta University, Tanta 31527, Egypt
4. Department of Pharmacognosy and Medicinal Plants, Faculty of Pharmacy (Boys), Al-Azhar University, Cairo 11884, Egypt; ismaila.hussein@azhar.edu.eg
5. Department of Pharmacognosy, Faculty of Pharmacy, Alsalam University, Tanta 3111, Egypt; drfatmaalzahraa1950@gmail.com
* Correspondence: engy.ali@pharm.tanta.edu.eg (E.E.); walaa.negm@pharm.tanta.edu.eg (W.A.N.)
† These authors contributed equally to this work.

Abstract: *Staphylococcus aureus* can cause a wide range of severe infections owing to its multiple virulence factors in addition to its resistance to multiple antimicrobials; therefore, novel antimicrobials are needed. Herein, we used *Gardenia thailandica* leaf extract (GTLE), for the first time for the biogenic synthesis of silver nanoparticles (AgNPs). The active constituents of GTLE were identified by HPLC, including chlorogenic acid (1441.03 µg/g) from phenolic acids, and quercetin-3-rutinoside (2477.37 µg/g) and apigenin-7-glucoside (605.60 µg/g) from flavonoids. In addition, the antioxidant activity of GTLE was evaluated. The synthesized AgNPs were characterized using ultraviolet-visible spectroscopy, Fourier-transform infrared spectroscopy, transmission and scanning electron microscopy (SEM), zeta potential, dynamic light scattering, and X-ray diffraction. The formed AgNPs had a spherical shape with a particle size range of 11.02–17.92 nm. The antimicrobial activity of AgNPs was investigated in vitro and in vivo against *S. aureus* clinical isolates. The minimum inhibitory concentration (MIC) of AgNPs ranged from 4 to 64 µg/mL. AgNPs significantly decreased the membrane integrity of 45.8% of the isolates and reduced the membrane potential by flow cytometry. AgNPs resulted in morphological changes observed by SEM. Furthermore, qRT-PCR was utilized to examine the effect of AgNPs on the gene expression of the efflux pump genes *nor*A, *nor*B, and *nor*C. The in vivo examination was performed on wounds infected with *S. aureus* bacteria in rats. AgNPs resulted in epidermis regeneration and reduction in the infiltration of inflammatory cells. Thus, GTLE could be a vital source for the production of AgNPs, which exhibited promising in vivo and in vitro antibacterial activity against *S. aureus* bacteria.

Keywords: AgNPs; antioxidant activity; flow cytometry; *Gardenia thailandica*; HPLC; infected wound; qRT-PCR

Citation: Attallah, N.G.M.; Elekhnawy, E.; Negm, W.A.; Hussein, I.A.; Mokhtar, F.A.; Al-Fakhrany, O.M. In Vivo and In Vitro Antimicrobial Activity of Biogenic Silver Nanoparticles against *Staphylococcus aureus* Clinical Isolates. *Pharmaceuticals* 2022, 15, 194. https://doi.org/10.3390/ph15020194

Academic Editor: Fu-Gen Wu

Received: 6 January 2022
Accepted: 1 February 2022
Published: 3 February 2022

Publisher's Note: MDPI stays neutral with regard to jurisdictional claims in published maps and institutional affiliations.

Copyright: © 2022 by the authors. Licensee MDPI, Basel, Switzerland. This article is an open access article distributed under the terms and conditions of the Creative Commons Attribution (CC BY) license (https://creativecommons.org/licenses/by/4.0/).

1. Introduction

Nanotechnology is a relatively novel discipline with massive applications, including those in the medical and pharmacological industries [1]. Recently, there is a growing interest in the usage of green-synthesized biocompatible silver nanoparticles (AgNPs) in various applications including antimicrobial products, anti-fungal preparations, drug delivery, the textile industry, and food packaging. Several chemical and physical methods for the synthesis of AgNPs have been reported, including sol-gel, chemical reduction, physical vapor deposition, thermal decomposition, and microwave irradiation [2,3]. Unfortunately, these techniques have many drawbacks such as the expense, use of high energy and/or hazardous chemicals, and production of toxic byproducts that are unsafe to the environment [4,5]. Besides, these toxic chemicals are attached to the end products, which considerably limits their application [2,6].

To overcome these limitations, it is crucial to find eco-friendly, easy to use, cost effective, and nontoxic alternative methods for the fabrication of AgNPs [4]. Recently, new methods based on green synthesis are emerging. These methods use eco-friendly compounds as reducing agents [1]. Plants and microorganisms are considered nontoxic biological reproducible resources that are safe for humans and the environment. They may be more suitable alternatives for the biosynthesis of AgNPs [7]. Plant extract nanoparticles are more favorable than microorganism-based nanoparticles since they do not require particular and complex processes such as culture management, isolation, and several purification steps [8]. In addition, using plants for the synthesis of nanoparticles has other advantages, such as the use of safer solvents, milder response conditions, more feasibility, and their various uses in surgical and pharmaceutical applications [3]. Due to the aforementioned limitations, researchers have developed green methods that employ various plant parts such as the leaf, peel, flower, fruit, and root. Numerous plant extract compounds (e.g., ascorbic acids, flavonoids, polyphenols, proteins, and terpenoids) play important roles in metal ion uptake, precursor salt reduction, and capping agents. Furthermore, several of them have antibacterial capabilities [9].

Gardenia thailandica Triveng. is a flowering plant native to Thailand. Gardenia species have a high medicinal potential and a long history of usage in traditional medicine to cure a variety of ailments such as jaundice, fever, hypertension, and skin ulcers. In addition, various Gardenia species have been linked to a variety of pharmacological effects, including anti-inflammatory, anti-viral, anti-cancer, and anti-apoptotic properties [10–13]. One of the most potential antimicrobials used in nanomedicine are AgNPs. AgNPs can interact with a microorganism's cell wall, producing reactive oxygen species that eventually cause cell death [9]. As a result, we can speculate that using *G. thailandica* extract will produce AgNPs with improved antimicrobial activity.

Staphylococcus aureus is a highly virulent pathogenic bacteria that can cause various clinical infections in humans. They are a major cause of infective endocarditis and bacteremia, in addition to skin and soft tissue, osteoarticular, pleuropulmonary, and device-associated infections [14]. Besides the various virulence factors they possess, antimicrobial resistance is widely spreading among these bacteria. Thus, new approaches should be studied to overcome different infections caused by *S. aureus* [15]. The green synthesized AgNPs could be a therapeutic alternative to the currently present antimicrobials.

In this study, we aimed to green synthesize AgNPs from *Gardenia thailandica* leaf extract (GTLE). Then, the produced AgNPs were characterized by different techniques. Furthermore, the antibacterial activity of the synthesized AgNPs was studied both in vitro and in vivo against *S. aureus* clinical isolates.

2. Results

2.1. High Performance Liquid Chromatographic Coupled with Diode Array Detector (HPLC-DAD) Analysis

The identification and quantification of phenolic compounds of GTLE was performed using the HPLC-DAD. Figure 1 displays the HPLC-DAD chromatogram for the identified flavonoids and phenolic compounds of GTLE. The abundant phenolic compounds were chlorogenic acid (1441.03 µg/g), while the major identified flavonoid compound was quercetin-3-rutinoside (2477.37 µg/g), as shown in Table 1.

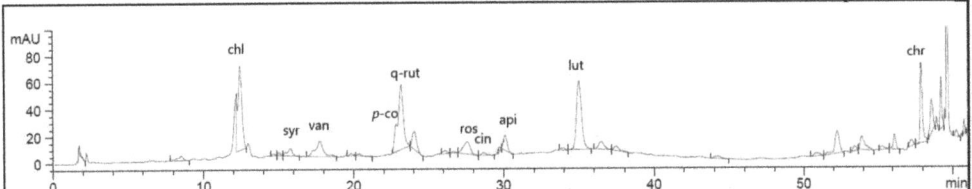

Figure 1. HPLC-DAD of GTLE (320 nm). Chl—chlorogenic acid; syr—syringic acid; van—vanillic acid; p-co—p-coumaric acid; q-rut—quercetin rutinoside; ros—rosmarinic acid; cin—cinnamic acid; api—apigenin; lut—luteolin; chr—chrysin.

Table 1. Chemical composition analysis of the phenolic and flavonoid compounds of GTLE by HPLC-DAD.

No	Retention Time (RT)	Compound	Concentration (µg/g) *
1	3.93	Gallic acid	ND
2	6.61	Protocatechuic acid	ND
3	9.91	p-hydroxybenzoic acid	ND
4	11.44	Gentisic acid	ND
5	12.18	Cateachin	ND
6	12.41	Chlorogenic acid	1441.03
7	13.33	Caffeic acid	ND
8	16.18	Syringic acid	10.09
9	17.69	Vanillic acid	44.17
10	20.26	Ferulic acid	ND
11	21.03	Sinapic acid	ND
12	22.26	p-coumaric	26.16
13	22.97	Quercetin-3-Rutinoside	2477.37
14	27.43	Rosmarinic acid	796.67
15	28.71	Apigenin-7-glucoside	605.60
16	30.04	Cinnamic acid	436.06
17	34.79	luteolin	753.18
18	39.50	Apigenin	ND
19	53.34	Kaempferol	ND
20	58.42	Chrysin	152.71

* ND stands for none detected.

2.2. Characterization of the Green-Synthesized AgNPs

2.2.1. Physical Observation

After 3 h of preservation in a cool and dark area, the physical appearance of the AgNO$_3$ solution changed to a dark solution after the addition of GTLE, indicating the chemical reduction reaction and synthesis of AgNPs.

2.2.2. UV-Vis Spectroscopy

UV-Vis spectroscopy was utilized as the first proof of nanoparticle formation owing to the selectivity of UV towards the formed nanoparticles. Since AgNPs have a characteristic optical reflectivity, they interact strongly with specific wavelengths of light. Because of the collective oscillation of electrons in AgNPs, free electrons produce a surface plasmon resonance (SPR) absorption band [16]. The absorption of AgNPs is controlled by the dielectric medium, chemical environment, shape of the particles, and particle size. The UV measurements of the produced AgNPs had an absorbance at 418 nm (Figure 2).

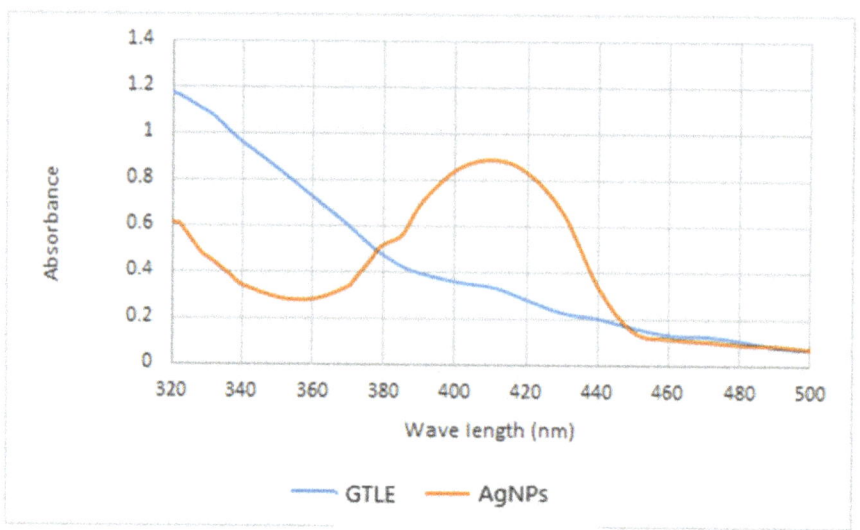

Figure 2. UV spectrum of the biosynthesized AgNPs by GTLE compared to GTLE.

2.2.3. Fourier-Transform Infrared (FTIR) Spectroscopy

The identity of the functional chemical groups of the GTLE involved in the reduction reaction to produce AgNPs was configurated by FTIR spectroscopy measurements. Peaks at 3417, 2926, 1632 cm^{-1} represent the functional groups as follows; OH, C aliphatic, and C=O of phenolic acids and flavonoids, while the polyphenols and aromatic compounds were represented by the peak at 1453 cm^{-1}. The secondary OH groups of GTLE were confirmed by the peak at 1080 cm^{-1} (Figure 3).

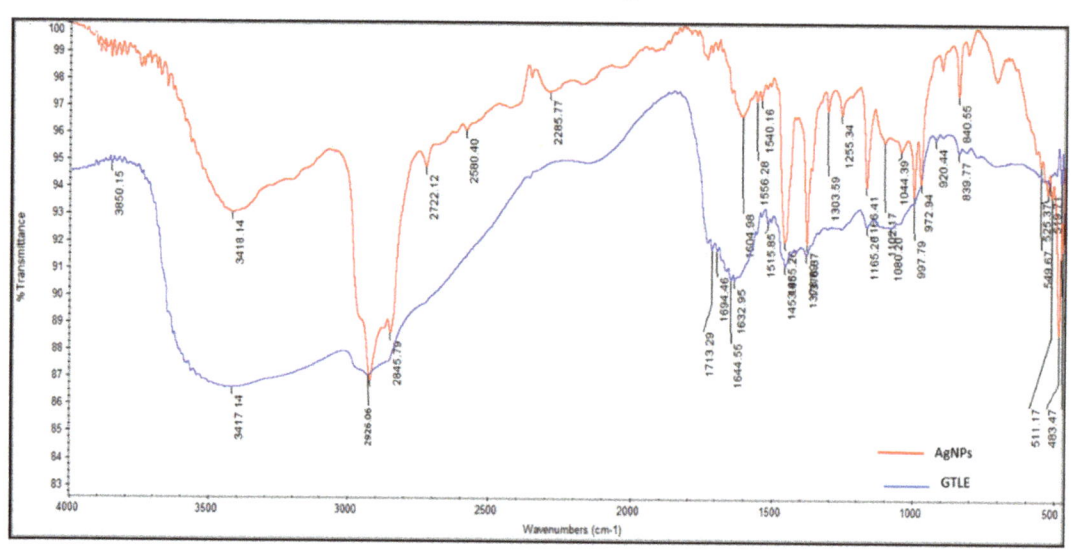

Figure 3. FTIR spectrum of the biosynthesized AgNPs by GTLE compared to GTLE.

2.2.4. High-Resolution Transmission Electron Microscope (HR-TEM)

The green-synthesized AgNPs using GTLE as a reducing agent were examined using HR-TEM, which revealed the formation of spherical shaped AgNPs with a particle size range of 11.02–17.92 nm and an average size of 14.24 nm (Figure 4). In addition, the selected area electron diffraction (SAED) pattern confirmed the crystalline nature of the formed AgNPs.

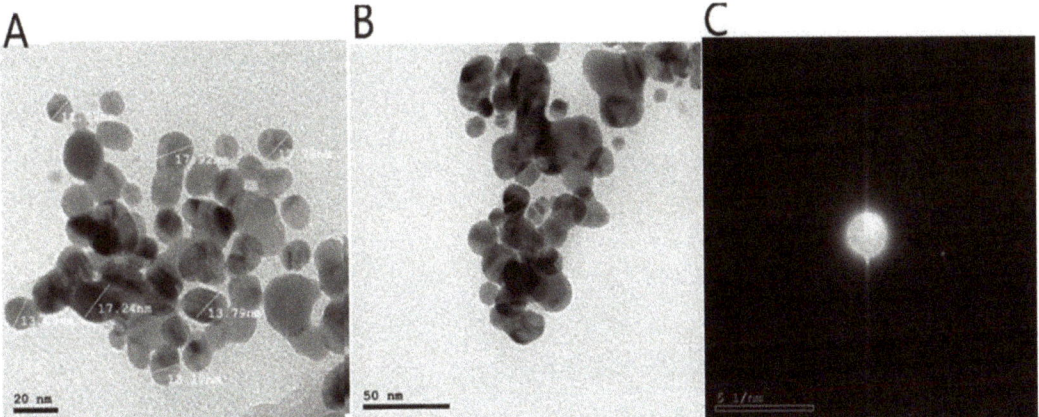

Figure 4. HR-TEM micrographs of the biosynthesized AgNPs using GTLE; (**A**): at 20 nm, (**B**); at 50 nm, (**C**): SAED confirmed the crystalline nature of the formed AgNPs.

2.2.5. Zeta Potential and Dynamic Light Scattering (DLS)

We used the zeta potential technique to evaluate the surface charge of the green-synthesized AgNPs. Herein, AgNPs had a zeta potential value of -6.54 ± 0.6 mV, where the negative charge highlighted the stability of the formed nanoparticles (Figure 5A).

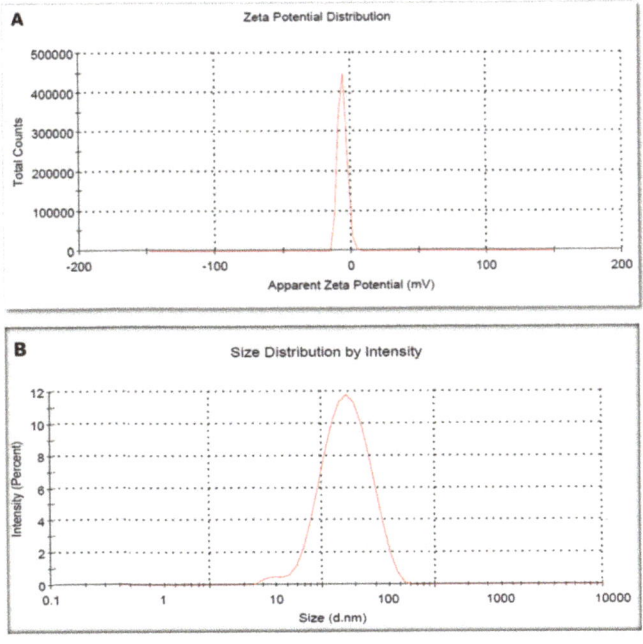

Figure 5. Zeta potential analysis (**A**) and DLS (**B**) of the biosynthesized AgNPs by GTLE.

Legend shells including the metallic shell of the formed nanoparticles were measured using the DLS technique; they had a size of 77.4 ± 1.88 nm (Figure 5B).

2.2.6. X-ray Diffraction (XRD)

The intense peaks were noticed at the 2θ scale of 38.26, 44.47, 64.71, and 77.73 corresponding to the (111), (200), (220), and (311) planes for silver, respectively (Figure 6).

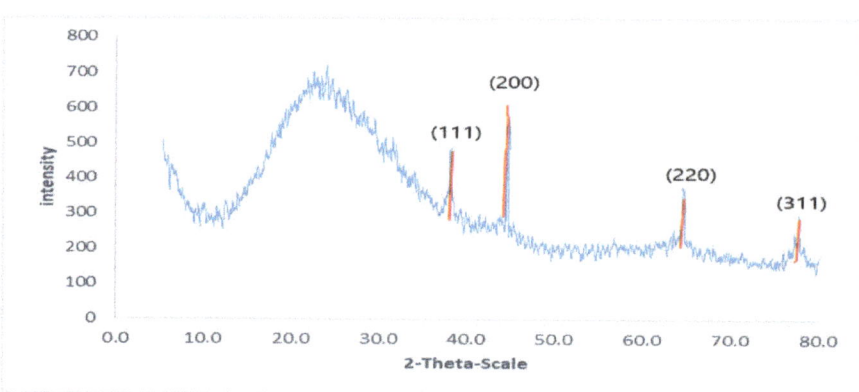

Figure 6. X-ray diffraction pattern of the biosynthesized AgNPs using GTLE.

2.2.7. Scanning Electron Microscope (SEM)

SEM is a useful tool for investigating an object's surface images. It can precisely illustrate the particle size, shape, and distribution of the tested material. In addition, it can determine the morphological appearance of the studied object and determine whether its size is at the micro- or nanoscale. SEM analysis of the biosynthesized AgNPs revealed that they are spherical in shape with a tendency to aggregate (Figure 7).

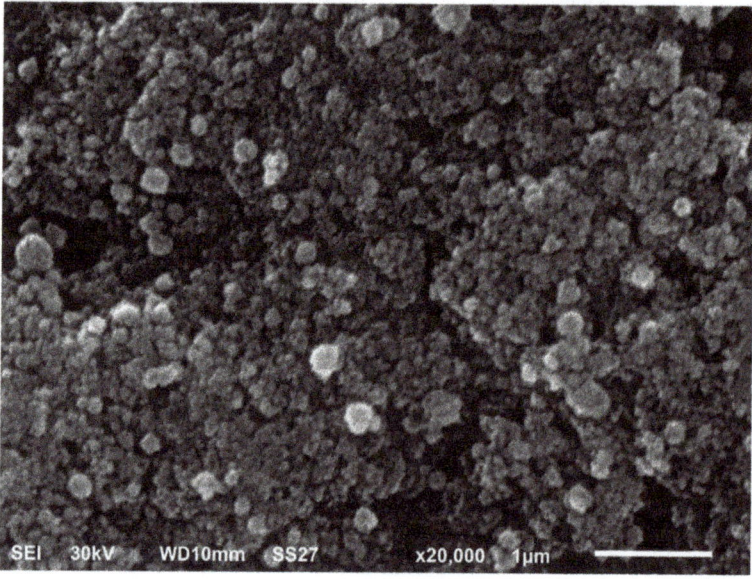

Figure 7. SEM of AgNPs biosynthesized using GTLE.

2.3. Total Content of Flavonoids and Polyphenolics

Total flavonoids were found to have a content of 162.98 mg/g equivalent to quercetin, while total polyphenols had a content of 287.89 mg/g equivalent to gallic acid. Findings indicate that *G. thailandica* possesses high contents of polyphenols and flavonoids.

2.4. Antioxidant Activity

The antioxidant activity of GTLE was investigated in this study using radical scavenging and metal-reducing assays. Radical scavenging assays used were 2,2-diphenyl-1-picrylhydrazyl (DPPH) and 2,2′-azino-bis(3-ethylbenzothiazoline-6-sulfonic acid) (ABTS) tests. GTLE exhibited antioxidant activity by DPPH and ABTS as (IC$_{50}$ 72.91 µg/mL) and (211.60 mg Trolox equivalents (TE)/mg), respectively. The metal-reducing assay used was the ferric reducing antioxidant power assay (FRAP) and the activity of GTLE was 70.95 mg TE/mg.

2.5. In Vitro Antibacterial Activity

2.5.1. In Vitro Susceptibility Testing

The green-synthesized AgNPs exhibited antibacterial activity against *S. aureus* clinical isolates, as they resulted in clear zones around the AgNPs discs by disc diffusion method. The broth microdilution method was utilized to identify the MIC values of AgNPs, and they ranged from 4 to 64 µg/mL. All the following tests were carried out after treatment of the tested isolates with 0.5 MIC values.

2.5.2. Time Kill Curve

The number of colony-forming units (CFU) per milliliter was reduced by more than 3 log units after incubation of *S. aureus* cells for 2 h with 4× MIC and 1 h with 8× MIC in 58.34% and 47.92% of the isolates, respectively. A representative example for the reduction of the CFU/mL is shown in Figure 8.

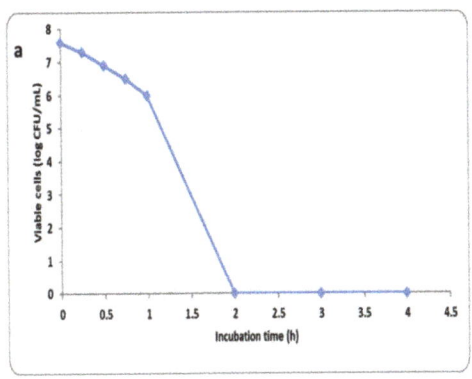
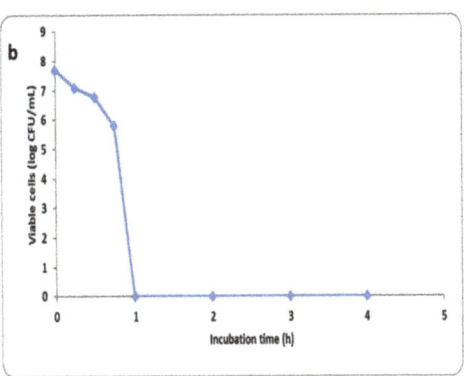

Figure 8. Time kill plot of (**a**) 4× MIC and (**b**) 8× MIC of the green synthesized AgNPs against *S. aureus* isolates.

2.5.3. Membrane Integrity and Permeability

We investigated the cell membrane integrity of *S. aureus* isolates after treatment with the green synthesized AgNPs (at concentrations equal to 0.5 MIC values) via detection of the release of the materials (DNA and RNA), which absorb at 260 nm, from the bacterial isolates. Herein, we found that the membrane integrity significantly decreased ($p < 0.05$) in 45.8% of the isolates after treatment with AgNPs. Figure 9a illustrates a representative example.

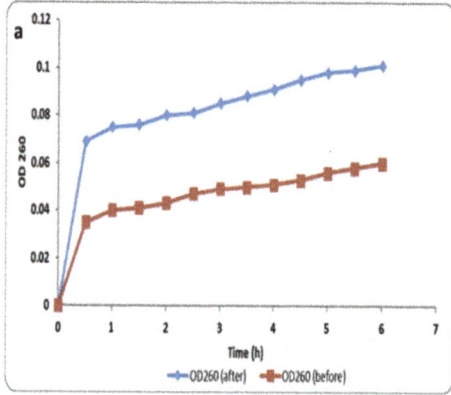 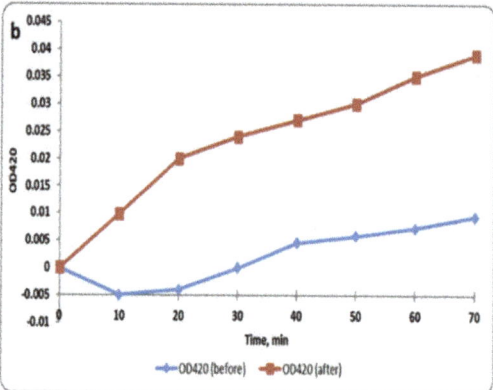

Figure 9. Line chart showing (**a**) the cell membrane integrity and (**b**) the membrane permeability of a representative *S. aureus* isolate before and after treatment with the green-synthesized AgNPs (at concentrations equal to 0.5 MIC values).

When the bacterial membrane permeability increases, O-nitrophenyl-β-galactopyranoside (ONPG) enters the bacterial cytoplasm in a large amount. In the cytoplasm, ONPG is broken down into O-nitrophenol (ONP) by a β-galactosidase enzyme that is present in the cytoplasm. Thus, the membrane permeability was tracked by monitoring the absorbance at OD_{420} (the yellow color of ONP can absorb at 420 nm) with time. The membrane permeability significantly increased ($p < 0.05$) in 56.25% of *S. aureus* isolates after treatment with AgNPs and an illustrative example is revealed in Figure 9b.

2.5.4. Membrane Depolarization

Membrane depolarization was determined in the tested isolates using DiBAC4(3) (bis-(1,3-dibutylbarbituric acid) trimethine oxonol) fluorescent stain. This is a membrane potential-sensitive stain that can enter the depolarized cell cytoplasm and bind to the intracellular proteins exhibiting an enhanced fluorescence. In the current investigation, we noticed that treatment with AgNPs exhibited a considerable reduction ($p < 0.05$) in the membrane potential in 35.42% of *S. aureus* isolates. A demonstrative example of the decrease in the membrane potential after AgNPs treatment is presented in Figure 10.

2.5.5. SEM Examination

The ultrastructural and morphological changes of *S. aureus* cells treated with the green-synthesized AgNPs were observed by SEM (Figure 11). The electron micrographs obtained by SEM revealed that the untreated cells had sphere-shaped, intact, smooth surfaces. On the other hand, the treated cells had a deformed and distorted shape.

2.5.6. Efflux Activity

The efflux activity of the tested *S. aureus* isolates was assessed by testing the capability of the cells to pump out ethidium bromide (EtBr) to the surrounding medium by the EtBr cartwheel method. Herein, we categorized the efflux activity of *S. aureus* isolates into three classes; negative, intermediate, and positive efflux activity as presented in Table 2. Eleven (22.92%) *S. aureus* isolates exhibited a reduction in their efflux activity when treated with the green-synthesized AgNPs. The efflux activity of these isolates changed from positive to either intermediate or negative.

2.5.7. Quantitative Real-Time PCR (qRT-PCR)

qRT-PCR was utilized for more in-depth exploration of the impact of the green-synthesized AgNPs on the efflux pump activity of the tested isolates. *S. aureus* bacteria

(n = 11) that displayed a decline in their efflux pump activity by the EtBr cartwheel method were selected for this assay. We found that the transcriptional levels of *nor*A, *nor*B, and *nor*C efflux pump genes decreased in 72.73%, 45.45%, and 18.18% of the tested isolates, respectively, after treatment with AgNPs. The fold changes mean values of *nor*A, *nor*B, and *nor*C efflux pump genes ranged from 0.11 to 0.47, respectively, as presented in Figure 12.

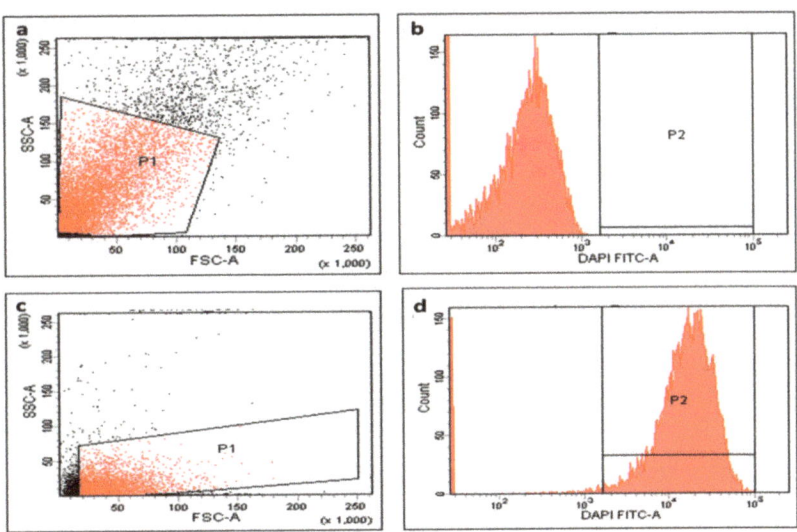

Figure 10. Flowcytometric chart (**a**) dot plot, (**b**) histogram (fluorescent gap = 67.1%) before treatment, (**c**) dot plot, and (**d**) histogram (fluorescent gap = 35.1%) after treatment of a representative *S. aureus* isolate with the green-synthesized AgNPs.

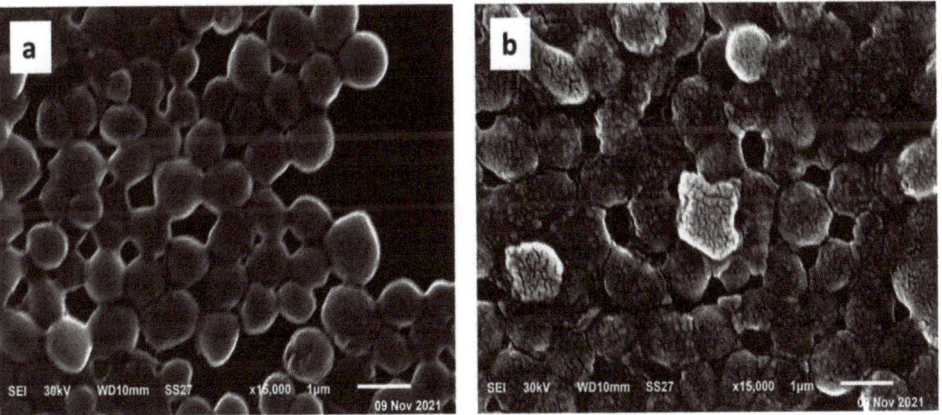

Figure 11. Scanning electron micrograph of a representative *S. aureus* isolate (**a**) before treatment and (**b**) after treatment with the green-synthesized AgNPs.

2.6. In Vivo Antibacterial Activity

The impact of AgNPs was investigated on macroscopic healing and skin histology following excisional wound healing as follows.

Table 2. Efflux activity of *S. aureus* isolates determined using EtBr cartwheel method, before and after treatment with AgNPs.

EtBr Conc. (mg/L) *	Number of Isolates (Before Treatment)	Number of Isolates (After Treatment)
≤0.5	5	7
1	10	12
1.5	11	13
2	8	13
2.5	14	3

* Concentration of EtBr at which *S. aureus* bacteria started to produce fluorescence. The isolates which emit fluorescence at 0.5 mg/L, 1–2 mg/L, and 2.5 mg/L lack efflux activity, have intermediate efflux activity, and have positive efflux activity, respectively.

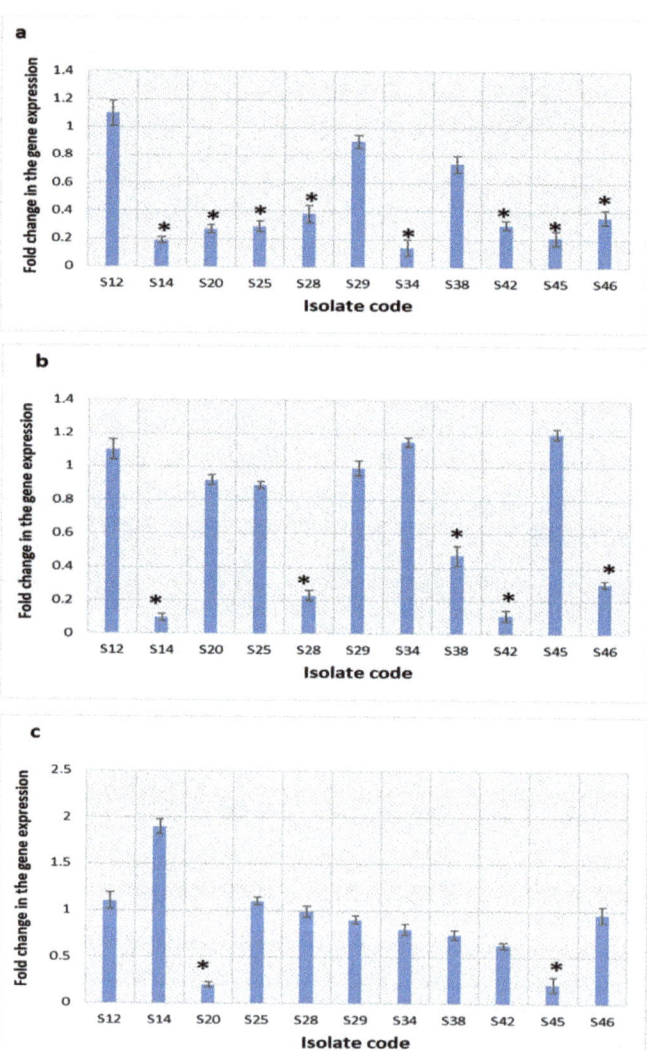

Figure 12. Bar charts showing the transcriptional level fold changes of (**a**) *nor*A, (**b**) *nor*B, and (**c**) *nor*C genes after treatment with the green-synthesized AgNPs. * represents a significant decrease in the fold change.

2.6.1. Macroscopic Healing

The rates of the macroscopic wound healing of the studied groups were inspected on days 0, 3, and 7, considering the day of the wound creation as day 0. Betadine™ and AgNPs groups demonstrated full and notable wound healing in comparison with the control group (Figure 13).

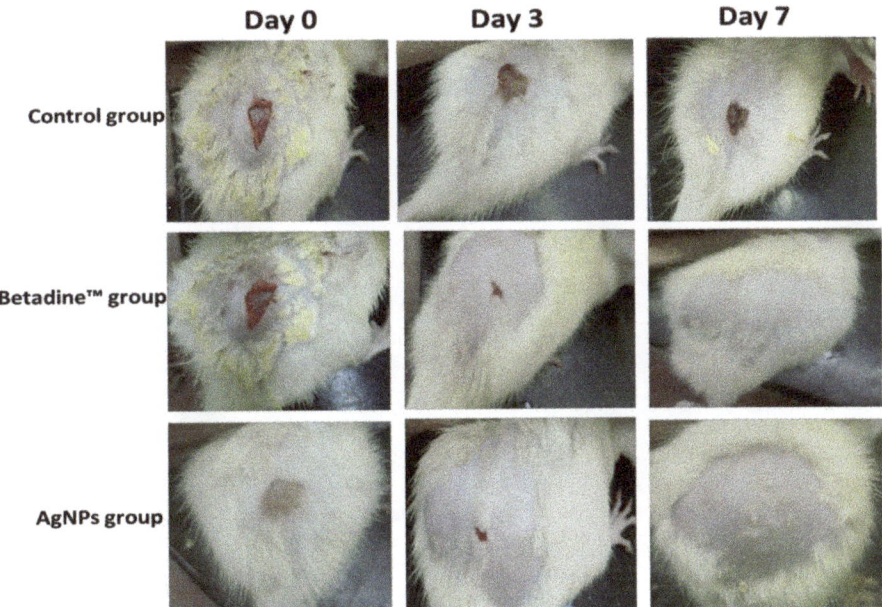

Figure 13. Macroscopic examination of the wound healing of the control, Betadine™, and AgNPs rat groups over the days 0, 3, and 7 starting from the day of wound creation (day 0).

Betadine™ and AgNPs groups exhibited significant wound healing on day 3 with wound healing percentages of 90.19 and 92.3, respectively, compared to the control group. Furthermore, they exhibited full wound healing on day 7 (with wound healing percentages of 99.02 and 99.23, respectively, in comparison with the control group (Figure 14a).

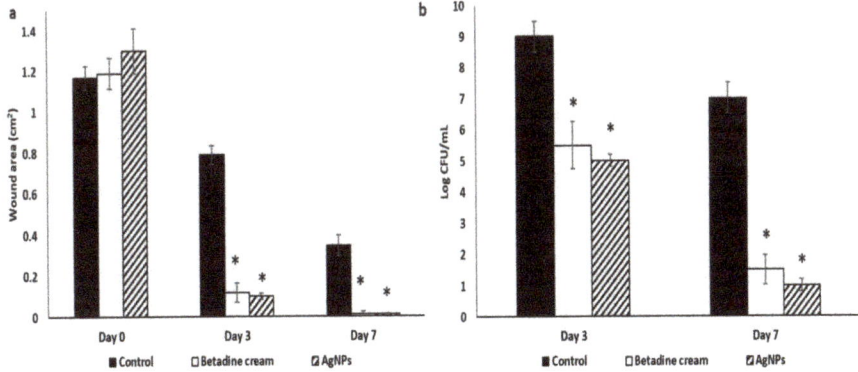

Figure 14. Wound characters on days 0, 3, and 7 in the different rat groups including (**a**) wound area, and (**b**) quantification of CFU/mL. * represents a significant decrease ($p < 0.05$).

In addition, both Betadine™ and AgNPs groups exhibited a significant decrease in CFU/mL in comparison with the control group (Figure 14b).

2.6.2. Histological Examination

The skin section of the wounds of the control group showed a wide area of epidermal loss and infiltration with inflammatory cells with granulation tissue formation (Figure 15a,d).

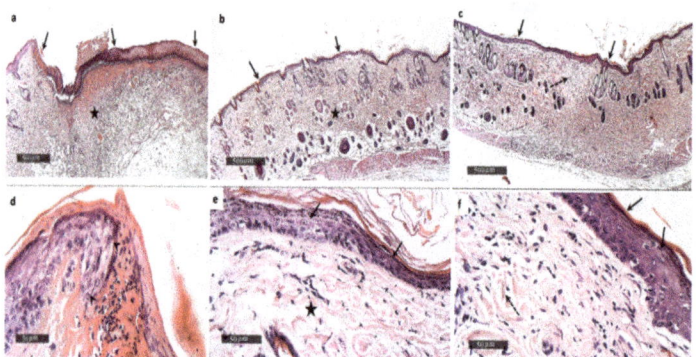

Figure 15. Histological examination of the wounds of rats on the seventh day. The control group (**a,d**) showed the presence of a wide area of epidermal loss in addition to ulceration in the wound gap (black arrow) with adjacent thickening in the epidermis (arrowhead). There was a highly cellular granulation tissue that is rich in inflammatory cells (black star). The Betadine™ group (**b,e**) showed an intact thin epidermal layer with intact keratinocytes and subcellular details (arrow). There was an intact layer of the dermis with a normal distribution of cellular elements and abundant well-organized collagen fibers (black star). The AgNPs group (**c,f**) showed efficient wound healing with complete epidermal re-epithelialization as well as wound closure (black arrow). There was more abundant fibroblastic activity with dermal granulation tissue rich in collagen (dashed arrow).

On the other hand, the skin section of the wounds of the Betadine™ group showed enhanced epidermal re-epithelialization and wound closure with more abundant fibroblastic activity and more collagen-rich dermal granulation tissue (Figure 15b,e). The section of the AgNPs treated group also exhibited complete wound healing with continuous epidermis and underlying fibrosis and collagenosis (Figure 15c,f).

3. Discussion

The widespread community and hospital-acquired infections caused by *S. aureus* isolates are a global consideration. In addition, these pathogenic bacteria are largely related to resistance to many commercially available antimicrobial agents [17]. Thus, many researchers have focused their studies on the exploration of new antimicrobial compounds against various types of pathogenic bacteria such as *S. aureus*. Natural products such as plants are showing promising antimicrobial activity with relatively low toxicity, low cost, and high bioavailability [18–23]. Herein, we used GTLE for the green synthesis of AgNPs. AgNPs drew our attention owing to their documented versatile activities in the literature, such as their antimicrobial and anti-inflammatory properties in addition to their wound healing promotion capability. AgNPs are currently utilized as interesting tools to face many emerging therapeutic challenges [24]. Despite their advantageous properties, the synthesis of AgNPs can be a high-cost process and a harmful approach. This is because of the utilization of chemical compounds and the possible production of certain harmful by-products [25–27]. Therefore, we decided to rely on the green synthesis of AgNPs to avoid these drawbacks. Many natural products such as plant extracts can be used in the green biosynthesis of AgNPs. In this case, the bioactive compounds of plants reduce silver to form AgNPs. In this way, we avoid the use of chemical reducers with their accompanying problems. The fundamental principle of the green methods is to utilize nontoxic biomolecules for the synthesis of nanoparticles via the reduction of metal ions in an aqueous solution. Most biomolecules such as DNA, proteins, and enzymes are quite expensive, easily decomposed, and vulnerable to being contaminated. On the other hand, many plant extracts are available, affordable, and stable against most environmental conditions (such as pH, temperature, and salt concentration) [5,28,29].

Nontoxic, environmentally friendly methods were used to synthesize bioinspired silver nanoparticles from GTLE. The quantification of the flavonoids and phenolic acids of *G. thailandica* was evaluated by the HPLC-DAD technique using 20 standard compounds. A few studies have documented the presence of these types of bioactive compounds in different *Gardenia* species. The HPLC-DAD analysis detected major phenolic compounds that are reported to possess antitumor, antibacterial, antioxidant, antidiabetic, and antihypercholesterolemic activities through different pathways [13]. Six phenolic acids were recognized (chlorogenic, rosmarinic, cinnamic, vanillic, *p*-coumaric, and syringic acid). In addition, four flavonoids were identified (quercetin-3-rutinoside, apigenin-7-glucoside, luteolin, and chrysin). The results of the HPLC-DAD analysis of *G. thailandica* revealed the presence of quercetin-3-rutinoside at a concentration of 2477.37 µg/g as the major flavonoid glycoside, while chlorogenic acid was the major phenolic acid in the extract at a concentration of 1441.03 µg/g, followed by rosmarinic acid at a concentration of 796.67 µg/g.

The antibacterial activity of the green-synthesized AgNPs was investigated both in vitro and in vivo. AgNPs exhibited antibacterial activity against *S. aureus* clinical isolates with MIC values that ranged from 4 to 64 µg/mL. Many studies have recorded that green-synthesized AgNPs have antibacterial activity against different pathogenic bacteria [30–33]. The short reproductive time of *S. aureus* bacteria is one of the principal reasons for the infectivity of such pathogenic bacteria [30]. Therefore, we investigated the impact of AgNPs on the time-kill curve of the tested *S. aureus* isolates and we found that the CFU/mL of *S. aureus* isolates was reduced by more than 3 log units after its incubation for 2 h with 4 × MIC and 1 h with 8 × MIC in 58.34% and 47.92% of the isolates, respectively.

As the bacterial cell membrane is an important target for several antimicrobials, we investigated the impact of the green-synthesized AgNPs on membrane characteristics including the membrane integrity, permeability, and depolarization. The bacterial cell membrane is considered to be a barrier with a selective permeability character, and the loss of this property can lead to cell death [34]. Herein, we investigated the membrane integrity of the tested bacteria before and after treatment with AgNPs by observing the leakage of materials absorbing 260 nm over time. We observed that treatment with AgNPs resulted in a massive reduction ($p < 0.05$) in the membrane integrity in 45.8% of the isolates. Many different techniques can be used for the evaluation of membrane permeability. In the current study, we used the ONPG method, which relies on the concept that when the bacteria are losing the ability to control their membrane permeability, the penetration of this compound increases [34]. Our results showed that the membrane permeability of the tested bacterial cells significantly increased ($p < 0.05$) in 56.25% of the isolates after treatment with AgNPs. Owing to the importance of the membrane potential in bacterial viability, we used DiBAC4, a fluorescent probe that enters the cell and links to the intracellular proteins when the membrane potential is lost. Here, the green-synthesized AgNPs resulted in a considerable reduction ($p < 0.05$) in membrane potential in 35.42% of *S. aureus* isolates.

SEM is widely utilized in microbiological research to study the different changes that occur in the ultrastructure and morphology of the bacterial cells when they are treated with antimicrobial agents [35]. Consequently, we used SEM in this study to explore the cell surface characters and external cell morphology to gain the benefit of the higher resolution of SEM when compared to light microscopes. Herein, we noticed that the AgNP-treated bacterial cells had a deformed and distorted shape in comparison with the non-treated ones.

The function of efflux pump proteins is to transfer harmful substances out of bacterial cells [26]. Therefore, efflux pumps are an important resistance mechanism to many antibiotics. In the current study, 22.92% of *S. aureus* isolates presented a reduction in their efflux activity after treatment with AgNPs. Efflux pumps in *S. aureus* bacteria are encoded by *nor*A, *nor*B, and *nor*C genes. For further elucidation of the effect of AgNPs on the efflux activity of the 11 *S. aureus* isolates that displayed a decline in their efflux activity by EtBr cartwheel assay, qRT-PCR was utilized. We noticed that treatment with AgNPs resulted in a substantial decrease in the expression of *nor*A, *nor*B, and *nor*C genes in 72.73%, 45.45%, and 18.18% of the tested isolates, respectively. Generally, metal nanoparticles could inhibit the efflux pump activity of bacteria by two mechanisms. The first possible mechanism is by direct binding to the efflux pumps' active site and the second mechanism is by disturbing the efflux kinetics [36].

The process of wound healing is associated with certain biological events such as re-epithelialization, fibroplasia in addition to extracellular matrix production. Many natural agents were found to produce satisfactory results in wound healing when compared to the chemical compounds, with the advantages of low cost and low toxicity [37]. Consequently, we used GTLE to synthesize AgNPs and investigated their effect on wound healing in rats with wounds infected with *S. aureus* isolates after seven days of treatment. The group treated with AgNPs exhibited notable wound healing when compared to the other groups. On the histological level, the AgNP-treated group displayed accelerated wound healing with complete epidermal re-epithelialization, abundant fibroblastic activity, formation of collagen-rich dermal granulation tissue, and minimal infiltration of inflammatory cells.

4. Materials and Methods

4.1. Plant Materials and Extract Preparation

Gardenia thailandica Tirveng. leaves were collected from a private garden on the Egypt Alexandria desert road. Esraa Ammar (Plant Ecology, Botany Department, Faculty of Science, Tanta University) confirmed the plant's identification. A voucher sample (PGA-GT-128-W) was maintained in the Tanta University Department of Pharmacognosy's herbarium. The powdered plant (650 g) was extracted with methanol using a maceration method (3×5 L). The extract was concentrated using a rotary evaporator to obtain a residue (7.89 g).

4.2. Drugs and Chemicals

All the chemicals and solvents used in this study were bought from Sigma-Aldrich (St. Louis, MO, USA) and were of high analytical quality.

4.3. HPLC-DAD of GTLE

An autosampler and a diode-array detector are included in the Agilent Technologies 1100 series liquid chromatography.

The analytical column was an Eclipse XDB-C18 (150 × 4.6 µm; 5 µm) with a C18 guard column (Phenomenex, Torrance, CA, USA). Acetonitrile (solvent A) and 2% acetic acid in water (Solvent B) made up the mobile phase. The flow rate was held constant at 0.8 mL/min for a total run of 70 min, and the gradient program was as follows: 100% B to 85% B in 30 min, 85% B to 50% B in 20 min, 50% B to 0% B in 5 min, and 0% B to 100% B in 5 min. The injection volume was 50 µL, and peaks for benzoic acid, cinnamic acid derivatives, and flavonoids were found simultaneously at 280, 320, and 360 nm, respectively. All samples were filtered using a 0.45 µm Acrodisc syringe filter (Gelman Laboratory, Michigan, USA) before injection. The peaks were identified using congruent retention durations and UV spectra, which were then compared to the standards.

4.4. Green Synthesis of AgNPs

One millimolar of an aqueous solution of silver nitrate ($AgNO_3$) was prepared and maintained in a cool dark area. For reduction of $Ag+$ ions, 10 mL of GTLE was added separately into 90 mL of an aqueous solution of 1 mM $AgNO_3$ and incubated overnight at room temperature in a dark area. The development of AgNPs was indicated by the production of a yellowish-brown color. The produced solutions were directly subjected to TEM and UV measurements. Centrifugation at 4000 rpm for 30 min was followed by a series of washing in distilled water and filtration to obtain pure AgNPs. The pure AgNPs were further characterized by FTIR, HR-TEM, XRD, zeta potential, and SEM [38–41].

4.5. Characterization of AgNPs

4.5.1. UV-Vis Spectroscopy

UV-Vis spectroscopy of the green-synthesized AgNPs was monitored using a UV–Vis spectrophotometer (Shimadzu, Kyoto, Japan) after dilution with distilled water.

4.5.2. FTIR

The different functional groups of the produced AgNPs were measured by FTIR spectrometer (Jasco, Tokyo, Japan) in the range of 4000–400 cm^{-1}.

4.5.3. HR-TEM

The morphology of the particles (shape and dimensions) in addition to SAED were examined by TEM. (JEOL-JEM-1011, Kyoto, Japan) and HR-TEM at 200 kV (JEOL-JEM-2100, Kyoto, Japan). Three milliliters of the sample were placed on the copper grid for TEM and HR-TEM examination and allowed to dry at room temperature for 15 min.

4.5.4. Zeta Potential and DLS

Particle size, homogeneity of distribution, and zeta potentials of AgNPs were examined using a zeta sizer nano ZN (Malvern Panalytical Ltd., England, UK). Before the measurements, an aliquot of nanoparticles was diluted with ultra-purified water and then sonicated for 15 min.

4.5.5. XRD

The XRD analysis was performed as a surface chemical analysis tool for the characterization of metal nanoparticles [42]. An XPERT-PRO-PANalytical Powder Diffractometer (PAN-alytical B.V., Almelo, The Netherlands) was used to perform XRD utilizing a monochromatic radiation source Cu-K α radiation (θ = 1.5406 Å) at 45 kV and 30 mA at ambient temperature. The silver nano-powder intensity data were gathered over a 2θ range of 4.01°–79.99°.

4.5.6. SEM

The morphology of the biosynthesized AgNPs was observed using SEM (TM1000, Hitachi, Chiyoda, Japan) as described previously [43].

4.6. Determination of the Total Content of Flavonoids and Polyphenols

The total flavonoid concentration was determined by colorimetric analysis of serial dilutions of the extract using the aluminum chloride technique and quercetin as a reference [44]. Using the Folin–Ciocalteu technique and gallic acid as a reference, the total content of polyphenols was determined [45]. The measured contents were expressed as mg/g equivalent of the corresponding standard for each method.

4.7. Antioxidant Activity of GTLE

4.7.1. The DPPH Radical Scavenging Capacity

The DPPH radical scavenging capacity of GTLE was evaluated according to the method of Boly et al. [46]. The decrease in DPPH color intensity was measured at 540 nm using the following equation:

$$Percentage\ inhibition = \frac{(Average\ absorbance\ of\ blank - average\ absorbance\ of\ the\ test)}{Average\ absorbance\ of\ blank} \times 100$$

The value of IC_{50} was calculated as previously described [47].

4.7.2. The ABTS Radical Scavenging Capacity

The assay was performed as previously reported [48]. Using the linear regression equation taken from the calibration curve, the results are presented as µM Trolox equivalents (TE)/ mg samples (linear dose-inhibition curve of Trolox).

4.7.3. FRAP Assay

The ferric reducing ability assay was conducted according to the method of Benzi et al. [49]. The result is expressed as µM TE/mg sample using the linear regression equation derived from the calibration curve (linear dose-response curve of Trolox).

4.8. In Vitro Antibacterial Activity

4.8.1. Bacterial Isolates

A total of 48 *S. aureus* clinical isolates were acquired from Tanta University Hospital. *S. aureus* isolates were microscopically examined and were biochemically identified as previously described [50]. *Staphylococcus aureus* (ATCC 29231) was utilized as a reference strain.

4.8.2. Susceptibility Testing

Disk Diffusion Method

The antimicrobial activity of AgNPs against *S. aureus* clinical isolates was performed using the Kirby–Bauer disk diffusion method [51]. Mueller–Hinton agar (MHA) (Merck, Germany) plates were inoculated with the bacterial isolates using sterile swabs. Sterile discs were thoroughly saturated with vancomycin and sterile water as positive and negative controls, respectively. In addition, a third disc was saturated with the green-synthesized AgNPs was added. Then, the disks were located on the MHA plates and incubated for 24 h at 37 °C. The formed inhibition zones were observed indicating antibacterial activity.

Minimum Inhibitory Concentration (MIC) Determination

The MIC values of the green-synthesized AgNPs were determined in a 96-well microdilution plate using the broth microdilution method [52]. The green-synthesized AgNP solution (500 µg/mL) was twofold diluted with each bacterial inoculum in 100 µL of MHB (10^6 CFU/mL). Each microtitration plate had a negative control (MHB only) and positive control (MHB containing bacteria). Each well of the microtitration plate was loaded with 30 µL of the resazurin solution and then the plates were incubated at 37 °C for 24 h. The variations in the color were detected.

4.8.3. Time Kill Curve

This was performed as previously reported [53]. Briefly, the green-synthesized AgNPs solution was diluted by MHB containing the bacterial suspensions to obtain a final concentration of 0× MIC, 0.5× MIC, 1× MIC, 2× MIC, 4× MIC, and 8× MIC for each bacterial isolate. The obtained cultures were then incubated in a shaking incubator at 37 °C. Aliquots of the cultures (100 µL) were distributed on the surface of MHA plates at 0, 0.25, 0.5, 1, 2, and 4 h. After incubation at 37 °C for 24 h, the colonies detected on the MHA plates were quantified in CFU/mL.

4.8.4. Membrane Integrity and Permeability

Membrane Integrity Assay

The effect of the green-synthesized AgNPs on the integrity of the cell membrane of the tested isolates was studied by monitoring the release of materials that have absorbance at 260 nm (A260) [54]. In brief, the optical density (OD) of the overnight bacterial cultures in nutrient broth was adjusted to be 0.4 at 630 nm. Then, the bacterial suspensions were centrifuged at $11,000 \times g$ for 10 min and the obtained pellets were resuspended in 0.5% NaCl solution and their absorbance was adjusted to 0.7 at 420 nm. The membrane integrity was assessed by checking the discharge of materials that have absorbance at 260 nm from the bacterial cytoplasm to the surrounding media over time using a UV/Vis spectrophotometer (SHIMADZU, Kyoto, Japan).

Membrane Permeability Assay

Membrane permeability was explored by quantifying the exit of a β-galactosidase enzyme from the bacterial cytoplasm using the substrate of the enzyme (ONPG) [55]. Briefly, 2% lactose was added to the overnight bacterial suspension in nutrient broth. This mixture was then centrifuged, and the obtained pellet was thoroughly rinsed using phosphate-buffered saline (PBS) and resuspended in NaCl solution (0.5%). Finally, each bacterial suspension (1.6 mL) was supplemented with 150 µL of ONPG solution (34 mM). The produced ONP was detected over time using an ELISA reader (Sunrise Tecan, Männedorf, Switzerland) to monitor the increase in absorbance at 420.

4.8.5. Membrane Depolarization

This test was carried out using DiBAC4(3), a fluorescent stain used for staining the tested bacterial cells (both treated and untreated with the green-synthesized AgNPs) [56]. A FACSVerse flow cytometer (BD Biosciences, Franklin Lakes, New Jersey, USA) was used to analyze the staining of the cells.

4.8.6. SEM

The morphological changes of the AgNPs treated *S. aureus* isolates in comparison with the non-treated ones were inspected by SEM (Hitachi, Chiyoda, Japan) as described by McDowell and Trump [57].

4.8.7. Efflux Activity

Efflux activity was tested by the EtBr cartwheel method [58] before and after treatment with AgNPs (at 0.5 MIC values) using the reference strain as a negative control. In brief, bacterial suspensions were inoculated as redial lines onto tryptic soy agar (TSA) plates using swabs and were incubated at 37 °C for 18 h. The TSA plates were supplied with EtBr (with concentrations that ranged from 0.5 to 2.5 mg/L). After incubation, the lowest EtBr concentrations that led to fluorescence production by the bacterial isolates were recorded by UV-Vis spectrophotometer (SHIMADZU, Kyoto, Japan). *S. aureus* isolates were then classified according to the recorded EtBr minimum concentrations as follows: isolates with no efflux activity are those that emitted fluorescence at an EtBr concentration of 0.5 mg/L, isolates with intermediate efflux activity are that emitted fluorescence at an EtBr concentration of 1–2.0 mg/L, and isolates with positive efflux activity are those that emitted fluorescence at an EtBr concentration of 2.5 mg/L.

4.8.8. qRT-PCR

We used qRT-PCR for detection of the expression levels of the genes encoding efflux pumps (*norA*, *norB*, and *norC*) in *S. aureus* isolates after treatment with AgNPs. In brief, total RNA was extracted from the pellets of overnight cultures of *S. aureus* isolates using the Purelink™ RNA Mini Kit (Thermo Scientific, Waltham, USA) according to the instructions of the manufacturer. The extracted RNA was then converted into cDNA by power™ cDNA synthesis kit (iNtRON Biotechnology, Seoul, Korea) as described by the manufacturer. Rotor-Gene Q 5plex (Qiagen, Hilden, Germany) was used for performing qRT-PCR for the calculation of the efflux pump gene expression fold changes. The sequences of the utilized primers in addition to the sequence of the housekeeping gene (16S rRNA) primer are presented in Table S1 [59,60]. The levels of the relative expression of the tested genes were quantified by the $2^{-\Delta\Delta Ct}$ method, considering the gene expression levels in the isolates before treatment to be 1 [61]. The statistically significant fold changes were those with two or more-fold changes (either increasing or decreasing) [62].

4.9. In Vivo Antibacterial Activity

4.9.1. Animals

We obtained 30 white albino male rats (190–210 g and 8 weeks old) from the animal house at the Faculty of Veterinary Medicine, Cairo University. All rats were kept in an environment with a constant temperature of 25 ± 2 °C and a pathogen-free atmosphere with a 12-h light/dark cycle. They were permitted free access to a standard pellet diet in addition to filtered water. The rats were given one week for acclimatization before being utilized in the research. The in vivo experiment performed in the current study followed the standards of the use of the laboratory animals authorized by the Faculty of Pharmacy Research Ethical Committee (Tanta University, Al Gharbiyah, Egypt) with approval number TP/RE/11-21-P-001.

4.9.2. Wound Model

The 30 rats were distributed into three groups, randomly, each with 10 rats. The groups were as follows: the control group (0.9% normal saline), Betadine™ ointment (Mundi pharma as standard drug) group, and AgNP group. The rats were anesthetized using diethyl ether; then a small area was carefully shaved on their backs. After that, full-thickness excisional wounds were created and infected with *S. aureus* bacteria (10^6 CFU/mL). For 7 days, the groups were administered the drugs topically on the surface of the wound [63].

4.9.3. Macroscopic Wound Healing

The day of the wound creation was considered as day 0, and the wound healing process was observed for 7 days starting from day 0. The wound images were taken on days 0, 3, and 7 using a digital camera. Wound areas (cm^2) were calculated on days 0, 3, and 7 for evaluation of the healing efficacy using a ruler beside the wounds [64]. The percentage of the wound healing was calculated using the following equation [65]:

$$Percentage\ of\ wound\ healing = \frac{wound\ area\ at\ day\ 0 - wound\ area\ at\ n\ th\ day}{wound\ area\ at\ day\ 0} \times 100$$

where n represents day 3 and day 7.

In addition, on days 3 and 7 post-wounding, wound tissues were excised and homogenized in PBS (10 mL), 10-fold serially diluted in MHB, and plated onto mannitol salt agar plates for CFU quantification after overnight incubation at 37 °C [66,67].

4.9.4. Histological Examination

The entire wound was isolated for histological examination at the end of the experiment, with a margin of about 5 mm of the surrounding intact skin. The skin sections were fixed using 10% formalin solution (pH 7.4) overnight. Then, they were processed using a series of alcohol and xylene grades. Following that, the tissues were inserted in paraffin wax at 65 °C. The blocks of tissues were cut into sections of 5 μm thickness, stained with hematoxylin and eosin (H&E), and finally viewed using a light microscope [63].

4.10. Statistical Analysis

All tests were accomplished in triplicate and the obtained results are presented as mean ± standard deviation (SD). Prism $8^®$ software (GraphPad, Inc., San Diego, CA, USA) was utilized to assess the statistical significance at $p < 0.05$.

5. Conclusions

According to our results, GTLE includes phenolic compounds. These substances have the ability to form and stabilize AgNPs. In conclusion, the green-synthesized AgNPs by GTLE exhibited good antibacterial activity both in vitro and in vivo against *S. aureus* clinical isolates. They significantly decreased ($p < 0.05$) the membrane integrity of 45.8% of the isolates. In addition, they reduced the membrane potential of 35.42% of isolates by flow cytometry. AgNPs also resulted in morphological and ultrastructural changes in the tested isolates, as revealed by SEM. Furthermore, they resulted in a significant reduction in efflux activity in 41.67% of the tested isolates. For more in-depth study of the impact of AgNPs on efflux activity, qRT-PCR was utilized to examine the relative gene expression of the efflux pump genes (*nor*A, *nor*B, and *nor*C). The in vivo examination was performed on wounds infected with *S. aureus* bacteria in rats. The group treated with AgNPs was characterized by epidermis regeneration and reduction in the infiltration of the inflammatory cells. Therefore, the green-synthesized AgNPs by GTLE could be a future alternative to chemical compounds and antimicrobials that are currently used to induce healing of wounds, especially infected ones.

Supplementary Materials: The following supporting information can be downloaded at: https://www.mdpi.com/article/10.3390/ph15020194/s1, Table S1. The sequences of the primers used in qRT-PCR.

Author Contributions: Conceptualization: W.A.N. and E.E.; Data curation, N.G.M.A. and I.A.H.; Formal analysis, N.G.M.A.; Funding acquisition, N.G.M.A.; Investigation, E.E., W.A.N., F.A.M. and O.M.A.-F.; Methodology, E.E., W.A.N., F.A.M., I.A.H. and O.M.A.-F.; Resources, F.A.M. and O.M.A.-F.; Writing—original draft, E.E., W.A.N., F.A.M. and O.M.A.-F.; Writing—review and editing, E.E., W.A.N., F.A.M., I.A.H. and O.M.A.-F. All authors have read and agreed to the published version of the manuscript.

Funding: This work was funded by Princess Nourah bint Abdulrahman University Researchers Supporting Project number (PNURSP2022R204), Princess Nourah bint Abdulrahman University, Riyadh, Saudi Arabia.

Institutional Review Board Statement: This study followed the standards of the use of the laboratory animals authorized by the Faculty of Pharmacy Research Ethical Committee (Tanta University, Egypt) with approval number TP/RE/11-21-P-001.

Informed Consent Statement: Not applicable.

Data Availability Statement: Data is contained within the article and supplementary materials.

Conflicts of Interest: The authors declare no conflict of interest.

References

1. Rodríguez-Luis, O.E.; Hernandez-Delgadillo, R.; Sánchez-Nájera, R.I.; Martínez-Castañón, G.A.; Niño-Martínez, N.; Navarro, M.D.C.S.; Ruiz, F.; Cabral-Romero, C. Green Synthesis of Silver Nanoparticles and Their Bactericidal and Antimycotic Activities against Oral Microbes. *J. Nanomater.* **2016**, *2016*, 9204573. [CrossRef]
2. Rajkumar, P.V.; Prakasam, A.; Rajeshkumar, S.; Gomathi, M.; Anbarasan, P.; Chandrasekaran, R. Green synthesis of silver nanoparticles using Gymnema sylvestre leaf extract and evaluation of its antibacterial activity. *South Afr. J. Chem. Eng.* **2020**, *32*, 1–4.
3. Ceylan, R.; Demirbas, A.; Ocsoy, I.; Aktumsek, A. Green synthesis of silver nanoparticles using aqueous extracts of three *Sideritis* species from Turkey and evaluations bioactivity potentials. *Sustain. Chem. Pharm.* **2021**, *21*, 100426. [CrossRef]
4. Huq, M.A. Green synthesis of silver nanoparticles using *Pseudoduganella eburnea* MAHUQ-39 and their antimicrobial mechanisms investigation against drug resistant human pathogens. *Int. J. Mol. Sci.* **2020**, *21*, 1510. [CrossRef] [PubMed]
5. He, X.; Gao, L.; Ma, N. One-Step Instant Synthesis of Protein-Conjugated Quantum Dots at Room Temperature. *Sci. Rep.* **2013**, *3*, 2825. [CrossRef]
6. Rolim, W.R.; Pelegrino, M.T.; de Araújo Lima, B.; Ferraz, L.S.; Costa, F.N.; Bernardes, J.S.; Rodigues, T.; Brocchi, M.; Seabra, A.B. Green tea extract mediated biogenic synthesis of silver nanoparticles: Characterization, cytotoxicity evaluation and antibacterial activity. *Appl. Surf. Sci.* **2019**, *463*, 66–74. [CrossRef]
7. Gomathi, M.; Prakasam, A.; Rajkumar, P.V.; Rajeshkumar, S.; Chandrasekaran, R.; Kannan, S. Phyllanthus reticulatus mediated synthesis and characterization of silver nanoparticles and its antibacterial activity against gram positive and gram negative pathogens. *Int. J. Res. Pharm. Sci.* **2019**, *10*, 3099–3106. [CrossRef]
8. Chiruvella, K.K.; Mohammed, A.; Dampuri, G.; Ghanta, R.G.; Raghavan, S.C. Phytochemical and Antimicrobial Studies of Methyl Angolensate and Luteolin-7-O-glucoside Isolated from Callus Cultures of *Soymida febrifuga*. *Int. J. Biomed. Sci. IJBS* **2007**, *3*, 269–278.

9. Garibo, D.; Borbón-Nuñez, H.A.; de León, J.N.D.; Mendoza, E.G.; Estrada, I.; Toledano-Magaña, Y.; Tiznado, H.; Ovalle-Marroquin, M.; Soto-Ramos, A.G.; Blanco, A.; et al. Green synthesis of silver nanoparticles using *Lysiloma acapulcensis* exhibit high-antimicrobial activity. *Sci. Rep.* **2020**, *10*, 12805. [CrossRef] [PubMed]
10. Zongram, O.; Ruangrungsi, N.; Palanuvej, C.; Rungsihirunrat, K. Leaf constant numbers of selected *Gardenia* species in Thailand. *J. Health Res.* **2017**, *31*, 69–75.
11. Phromnoi, K.; Reuter, S.; Sung, B.; Limtrakul, P.; Aggarwal, B.B. A Dihydroxy-pentamethoxyflavone from *Gardenia obtusifolia* suppresses proliferation and promotes apoptosis of tumor cells through modulation of multiple cell signaling pathways. *Anticancer Res.* **2010**, *30*, 3599–3610. [PubMed]
12. Kongkum, N.; Tuchinda, P.; Pohmakotr, M.; Reutrakul, V.; Piyachaturawat, P.; Jariyawat, S.; Suksen, K.; Akkarawongsapat, R.; Kasisit, J.; Napaswad, C. Cytotoxic, Antitopoisomerase IIα, and Anti-HIV-1 Activities of Triterpenoids Isolated from Leaves and Twigs of *Gardenia carinata*. *J. Nat. Prod.* **2013**, *76*, 530–537. [CrossRef]
13. Tuchinda, P.; Saiai, A.; Pohmakotr, M.; Yoosook, C.; Kasisit, J.; Napaswat, C.; Santisuk, T.; Reutrakul, V. Anti-HIV-1 Cycloartanes from Leaves and Twigs of *Gardenia thailandica*. *Planta Med.* **2004**, *70*, 366–370.
14. Tong, S.Y.C.; Davis, J.S.; Eichenberger, E.; Holland, T.L.; Fowler, V.G., Jr. *Staphylococcus aureus* infections: Epidemiology, pathophysiology, clinical manifestations, and management. *Clin. Microbiol. Rev.* **2015**, *28*, 603–661. [CrossRef] [PubMed]
15. Derakhshan, S.; Navidinia, M.; Haghi, F. Antibiotic susceptibility of human-associated *Staphylococcus aureus* and its relation to *agr* typing, virulence genes, and biofilm formation. *BMC Infect. Dis.* **2021**, *21*, 1–10. [CrossRef] [PubMed]
16. Noginov, M.A.; Zhu, G.; Bahoura, M.; Adegoke, J.; Small, C.; Ritzo, B.A.; Drachev, V.P.; Shalaev, V.M. The effect of gain and absorption on surface plasmons in metal nanoparticles. *Appl. Phys. B* **2007**, *86*, 455–460. [CrossRef]
17. McNeilly, O.; Mann, R.; Hamidian, M.; Gunawan, C. Emerging Concern for Silver Nanoparticle Resistance in Acinetobacter baumannii and Other Bacteria. *Front. Microbiol.* **2021**, *12*, 652863. [CrossRef] [PubMed]
18. Stan, D.; Enciu, A.-M.; Mateescu, A.L.; Ion, A.C.; Brezeanu, A.C.; Tanase, C. Natural Compounds with Antimicrobial and Antiviral Effect and Nanocarriers Used for Their Transportation. *Front. Pharmacol.* **2021**, *12*, 2405. [CrossRef]
19. Ekrikaya, S.; Yılmaz, E.; Celik, C.; Demirbuga, S.; Ildiz, N.; Demirbas, A.; Ocsoy, I. Investigation of ellagic acid rich-berry extracts directed silver nanoparticles synthesis and their antimicrobial properties with potential mechanisms towards *Enterococcus faecalis* and *Candida albicans*. *J. Biotechnol.* **2021**, *341*, 155–162. [CrossRef] [PubMed]
20. Demirbaş, A.; Yılmaz, V.; Ildiz, N.; Baldemir, A.; Ocsoy, I. Anthocyanins-rich berry extracts directed formation of Ag NPs with the investigation of their antioxidant and antimicrobial activities. *J. Mol. Liq.* **2017**, *248*, 1044–1049. [CrossRef]
21. Negm, W.A.; Ibrahim, A.E.-R.S.; El-Seoud, K.A.; Attia, G.I.; Ragab, A.E. A new cytotoxic and antioxidant Amentoflavone Monoglucoside from *Cycas revoluta* Thunb growing in Egypt. *J. Pharm. Sci. Res.* **2016**, *8*, 343–350.
22. Negm, W.A.; El-Seoud, K.A.A.; Kabbash, A.; Kassab, A.A.; El-Aasr, M. Hepatoprotective, cytotoxic, antimicrobial and antioxidant activities of *Dioon spinulosum* leaves Dyer Ex Eichler and its isolated secondary metabolites. *Nat. Prod. Res.* **2020**, *35*, 5166–5176. [CrossRef] [PubMed]
23. Elmongy, E.I.; Negm, W.A.; Elekhnawy, E.; El-Masry, T.A.; Attallah, N.G.M.; Altwaijry, N.; Batiha, G.E.-S.; El-Sherbeni, S.A. Antidiarrheal and Antibacterial Activities of Monterey Cypress Phytochemicals: In Vivo and In Vitro Approach. *Molecules* **2022**, *27*, 346. [CrossRef]
24. Ferdous, Z.; Nemmar, A. Health Impact of Silver Nanoparticles: A Review of the Biodistribution and Toxicity Following Various Routes of Exposure. *Int. J. Mol. Sci.* **2020**, *21*, 2375. [CrossRef]
25. Turek, D.; van Simaeys, D.; Johnson, J.; Ocsoy, I.; Tan, W. Molecular recognition of live methicillin-resistant *staphylococcus aureus* cells using DNA aptamers. *World J. Transl. Med.* **2013**, *2*, 67–74. [CrossRef] [PubMed]
26. Some, S.; Bulut, O.; Biswas, K.; Kumar, A.; Roy, A.; Sen, I.K.; Mandal, A.; Franco, O.L.; Ince, I.A.; Neog, K.; et al. Effect of feed supplementation with biosynthesized silver nanoparticles using leaf extract of *Morus indica* L. V1 on *Bombyx mori* L. (Lepidoptera: Bombycidae). *Sci. Rep.* **2019**, *9*, 14839. [CrossRef]
27. Some, S.; Sarkar, B.; Biswas, K.; Jana, T.K.; Bhattacharjya, D.; Dam, P.; Mondal, R.; Kumar, A.; Deb, A.K.; Sadat, A.; et al. Biomolecule functionalized rapid one-pot green synthesis of silver nanoparticles and their efficacy toward the multidrug resistant (MDR) gut bacteria of silkworms (*Bombyx mori*). *RSC Adv.* **2020**, *10*, 22742–22757. [CrossRef]
28. Ocsoy, I.; Paret, M.L.; Ocsoy, M.A.; Kunwar, S.; Chen, T.; You, M.; Tan, W. Nanotechnology in Plant Disease Management: DNA-Directed Silver Nanoparticles on Graphene Oxide as an Antibacterial against *Xanthomonas perforans*. *ACS Nano* **2013**, *7*, 8972–8980. [CrossRef]
29. Strayer, M.A.L.; Ocsoy, I.; Tan, W.; Jones, J.B.; Paret, M.L. Low Concentrations of a Silver-Based Nanocomposite to Manage Bacterial Spot of Tomato in the Greenhouse. *Plant Dis.* **2016**, *100*, 1460–1465. [CrossRef]
30. Loo, Y.Y.; Rukayadi, Y.; Nor-Khaizura, M.-A.-R.; Kuan, C.H.; Chieng, B.W.; Nishibuchi, M.; Radu, S. In Vitro Antimicrobial Activity of Green Synthesized Silver Nanoparticles Against Selected Gram-negative Foodborne Pathogens. *Front. Microbiol.* **2018**, *9*, 1555. [CrossRef] [PubMed]
31. Salayová, A.; Bedlovičová, Z.; Daneu, N.; Baláž, M.; Bujňáková, Z.L.; Balážová, Ľ.; Tkáčiková, Ľ. Green Synthesis of Silver Nanoparticles with Antibacterial Activity Using Various Medicinal Plant Extracts: Morphology and Antibacterial Efficacy. *Nanomaterials* **2021**, *11*, 1005. [CrossRef] [PubMed]
32. Senthil, B.; Devasena, T.; Prakash, B.; Rajasekar, A. Non-cytotoxic effect of green synthesized silver nanoparticles and its antibacterial activity. *J. Photochem. Photobiol. B Biol.* **2017**, *177*, 1–7. [CrossRef] [PubMed]

33. Csakvari, A.C.; Moisa, C.; Radu, D.G.; Olariu, L.M.; Lupitu, A.I.; Panda, A.O.; Pop, G.; Chambre, D.; Socoliuc, V.; Copolovici, L.; et al. Green Synthesis, Characterization, and Antibacterial Properties of Silver Nanoparticles Obtained by Using Diverse Varieties of *Cannabis sativa* Leaf Extracts. *Molecules* **2021**, *26*, 4041. [CrossRef] [PubMed]
34. Bouyahya, A.; Abrini, J.; Dakka, N.; Bakri, Y. Essential oils of *Origanum compactum* increase membrane permeability, disturb cell membrane integrity, and suppress quorum-sensing phenotype in bacteria. *J. Pharm. Anal.* **2019**, *9*, 301–311. [CrossRef]
35. Torres, M.R.; Slate, A.J.; Ryder, S.F.; Akram, M.; Iruzubieta, C.J.C.; Whitehead, K.A. Ionic gold demonstrates antimicrobial activity against *Pseudomonas aeruginosa* strains due to cellular ultrastructure damage. *Arch. Microbiol.* **2021**, *203*, 3015–3024. [CrossRef] [PubMed]
36. Gupta, D.; Singh, A.; Khan, A.U. Nanoparticles as Efflux Pump and Biofilm Inhibitor to Rejuvenate Bactericidal Effect of Conventional Antibiotics. *Nanoscale Res. Lett.* **2017**, *12*, 454. [CrossRef]
37. Ibrahim, N.; Wong, S.K.; Mohamed, I.N.; Mohamed, N.; Chin, K.-Y.; Ima-Nirwana, S.; Shuid, A.N. Wound Healing Properties of Selected Natural Products. *Int. J. Environ. Res. Public Health* **2018**, *15*, 2360. [CrossRef]
38. Aziz, S.B.; Abdullah, O.G.; Saber, D.R.; Rasheed, M.A.; Ahmed, H.M. Investigation of metallic silver nanoparticles through UV-Vis and optical micrograph techniques. *Int. J. Electrochem. Sci.* **2017**, *12*, 363–373. [CrossRef]
39. Edison, T.N.J.I.; Atchudan, R.; Kamal, C.; Lee, Y.R. *Caulerpa racemosa*: A marine green alga for eco-friendly synthesis of silver nanoparticles and its catalytic degradation of methylene blue. *Bioprocess Biosyst. Eng.* **2016**, *39*, 1401–1408. [CrossRef]
40. Edison, T.N.J.I.; Atchudan, R.; Karthik, N.; Balaji, J.; Xiong, D.; Lee, Y.R. Catalytic degradation of organic dyes using green synthesized N-doped carbon supported silver nanoparticles. *Fuel* **2020**, *280*, 118682. [CrossRef]
41. Edison, T.N.J.I.; Atchudan, R.; Lee, Y.R. Optical Sensor for Dissolved Ammonia Through the Green Synthesis of Silver Nanoparticles by Fruit Extract of *Terminalia chebula*. *J. Clust. Sci.* **2016**, *27*, 683–690. [CrossRef]
42. Baer, D.R.; Gaspar, D.J.; Nachimuthu, P.; Techane, S.D.; Castner, D.G. Application of surface chemical analysis tools for characterization of nanoparticles. *Anal. Bioanal. Chem.* **2010**, *396*, 983–1002. [CrossRef] [PubMed]
43. Zhang, X.-F.; Liu, Z.-G.; Shen, W.; Gurunathan, S. Silver Nanoparticles: Synthesis, Characterization, Properties, Applications, and Therapeutic Approaches. *Int. J. Mol. Sci.* **2016**, *17*, 1534. [CrossRef]
44. Kiranmai, M.; Kumar, C.B.M.; Ibrahim, M. Comparison of total flavanoid content of *Azadirachta indica* root bark extracts prepared by different methods of extraction. *Res. J. Pharm. Biol. Chem. Sci.* **2011**, *2*, 254–261.
45. Attard, E. A rapid microtitre plate Folin-Ciocalteu method for the assessment of polyphenols. *Open Life Sci.* **2013**, *8*, 48–53. [CrossRef]
46. Boly, R.; Lamkami, T.; Lompo, M.; Dubois, J.; Guissou, I. DPPH free radical scavenging activity of two extracts from *Agelanthus dodoneifolius* (Loranthaceae) leaves. *Int. J. Toxicol. Pharmacol. Res.* **2016**, *8*, 29–34.
47. Chen, Z.; Bertin, R.; Froldi, G. EC50 estimation of antioxidant activity in DPPH assay using several statistical programs. *Food Chem.* **2013**, *138*, 414–420. [CrossRef] [PubMed]
48. Arnao, M.B.; Cano, A.; Acosta, M. The hydrophilic and lipophilic contribution to total antioxidant activity. *Food Chem.* **2001**, *73*, 239–244. [CrossRef]
49. Benzie, I.F.F.; Strain, J.J. The ferric reducing ability of plasma (FRAP) as a measure of "antioxidant power": The FRAP assay. *Anal. Biochem.* **1996**, *239*, 70–76. [CrossRef]
50. MacFaddin, J.F. *Biochemical Tests for Identification of Medical Bacteria*; Williams and Wilkins: Philadelphia, PA, USA, 2000; p. 113.
51. Bauer, A.W.; Kirby, W.M.; Sherris, J.C.; Truck, M. Antibiotic Susceptibility Testing by a Standardized Single Disk Method. *Am. Clin. Pathol.* **1966**, *45*, 493–496. [CrossRef]
52. Weinstein, M.P.; Lewis, J.S., II; Bobenchik, A.M.; Campeau, S.; Cullen, S.K.; Galas, M.F.; Gold, H.; Humphries, R.M.; Kirn, T.J., Jr.; Limbago, B.; et al. *M100 Performance Standards for Antimicrobial Susceptibility Testing*; Clinical and Laboratory Standards Institute: Wayne, PA, USA, 2020.
53. Lau, K.Y.; Zainin, N.S.; Abas, F.; Rukayadi, Y. Antibacterial and sporicidal activity of *Eugenia polyantha* Wight against *Bacillus cereus* and *Bacillus subtilis*. *Int. J. Curr. Microbiol. Appl. Sci.* **2014**, *3*, 499–510.
54. Negm, W.A.; El-Aasr, M.; Kamer, A.A.; Elekhnawy, E. Investigation of the Antibacterial Activity and Efflux Pump Inhibitory Effect of *Cycas thouarsii* R.Br. Extract against *Klebsiella pneumoniae* Clinical Isolates. *Pharmaceuticals* **2021**, *14*, 756. [CrossRef] [PubMed]
55. Alotaibi, B.; Negm, W.A.; Elekhnawy, E.; El-Masry, T.A.; Elseady, W.S.; Saleh, A.; Alotaibi, K.N.; El-Sherbeni, S.A. Antibacterial, Immunomodulatory, and Lung Protective Effects of *Boswelliadalzielii* Oleoresin Ethanol Extract in Pulmonary Diseases: In Vitro and In Vivo Studies. *Antibiotics* **2021**, *10*, 1444. [CrossRef]
56. Elekhnawy, E.; Sonbol, F.; Abdelaziz, A.; Elbanna, T. An investigation of the impact of triclosan adaptation on *Proteus mirabilis* clinical isolates from an Egyptian university hospital. *Braz. J. Microbiol.* **2021**, *52*, 927–937. [CrossRef] [PubMed]
57. McDowell, E.M.; Trump, B.F. Histologic fixatives suitable for diagnostic light and electron microscopy. *Arch. Pathol. Lab. Med.* **1976**, *100*, 405–414. [PubMed]
58. Attallah, N.G.M.; Negm, W.A.; Elekhnawy, E.; Elmongy, E.I.; Altwaijry, N.; El-Haroun, H.; El-Masry, T.A.; El-Sherbeni, S.A. Elucidation of Phytochemical Content of *Cupressus macrocarpa* Leaves: In Vitro and In Vivo Antibacterial Effect against Methicillin-Resistant *Staphylococcus aureus* Clinical Isolates. *Antibiotics* **2021**, *10*, 890. [CrossRef] [PubMed]

59. El-Hamid, M.I.A.; El-Naenaeey, E.-S.Y.; Kandeel, T.M.; Hegazy, W.A.H.; Mosbah, R.A.; Nassar, M.S.; Bakhrebah, M.A.; Abdulaal, W.H.; Alhakamy, N.A.; Bendary, M.M. Promising antibiofilm agents: Recent breakthrough against biofilm producing methicillin-resistant *Staphylococcus aureus*. *Antibiotics* **2020**, *9*, 667. [CrossRef]
60. Kwak, Y.G.; Truong-Bolduc, Q.C.; Kim, H.B.; Song, K.-H.; Kim, E.S.; Hooper, D.C. Association of norB overexpression and fluoroquinolone resistance in clinical isolates of *Staphylococcus aureus* from Korea. *J. Antimicrob. Chemother.* **2013**, *68*, 2766–2772. [CrossRef]
61. Livak, K.J.; Schmittgen, T.D. Analysis of relative gene expression data using real-time quantitative PCR and the $2^{-\Delta\Delta CT}$ method. *Methods* **2001**, *25*, 402–408. [CrossRef]
62. Attallah, N.G.M.; Negm, W.A.; Elekhnawy, E.; Altwaijry, N.; Elmongy, E.I.; El-Masry, T.A.; Alturki, E.A.; Yousef, D.A.; Shoukheba, M.Y. Antibacterial Activity of *Boswellia sacra* Flueck. Oleoresin Extract against *Porphyromonas gingivalis* Periodontal Pathogen. *Antibiotics* **2021**, *10*, 859. [CrossRef]
63. Diniz, F.R.; Maia, R.C.A.P.; Andrade, L.R.; Andrade, L.N.; Chaud, M.V.; da Silva, C.F.; Corrêa, C.B.; de Albuquerque Junior, R.L.C.; da Costa, L.P.; Shin, S.R.; et al. Silver Nanoparticles-Composing Alginate/Gelatine Hydrogel Improves Wound Healing In Vivo. *Nanomaterials* **2020**, *10*, 390. [CrossRef] [PubMed]
64. Cheng, K.-Y.; Lin, Z.-H.; Cheng, Y.-P.; Chiu, H.-Y.; Yeh, N.-L.; Wu, T.-K.; Wu, J.-S. Wound healing in streptozotocin-induced diabetic rats using atmospheric-pressure argon plasma jet. *Sci. Rep.* **2018**, *8*, 12214. [CrossRef] [PubMed]
65. Alsenani, F.; Ashour, A.M.; Alzubaidi, M.A.; Azmy, A.F.; Hetta, M.H.; Abu-Baih, D.H.; Elrehany, M.A.; Zayed, A.; Sayed, A.M.; Abdelmohsen, U.R.; et al. Wound Healing Metabolites from Peters' Elephant-Nose Fish Oil: An In Vivo Investigation Supported by In Vitro and In Silico Studies. *Mar. Drugs* **2021**, *19*, 605. [CrossRef]
66. Krausz, A.E.; Adler, B.L.; Cabral, V.; Navati, M.; Doerner, J.; Charafeddine, R.A.; Chandra, D.; Liang, H.; Gunther, L.; Clendaniel, A.; et al. Curcumin-encapsulated nanoparticles as innovative antimicrobial and wound healing agent. *Nanomed. Nanotechnol. Biol. Med.* **2015**, *11*, 195–206. [CrossRef] [PubMed]
67. Elekhnawy, E.A.; Sonbol, F.I.; Elbanna, T.E.; Abdelaziz, A.A. Evaluation of the impact of adaptation of *Klebsiella pneumoniae* clinical isolates to benzalkonium chloride on bioflm formation. *Egypt. J. Med. Hum. Genet.* **2021**, *22*, 1–6. [CrossRef]

Article

Azeotropic Distillation-Induced Self-Assembly of Mesostructured Spherical Nanoparticles as Drug Cargos for Controlled Release of Curcumin

Long Chen [1], Xin Fu [2], Mei Lin [3] and Xingmao Jiang [1],*

[1] Key Laboratory for Green Chemical Process of Ministry of Education, School of Chemical Engineering & Pharmacy, Wuhan Institute of Technology, Wuhan 430205, China; chenlong0729@126.com
[2] Nanjing Zhongwei Biomaterials Research Institute Co., Ltd., Nanjing 210008, China; tiger6105180@163.com
[3] Institute of Clinical Medicine, Taizhou People's Hospital Affiliated to Nantong University, Taizhou 225300, China; l_mei@163.com
* Correspondence: jxm@wit.edu.cn

Abstract: Methods of large-scale controllable production of uniform monodispersed spherical nanoparticles have been one of the research directions of scientists in recent years. In this paper, we report an azeotropic distillation-induced evaporation self-assembly method as a universal method, and monodispersed hydrophobic ordered mesoporous silica nanospheres (MHSs) were successfully synthesized by this method, using triethoxymethylsilane (MTES) as the silica precursor and hexadecyl trimethyl ammonium bromide (CTAB) as the template. SEM and TEM images showed good monodispersity, sphericity, and uniform diameter. Meanwhile, SAXS and N_2 adsorption–desorption measurements demonstrated a highly ordered lamellar mesostructure with a large pore volume. The model drug, curcumin was successfully encapsulated in MHSs for drug delivery testing, and their adsorption capacity was 3.45 mg g^{-1}, which greatly improved the stability of curcumin. The release time when net release rate of curcumin reached 50% was extended to 6 days.

Keywords: drug delivery; curcumin; azeotropic distillation; self-assembly; hydrophobic; mesoporous silica nanospheres

Citation: Chen, L.; Fu, X.; Lin, M.; Jiang, X. Azeotropic Distillation-Induced Self-Assembly of Mesostructured Spherical Nanoparticles as Drug Cargos for Controlled Release of Curcumin. *Pharmaceuticals* 2022, *15*, 275. https://doi.org/10.3390/ph15030275

Academic Editor: Fu-Gen Wu

Received: 11 January 2022
Accepted: 19 February 2022
Published: 23 February 2022

Publisher's Note: MDPI stays neutral with regard to jurisdictional claims in published maps and institutional affiliations.

Copyright: © 2022 by the authors. Licensee MDPI, Basel, Switzerland. This article is an open access article distributed under the terms and conditions of the Creative Commons Attribution (CC BY) license (https://creativecommons.org/licenses/by/4.0/).

1. Introduction

In recent decades, drug delivery controlled by carrier systems has been demonstrated to have successful applications in the diagnosis and treatment of various diseases [1–3]. Mesostructured spherical nanoparticles are promising intracellular delivery systems for anticancer, immunomodulatory drugs and cell activity modulators, etc. [4,5]. The cellular uptake of nanoparticles by living cells is strongly size-dependent [6]. Small nanoparticle size (≈50 nm) is most efficient for the intracellular delivery [1,7]. The development of a suitable nanostructured carrier system with good biocompatibility and selective delivery of drugs to target cells is the central problem of nanomedicine. Curcumin is a natural bioactive substance, which has been of great interest to researchers due to its wide range of biological activities and pleiotropic therapeutic potential such as antioxidant, anti-inflammatory [8–12], antibacterial, antifungal, antiviral, antiprotozoal, and antiparasitic activities [11,13–15], but its application has been strictly limited because of its poor solubility in water, short half-life, low bioavailability, and pharmacokinetic profile.

Compared to general organic carriers such as liposomes [16], micelles [17], PLGA [18], cyclodextrin [19], viruses, etc., mesoporous silica nanoparticles have the significant advantage of sustained release profile [20,21], good biocompatibility, and large drug loading capacity, which is largely depending on their tunable surface chemistry and particle size, uniform pore size, high surface area. Therefore, mesoporous silica nanoparticles have attracted great research attention as cargos for delivery and controlled release of various drugs. Tang and co-workers found that mesoporous silica nanospheres modified by hydrophobic groups showed enhanced hydrothermal stability and delayed release of hydrophobic drug [22]. Highly monodispersed mesoporous silica nanospheres are needed to control the delivery rate [23]. While the method of large-scale controllable production of uniform monodispersed spherical nanoparticles has been one of the research directions of scientists in recent years. Brinker and his co-workers succeeded in fabricating mesoporous silica nanospheres by developed aerosol-assisted evaporation-induced self-assembly (EISA) progress [24]. The monodispersed spherical silica nanoparticles can be obtained after a few seconds of EISA progress, which greatly improves the production efficiency. Various templates (CTAB, Brij-56, Brij-58, P123) have been used to control the pore size and mesostructures of silica [25]. Then, they synthesized protocells by fusion of lipid bilayers to the mesoporous silica nanospheres, and this structure was prominent for drug delivery [1,3,5,26–29]. However, mesoporous silica nanospheres made using a commercial atomizer (Model 3076, TSI, Inc., St Paul, MN, USA) have a wide size distribution (~50–1000 nm), requiring separation. Although the electrospray aerosol generator (Model 3480, TSI, Inc., St Paul, MN, USA) can produce high concentrations of monodisperse submicron particles with diameters ranging from 2 nm to 100 nm [30], its yield is too low (Liquid flow rate: 50 to 100 nL min^{-1}). Large-scale controllable production of MHSs with uniform particle size by aerosol-induced EISA is still a challenge.

Here, we report a facile synthesis method based on azeotropic distillation-induced self-assembly to prepare MHSs with good monodispersity and controllable uniform size. This method is easy to operate, and can solve the deficiencies in the synthesis of MHSs by aerosol-assisted EISA, such as wide size distribution, small specific surface area and pore volume. In addition, it can be used for the synthesis of many functional nanomaterials, and can be effectively applied to large-scale industrial production. In this report, the water, ethanol, benzene and CTAB form a stable reverse microemulsion under stirring and MTES has enough time to be hydrolyzed after adding to the system. The rising temperature causes the solvent to evaporate as an azeotrope, then the water phase and benzene will be condensed, the water phase can be separated by a water separator, and the benzene can reflux back to the reaction system. Each MHS can be formed from one single aqueous droplet in the microemulsion after evaporation and removal of water. The usual aerosol-assisted EISA is not able to synthesis of hydrophobic ordered porous silica with MTES, due to the fast evaporation and hydrolysis rate which cannot match with slow self-assembly rate of MTES. As a result, the relatively slow evaporation rate of azeotropic distillation is necessary for continuous self-assembly of MTES into highly ordered porous structures.

2. Results

2.1. Characterization of MHS Samples

Field emission scanning electron microscopy (FESEM) and transmission electron microscopy (TEM) images of the washed and dried MHS samples are shown in Figure 1. The MHS samples are uniform in size and spherical in shape (Figure 1a,b,d), and their particle size could be adjusted by the amount of CTAB/MTES mole ratio (Table 1). The particle size of MHS-1 was about 30 nm, while the particle size of MHS-2 is about 132 nm, because of the CTAB/MTES mole ratio of MHS-2 is lower. From the HRTEM image of MHS-1 (Figure 1c), CTAB produces particles exhibiting a highly ordered lamellar mesostructure.

To determine the pore ordering of the synthesized MHS samples, Small Angle X-ray Scattering (SAXS) analysis was performed from 0.8° to 12° (Figure 2). MHS-1 and MHS-2 showed sharp peaks at a low angle (2θ = 3.41°) and two weak peaks at higher angles (2θ = 6.81° and 10.18°) corresponding to the (100), (200), and (300) planes, respectively. The SAXS curve of MHS-1 shows that the washed and dried particles have an ordered lamellar structure. Compared to MHS-1, due to the decrease of CTAB/MTES mole ratio, it is difficult for silica-surfactant liquid-crystalline mesophase to grow in an orderly fashion in the process of self-assembly induced by azeotropic distillation, which leads the structural order of MHS-2 to decrease.

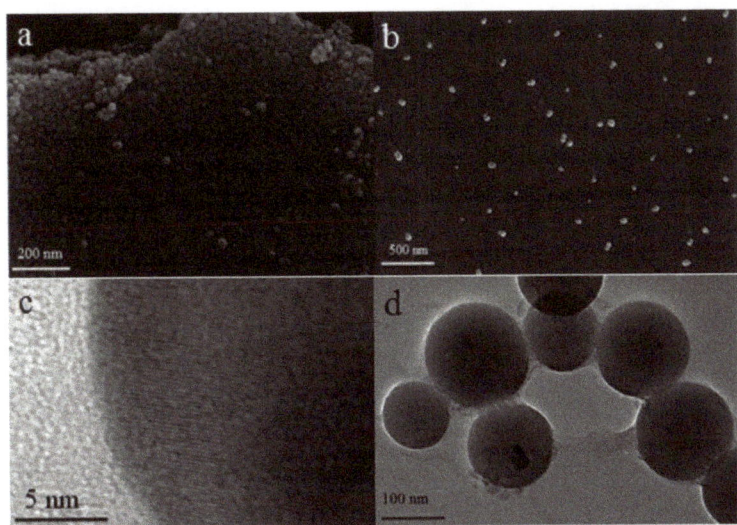

Figure 1. (**a,b**) FESEM images of MHS-1 sample (original magnification = 20,000×), (**c**) HRTEM image of MHS-1 sample, (**d**) TEM image of MHS-2 sample.

Table 1. Physicochemical characteristics of MHS samples [a].

Sample	C/M MR	MD (nm)	PV (cm^3 g^{-1})	SA$_{BET}$ (m^2 g^{-1})	PD (nm)
MHS-1	2.75	30	1.208	477.002	4.3110
MHS-2	0.183	132	0.654	257.452	12.4022

[a] C/M MR: CTAB/MTES mole ratio, MD: Mean diameter, PV: Pore volume, SA$_{BET}$: BET surface area, PD: Pore diameter.

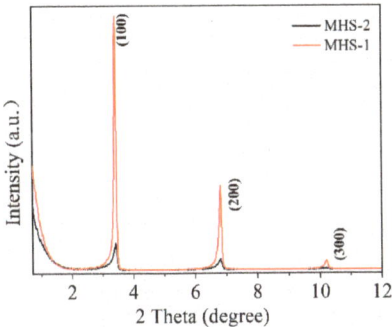

Figure 2. Small angle X-ray scattering (SAXS) curves of MHS-1 and MHS-2.

Figure 3 shows the N_2 adsorption–desorption isotherms and the pore size distribution curve for washed and dried MHS-1 and MHS-2 samples. The samples were outgassed at 250 °C for 10 h before measurements. The MHS-1 (Figure 3a) exhibited a typical type IV isotherm with a H4 hysteresis loop, giving a large pore volume (1.208 cm^3 g^{-1}) and a narrow pore size distribution (centered at 4.3110 nm) (Table 1). In Figure 3b, MHS-2 showed a type III isotherm with a H3 hysteresis loop, indicating that the pore size is non-uniform, and compared with MHS-1, its pore volume (0.654 cm^3 g^{-1}) and the surface area (257.452 m^2 g^{-1}) drastically decreased (Table 1). This was attributed to the fact that the CTAB/MTES mole ratio decreased, and the structural order of MHS-2 decreased. This is consistent with the results of SAXS analysis. All results mentioned above further demonstrate that MHS-1 with a smaller size, uniform pore size and higher specific area is more suitable for hydrophobic drug storage and release.

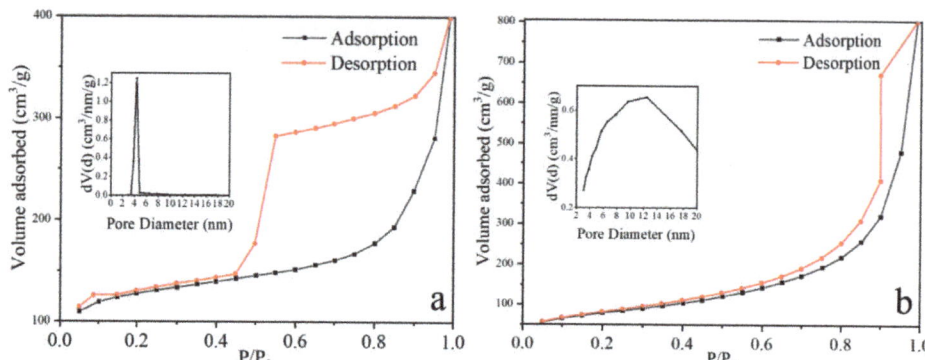

Figure 3. N_2 adsorption–desorption isotherms corresponding pore size distribution curve of (**a**) MHS-1 sample and (**b**) MHS-2 sample.

The surface properties of washed MHS-1 and MHS-2 samples with different heat treatment temperatures were measured by the contact angle test in Figure 4. Tablets of MHSs were prepared using a cylindrical stainless-steel die with a diameter of 1 cm. A pressure of 15 bar was applied for 10 min using a manual hydraulic press. The contact angle results of washed MHS-1 and MHS-2 dried at 60 °C were 121.2° and 120.5° (Figure 4a,c), which showed that the silica has good hydrophobic property. The contact angle of the particles increased to 126.3° and 129.9° (Figure 4b,d) after increasing drying temperature to 300 °C, indicating that the free water and hydroxyl groups on the surface of MHS samples decreased with the increase of temperature. The thermal, chemical and physical properties of MHS-1 and MHS-2 were measured at different atmosphere by simultaneous thermogravimetry and differential scanning calorimetry (TG-DSC) in N_2 (Figure 5a,c) or air (Figure 5b,d) at a constant heating rate of 10 °C min^{-1} in the temperature range between 30 °C and 800 °C. Compared with the data under N_2 conditions, one strong exothermic peak appeared at 420 °C under air conditions, suggesting oxidation of methyl groups by oxygen. This result demonstrated a high thermal stability for MHSs. The mesopores were covered by lipophilic-CH_3 groups, enabling curcumin molecules to easily enter and stay in the pores.

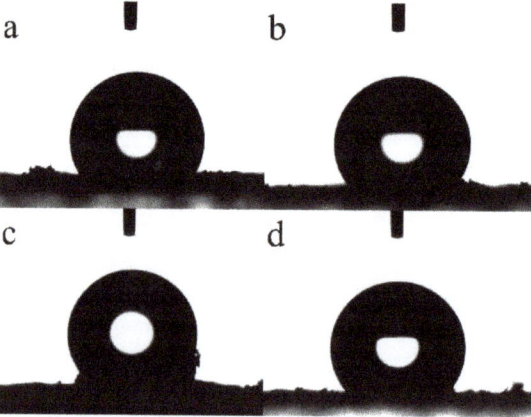

Figure 4. Contact angle measurements of MHS-1 and MHS-2 samples were dried at (**a,c**) 60 °C and (**b,d**) 300 °C.

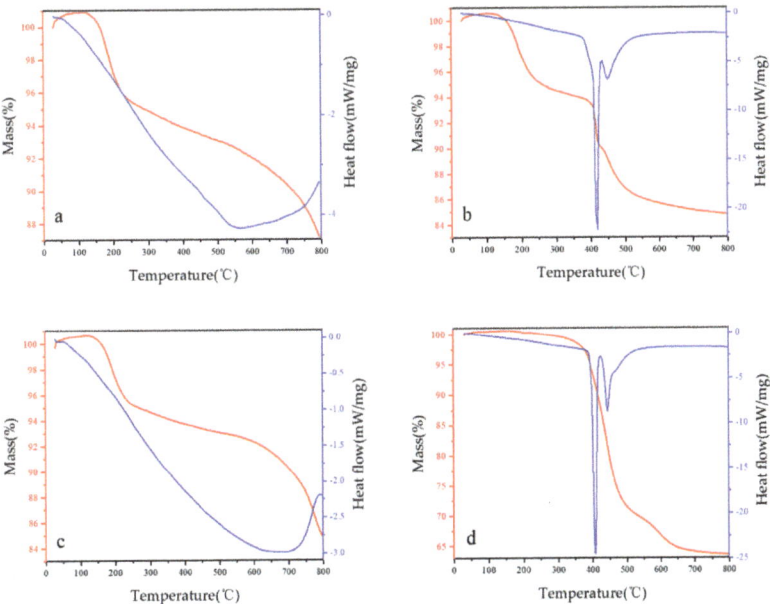

Figure 5. TG-DSC curves for MHS-1 and MHS-2 calcined under (**a,c**) N_2 or (**b,d**) air with a heating rate of 10 °C min^{-1}.

2.2. Adsorption and Release Experiment of Curcumin

After loading with curcumin, MHS-1 and MHS-2 were accordingly marked as MHSAC-1 and MHSAC-2, respectively. Figure 6 showed the FTIR spectra of MHS-1, MHSAC-1, MHS-2, MHSAC-2 and curcumin samples. The FTIR peaks in the ranges of 1026–1125 cm^{-1}, 772–801 cm^{-1} and 434–441 cm^{-1} correspond to the asymmetric, symmetric stretching and bending modes of the Si-O-Si [31], respectively (I and III in Figure 6). A sharp peak appearing at 2974 cm^{-1} can be assigned to -CH$_3$ group and an absorption peak at 1275 cm^{-1} belongs to Si–CH$_3$ stretching vibrations [31]. The FTIR peaks at 3468 cm^{-1} belong to OH groups of Si-OH on the surface of the MHS samples. In the FTIR spectra for curcumin (V in Figure 6), a sharp peak at 3511 cm^{-1} is assigned to the phenolic O-H stretching with a broad band at a range from 3100–3400 cm^{-1}, which is due to the -OH group (in enol form). The strong peak at 1627 cm^{-1} is associated with mixed C=O and C=C species of curcumin. Another strong band at 1603 cm^{-1} is attributed to the symmetric aromatic ring stretching vibrations C=C ring. The 1509 cm^{-1} peak is assigned to the C=O, and the C-O-C stretching peak of ether at 1027 cm^{-1} [32]. After curcumin was adsorbed, a new absorption band belonging to the heptadiene-dione chromophore group of curcumin appeared in the range of 1429–1627 cm^{-1} (II and IV in Figure 6). Other FTIR peaks belonging to MHS samples had no obvious shift. These results suggested that curcumin molecule had been adsorbed to hydrophobic silica [33]. It was also found that the absorption band intensity at 1429–1627 cm^{-1} of MHSAC-1 sample was much stronger than that of MHSAC-2 sample as a result of more curcumin loaded in MHSAC-1. All these results demonstrate that curcumin was successfully encapsulated in the as-synthesized MHS-1 samples.

Curcumin was encapsulated in hydrophobic mesopores by repeated heating and cooling of curcumin solution (V_{water}:$V_{ethanol}$ = 1:1) and MHS samples [34]. The remaining curcumin solutions after removal of MHSAC-1 and MHSAC-2, were accordingly marked as C-1 and C-2, respectively. Figure 7a shows the UV-vis spectra of 3.94 mg L^{-1} curcumin solution, C-1 and C-2. A sharp absorption peak of curcumin in a mixed solution of water and ethanol (V_{water}:$V_{ethanol}$ = 1:1) appeared at 432 nm. A series of curcumin solutions with different concentration were prepared, and their absorbance at 432 nm was measured to obtain the standard curve of curcumin solution (Figure 7b). After curcumin was adsorbed, the intensity of the absorption peaks for C-1 and C-2 decreased as a result of attachment of curcumin to the MHS-1 and MHS-2. C-1 exhibited a much lower absorption peak than C-2, indicating that higher adsorption capacity of MHS-1 due to larger specific area and pore volume under the same conditions. After calculation, the curcumin adsorption capacity of MHS-1 was 3.45 mg g^{-1}, while it was only 0.91 mg g^{-1} for MHS-2. It was proved that a mesoporous carrier with a high specific surface area, large pore volume, and appropriate pore size (larger than the kinetic diameter of the drug) would be beneficial for improving the adsorption capacity.

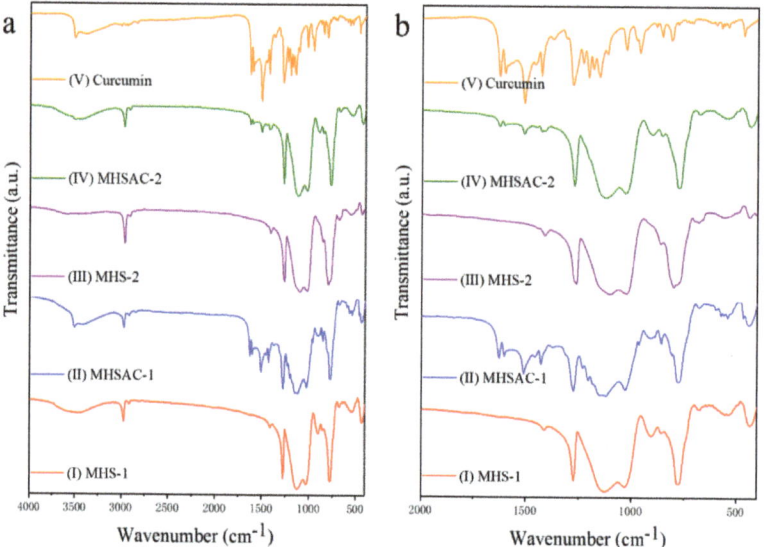

Figure 6. FTIR spectra for MHS-1, MHSAC-1, MHS-2, MHSAC-2 and curcumin: (**a**) range from 4000 cm^{-1} to 400 cm^{-1}, (**b**) range from 2000 cm^{-1} to 400 cm^{-1}.

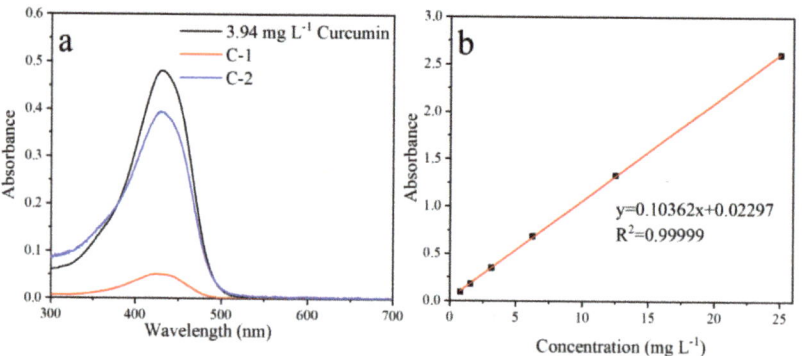

Figure 7. (**a**) UV-vis spectra of 3.94 mg L^{-1} curcumin mixed solution (V_{water}:$V_{ethanol}$ = 1:1), C-1 and C-2, (**b**) standard curve of curcumin.

As shown in Figure 8a, the in vitro sustained release process of curcumin was investigated by means of UV-vis spectroscopy. Curcumin was released from MHSAC-1 for 21 days in phosphate-buffered saline (PBS, water, pH = 7.4). Figure 8b shows the in vitro release kinetics of curcumin from MHSAC-1 samples in PBS (water, pH = 7.4), which was calculated according to the standard curve of curcumin in Figure 7b. As shown, 70.6%

of absorbed curcumin released slowly from MHSAC-1 samples in PBS, which lasted for 21 days. On about the sixth day, the net release rate of curcumin just reached 50% and then then rate of release became slow. It was important to extend the release time for practical controlled release. These results proved that the MHS nanospheres with a small size, larger specific area and pore volume (MHSAC-1) displayed a sustained release of curcumin and have great application potential in the study on the controllable slow release of hydrophobic drugs.

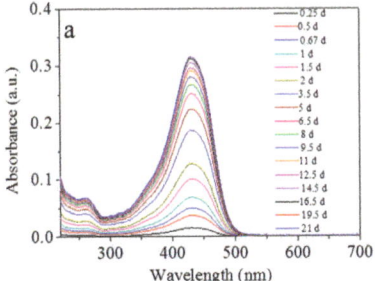
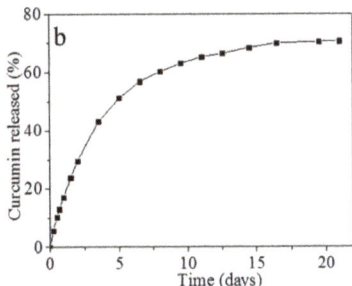

Figure 8. (a) UV-vis spectra of curcumin PBS (water, pH = 7.4) at different release times from MHSAC-1. (b) The curcumin release curve of MHSAC-1 samples in PBS (water, pH = 7.4).

3. Discussion

Usually, because the atomization process is uncontrollable, the diameter distribution of aerosol droplets obtained by commercial atomizers (such as Model 3076, TSI, Inc., St Paul, MN, USA) is wide. Therefore, it is difficult to achieve uniformity of as-synthesized particle size. Some scholars have added a microwave radiation zone at the back end of commercial atomizers to cause aerosol droplets to break up due to overheating under microwave radiation. Due to the surface tension of droplets, it will make the aerosol droplets in the carrier gas maintain the maximum total contact area and minimize the Gibbs free energy, so that the particle size of the droplets entering the drying zone tends to be uniform.

The self-assembly induced by azeotropic distillation is designed based on this principle. On the one hand, it has some similarities with aerosol-assisted self-assembly, beginning with a homogeneous solution of soluble silica and surfactant prepared in ethanol/water solvent with $c_0 \ll$ critical micelle concentration (cmc). CTAB as a stabilizer [35] can disperse the water phase into small droplets with uniform size in the oil phase. Additionally, due to the surface tension of the droplets, this stirred and heated system tends to maintain the maximum total contact area, which can cause the Gibbs free energy to be the lowest. The hydrolysis of MTES in the droplets is controlled, which limits the excessive growth of silica particles. With the slow separation of the water phase in the azeotrope and the reflux of benzene, the evaporation of the solvent creates a radial gradient of surfactant concentration from surface of each droplet to inside that steepens in time [36]. As the surfactant on the droplet surface first reaches the critical micelle concentration (cmc) [24], the ordered silica-surfactant liquid-crystalline mesophase grows radially inward from the surface. Finally, as the solvent continues to decrease, the silica-surfactant mesophase dries and shrinks to a sphere. For all the nanoparticles that can be synthesized by aerosol-assisted self-assembly, the self-assembly induced by azeotropic distillation can be applied, and the uniformity of particles can be ensured.

On the other hand, it is also different from aerosol-assisted self-assembly. Azeotropic distillation can adjust the evaporation rate of azeotrope by controlling the temperature of the system. In a typical synthesis, it takes about 0.5 h to get 0.5 mL of condensed azeotrope, and all ethanol and water in the system can be separated for as long as several hours (It only takes a few seconds for ethanol and water to evaporate in the process of aerosol-assisted self-assembly). This ensures that there is sufficient time for ordered self-assembly of MTES and CTAB liquid crystal mesophase. Therefore, this method can also be used to synthesize some nanoparticles with special morphology, such as cube shape [37] and rod shape. Of course, the self-assembly process induced by azeotropic distillation also has some shortcomings. The system is an inverse microemulsion system which depends on stirring to achieve uniform dispersion. Stirring speeds that are too violent or too slow may lead to irregular morphology of the final particles.

The MHSs prepared by self-assembly induced by azeotropic distillation show a good ability to control the sustained release of hydrophobic drugs. The release time at which the net release rate of curcumin reached 50% was extended to 6 days, which was much slower than curcumin-conjugated silica nanoparticles (7 h) [33], L- methionine encapsulated by hollow mesoporous silica nanoparticles (50 min) [34], and curcumin loaded by mesoporous silica nanoparticles (functionalized by 3-aminopropyltriethoxyorthosilane) (50 h) [38]. These results prove that this material can also be expected to be used in the encapsulation of other fat-soluble drugs, such as taxol, methotrexate, doxorubicin, etc.

4. Materials and Methods

4.1. Reagents and Instruments

All reagent-grade chemicals were used as received without further purification, and ultra nanopure distilled water (18.25 MΩ·cm) was used in all experiments. Curcumin, triethoxymethylsilane (MTES) and hexadecyl trimethyl ammonium bromide (CTAB) were purchased from Shanghai Aladdin Biochemical Technology Co., Ltd. (Shanghai, China). Ethanol, isopropyl alcohol, and benzene were bought from Sinopharm Chemical Reagent Co., Ltd. (Shanghai, China).

X-ray powder diffraction (XRD) patterns were recorded on a Rigaku Ultima IV powder diffractometer with a Cu Kα radiation source (λ = 1.5406 Å, 40 kV, 100 mA). Field emission scanning electron microscopy (FESEM, SIGMA Zeiss, Germany) and transmission electron microscopy (TEM, FEI Tecnai 30, 300 kV, Philips) were applied for characterization of the morphology of the samples. Fourier transform infrared spectroscopy (FTIR) were measured with a Tensor-II spectrometer (Bruker Co., Germany) by averaging 64 scans with a spatial resolution of 4 cm^{-1}. UV-vis absorption spectra were measured by Shimadzu UV-3600 UV-Vis-NIR spectrophotometer. The contact angles were measured by the XG-CAMA static contact angle tester. Thermal behavior of the samples was analyzed by thermogravimetry and differential scanning calorimetry (TG/DSC) (NETZSCH STA 409 PC, Germany). Labsys Evo simultaneous thermal analyzer was used to test the thermal stability of mesoporous hydrophobic silica. BET-surface area was measured by N_2 adsorption–desorption at liquid nitrogen temperature using an Autosorb-iQ2-MP (Quantachrome) gas sorption system. Specific surface areas were calculated using the Brunauer–Emmett–Teller (BET) model, and the pore size distributions were evaluated from the adsorption branches of the nitrogen isotherms using the Barrett–Joyner–Halenda (BJH) model.

4.2. Synthesis of MHS Samples

MHS samples were synthesized by azeotropic distillation-induced self-assembly, as defined in Scheme 1. In a typical synthesis, MHS-1 was prepared as follows: 1 g CTAB was dissolved in a solution of 7.5 mL deionized water, 18.5 mL ethanol and 74 mL benzene. The mixture was poured into a 250 mL three-necked flask mounted with a Dean-Stark trap followed by stirring at room temperature for 30 min to form a reverse microemulsion. Then 0.2 mL MTES was added into the above solution, which was kept at 45 °C for 2 h. With the hydrolysis of MTES molecules and formation of silanols, silica species became more hydrophilic and enriched in the aqueous droplets as a result of phase equilibrium. With constant hydrolysis of MTES and the enhanced hydrophilic of silica precursors, silica species continuously diffused from the benzene phase into water–ethanol droplets due to decreased solubility. Evaporation of water and ethanol from the droplets by azeotropic distillation at 64.9 °C (azeotrope composition: ethanol 18.5%, benzene 74% and water 7.5%) increased the concentrations of the silica species and CTAB, which led to self-assembly of the micelles into liquid crystalline mesophase. Meanwhile, hydrophilic inorganic precursors were also condensed into ordered porous silica with the liquid crystal as templates. The solution was then heated to 115 °C to remove the solvent. Subsequently, the samples were cooled to room temperature, collected and washed with a solution (deionized water and isopropyl alcohol, the volume ratio of 1: 1) to remove the CTAB, and further dried at 60 °C for 6 h. The preparation process for MHS-2 is the same as for MHS-1 except for adding 3 mL MTES instead of 0.2 mL MTES.

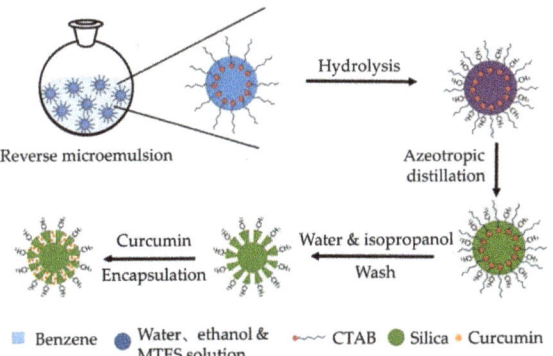

Scheme 1. Schematic diagram of the azeotropic distillation assisted route for formation of mesoporous hydrophobic silica nanospheres.

4.3. Adsorption and Release Experiment of Curcumin

MHSAC-1 was prepared as follows: 0.1 g MHS-1 was dissolved in 100 mL 3.94 mg L^{-1} curcumin solution (V_{water}:$V_{ethanol}$ = 1:1), followed by repeated heating and cooling several times until the color of the solution did not change. The samples were collected by centrifugation at 13000 rpm for 10 min and dried at 60 °C for 1 h. The preparation process for MHSAC-2 is the same as for MHSAC-1. The in vitro release kinetics of curcumin from MHSAC was as follows: Sixteen MHSAC samples of 0.01 g each were dissolved in 10 mL phosphate-buffered saline (PBS, water, pH 7.4). Then these samples were stirred at 100 rpm with a magnetic stirrer at 37 °C. After a period of time, the solution was centrifuged at 7000 rpm for 3 min to separate curcumin from the bottom of phosphate-buffered saline. The separated curcumin was dissolved in 10 mL mixed solution (V_{water}:$V_{ethanol}$ = 1:1, pH 7.0) then the UV-vis absorption spectra were recorded.

5. Conclusions

In summary, ordered mesoporous hydrophobic silica nanoparticles (MHSs) with a uniform size were successfully one-step synthesized by an azeotropic distillation-assisted method with MTES as precursor and CTAB as an ordered mesoporous template. The obtained MHSs exhibited high monodispersity, good sphericity, and large pore volume, with a highly ordered lamellar mesostructure, while the particle size can also be adjusted. This method solved the deficiencies in the synthesis of MHSs by aerosol-assisted EISA, such as wide size distribution, and small specific surface area and pore volume. Curcumin was successfully encapsulated in MHSs, and their adsorption capacity was 3.45 mg g^{-1}, greatly improving the stability of curcumin. The release time after which net release rate of curcumin reached 50% was extended to 6 days. Curcumin can be released slowly from MHSs, guaranteeing that curcumin has enough time to reach and inhibit cancer cells, bacteria, fungus, etc. MHSs also have great application potential in the study on the encapsulation of other hydrophobic drugs for drug delivery such as taxol, methotrexate, doxorubicin, etc.

6. Patents

There was a patent (Patent number: CN104876230B) resulting from the work reported in this manuscript and it was licensed.

Author Contributions: Conceptualization, L.C. and X.J.; methodology, L.C., X.F., M.L. and X.J.; investigation, L.C., X.F., M.L. and X.J.; software, resources and visualization, L.C. and X.F.; literature search, L.C. and X.F.; writing—original draft preparation, L.C. and X.F.; writing—review and editing, L.C., X.F., M.L. and X.J.; funding acquisition, M.L. and X.J. All authors have read and agreed to the published version of the manuscript.

Funding: This research was financially supported by the National Natural Science Foundation of China (NO.21878237) and the Graduate Innovative Fund of Wuhan Institute of Technology (NO. CX2020018).

Institutional Review Board Statement: Not applicable.

Informed Consent Statement: Not applicable.

Data Availability Statement: Data is contained within the article.

Acknowledgments: We record our sincere thanks for the measurement provided by Key Laboratory for Green Chemical Process of Ministry of Education, School of Chemical Engineering & Pharmacy, Wuhan Institute of Technology, Wuhan, China. In addition, also many thanks for the curcumin provided by Institute of Clinical Medicine, Taizhou People's Hospital Affiliated to Nantong University, Taizhou, Jiangsu 225300, China.

Conflicts of Interest: The authors declare no conflict of interest.

References

1. Ashley, C.E.; Carnes, E.C.; Phillips, G.K.; Padilla, D.; Durfee, P.N.; Brown, P.A.; Hanna, T.N.; Liu, J.; Phillips, B.; Carter, M.B.; et al. The targeted delivery of multicomponent cargos to cancer cells by nanoporous particle-supported lipid bilayers. *Nat. Mater.* **2011**, *10*, 389–397. [CrossRef] [PubMed]
2. Langer, R. Polymer-controlled drug delivery systems. *Acc. Chem. Res.* **1993**, *26*, 537–542. [CrossRef]
3. Liu, J.; Stace-Naughton, A.; Jiang, X.; Brinker, C.J. Porous nanoparticle supported lipid bilayers (protocells) as delivery vehicles. *J. Am. Chem. Soc.* **2009**, *131*, 1354–1355. [CrossRef]
4. Tarn, D.; Ashley, C.E.; Xue, M.; Carnes, E.C.; Zink, J.I.; Brinker, C.J. Mesoporous silica nanoparticle nanocarriers: Biofunctionality and biocompatibility. *Acc. Chem. Res.* **2013**, *46*, 792–801. [CrossRef] [PubMed]
5. Liu, J.; Stace-Naughton, A.; Brinker, C.J. Silica nanoparticle supported lipid bilayers for gene delivery. *Chem. Commun.* **2009**, 5100–5102. [CrossRef]
6. Zhang, S.; Li, J.; Lykotrafitis, G.; Bao, G.; Suresh, S. Size-dependent endocytosis of nanoparticles. *Adv. Mater.* **2009**, *21*, 419–424. [CrossRef]
7. Jiang, W.; Kim, B.Y.; Rutka, J.T.; Chan, W.C. Nanoparticle-mediated cellular response is size-dependent. *Nat. Nanotechnol.* **2008**, *3*, 145–150. [CrossRef]
8. Pan, Y.; Chen, Y.; Li, Q.; Yu, X.; Wang, J.; Zheng, J. The synthesis and evaluation of novel hydroxyl substituted chalcone analogs with in vitro anti-free radicals pharmacological activity and in vivo anti-oxidation activity in a free radical-injury Alzheimer's model. *Molecules* **2013**, *18*, 1693–1703. [CrossRef]
9. Hatcher, H.; Planalp, R.; Cho, J.; Torti, F.; Torti, S. Curcumin: From ancient medicine to current clinical trials. *Cell. Mol. Life Sci.* **2008**, *65*, 1631–1652. [CrossRef]
10. Liu, Z.; Tang, L.; Zou, P.; Zhang, Y.; Wang, Z.; Fang, Q.; Jiang, L.; Chen, G.; Xu, Z.; Zhang, H.; et al. Synthesis and biological evaluation of allylated and prenylated mono-carbonyl analogs of curcumin as anti-inflammatory agents. *Eur. J. Med. Chem.* **2014**, *74*, 671–682. [CrossRef]
11. Zorofchian Moghadamtousi, S.; Abdul Kadir, H.; Hassandarvish, P.; Tajik, H.; Abubakar, S.; Zandi, K. A review on antibacterial, antiviral, and antifungal activity of curcumin. *BioMed Res. Int.* **2014**, *2014*, 186864. [CrossRef] [PubMed]
12. Hewlings, S.J.; Kalman, D.S. Curcumin: A review of its effects on human health. *Foods* **2017**, *6*, 92. [CrossRef]
13. Praditya, D.; Kirchhoff, L.; Brüning, J.; Rachmawati, H.; Steinmann, J.; Steinmann, E. Anti-infective properties of the golden spice curcumin. *Front. Microbiol.* **2019**, *10*, 912. [CrossRef] [PubMed]
14. Rai, M.; Ingle, A.P.; Pandit, R.; Paralikar, P.; Anasane, N.; Santos, C.A.D. Curcumin and curcumin-loaded nanoparticles: Antipathogenic and antiparasitic activities. *Expert Rev. Anti-Infect. Ther.* **2020**, *18*, 367–379. [CrossRef] [PubMed]
15. Adamczak, A.; Ożarowski, M.; Karpiński, T.M. Curcumin, a natural antimicrobial agent with strain-specific activity. *Pharmaceuticals* **2020**, *13*, 153. [CrossRef] [PubMed]

16. Akbarzadeh, A.; Rezaei-Sadabady, R.; Davaran, S.; Joo, S.W.; Zarghami, N.; Hanifehpour, Y.; Samiei, M.; Kouhi, M.; Nejati-Koshki, K. Liposome: Classification, preparation, and applications. *Nanoscale Res. Lett.* **2013**, *8*, 102. [CrossRef]
17. Yoo, J.-W.; Irvine, D.J.; Discher, D.E.; Mitragotri, S. Bio-inspired, bioengineered and biomimetic drug delivery carriers. *Nat. Rev. Drug Discov.* **2011**, *10*, 521–535. [CrossRef]
18. Yallapu, M.M.; Gupta, B.K.; Jaggi, M.; Chauhan, S.C. Fabrication of curcumin encapsulated PLGA nanoparticles for improved therapeutic effects in metastatic cancer cells. *J. Colloid Interface Sci.* **2010**, *351*, 19–29. [CrossRef]
19. Yallapu, M.M.; Jaggi, M.; Chauhan, S.C. β-Cyclodextrin-curcumin self-assembly enhances curcumin delivery in prostate cancer cells. *Colloids Surf. B Biointerfaces* **2010**, *79*, 113–125. [CrossRef]
20. Choi, E.; Lu, J.; Tamanoi, F.; Zink, J.I. Drug Release from Three-Dimensional Cubic Mesoporous Silica Nanoparticles Controlled by Nanoimpellers. *Z. Anorg. Allg. Chem.* **2014**, *640*, 588–594. [CrossRef]
21. Li, Z.; Barnes, J.C.; Bosoy, A.; Stoddart, J.F.; Zink, J.I. Mesoporous silica nanoparticles in biomedical applications. *Chem. Soc. Rev.* **2012**, *41*, 2590–2605. [CrossRef] [PubMed]
22. Tang, Q.; Xu, Y.; Wu, D.; Sun, Y.; Wang, J.; Xu, J.; Deng, F. Studies on a new carrier of trimethylsilyl-modified mesoporous material for controlled drug delivery. *J. Control. Release* **2006**, *114*, 41–46. [CrossRef] [PubMed]
23. Rama Rao, G.V.; López, G.P.; Bravo, J.; Pham, H.; Datye, A.K.; Xu, H.F.; Ward, T.L. Monodisperse mesoporous silica microspheres formed by evaporation-induced self assembly of surfactant templates in aerosols. *Adv. Mater.* **2002**, *14*, 1301–1304. [CrossRef]
24. Brinker, C.J.; Lu, Y.; Sellinger, A.; Fan, H. Evaporation-induced self-assembly: Nanostructures made easy. *Adv. Mater.* **1999**, *11*, 579–585. [CrossRef]
25. Lu, Y.; Fan, H.; Stump, A.; Ward, T.L.; Rieker, T.; Brinker, C.J. Aerosol-assisted self-assembly of mesostructured spherical nanoparticles. *Nature* **1999**, *398*, 223–226. [CrossRef]
26. Durfee, P.N.; Lin, Y.-S.; Dunphy, D.R.; Muñiz, A.E.J.; Butler, K.S.; Humphrey, K.R.; Lokke, A.J.; Agola, J.O.; Chou, S.S.; Chen, I.-M.; et al. Mesoporous silica nanoparticle-supported lipid bilayers (protocells) for active targeting and delivery to individual leukemia cells. *ACS Nano* **2016**, *10*, 8325–8345. [CrossRef]
27. Butler, K.S.; Durfee, P.N.; Theron, C.; Ashley, C.E.; Carnes, E.C.; Brinker, C.J. Protocells: Modular mesoporous silica nanoparticle-supported lipid bilayers for drug delivery. *Small* **2016**, *12*, 2173–2185. [CrossRef]
28. Villegas, M.R.; Baeza, A.; Noureddine, A.; Durfee, P.N.; Butler, K.S.; Agola, J.O.; Brinker, C.J.; Vallet-Regí, M. Multifunctional protocells for enhanced penetration in 3D extracellular tumoral matrices. *Chem. Mater.* **2018**, *30*, 112–120. [CrossRef]
29. Dengler, E.C.; Liu, J.; Kerwin, A.; Torres, S.; Olcott, C.M.; Bowman, B.N.; Armijo, L.; Gentry, K.; Wilkerson, J.; Wallace, J.; et al. Mesoporous silica-supported lipid bilayers (protocells) for DNA cargo delivery to the spinal cord. *J. Control. Release* **2013**, *168*, 209–224. [CrossRef]
30. Kelly, J.T.; Asgharian, B.; Kimbell, J.S.; Wong, B.A. Particle deposition in human nasal airway replicas manufactured by different methods. Part II: Ultrafine particles. *Aerosol Sci. Technol.* **2004**, *38*, 1072–1079. [CrossRef]
31. Tao, C.; Yang, K.; Zou, X.; Yan, H.; Yuan, X.; Zhang, L.; Jiang, B. Double-layer tri-wavelength hydrophobic antireflective coatings derived from methylated silica nanoparticles and hybrid silica nanoparticles. *J. Sol-Gel Sci. Technol.* **2018**, *86*, 285–292. [CrossRef]
32. Hasan, M.; Messaoud, G.B.; Michaux, F.; Tamayol, A.; Kahn, C.J.; Belhaj, N.; Linder, M.; Arab-Tehrany, E. Chitosan-coated liposomes encapsulating curcumin: Study of lipid–polysaccharide interactions and nanovesicle behavior. *RSC Adv.* **2016**, *6*, 45290–45304. [CrossRef]
33. Gangwar, R.K.; Tomar, G.B.; Dhumale, V.A.; Zinjarde, S.; Sharma, R.B.; Datar, S. Curcumin conjugated silica nanoparticles for improving bioavailability and its anticancer applications. *J. Agric. Food Chem.* **2013**, *61*, 9632–9637. [CrossRef]
34. Jiang, X.; Ward, T.L.; Cheng, Y.-S.; Liu, J.; Brinker, C.J. Aerosol fabrication of hollow mesoporous silica nanoparticles and encapsulation of L-methionine as a candidate drug cargo. *Chem. Commun.* **2010**, *46*, 3019–3021. [CrossRef]
35. Chen, F.; Xu, G.-Q.; Hor, T.A. Preparation and assembly of colloidal gold nanoparticles in CTAB-stabilized reverse microemulsion. *Mater. Lett.* **2003**, *57*, 3282–3286. [CrossRef]
36. Jayanthi, G.; Zhang, S.; Messing, G.L. Modeling of solid particle formation during solution aerosol thermolysis: The evaporation stage. *Aerosol Sci. Technol.* **1993**, *19*, 478–490. [CrossRef]
37. Min, J.; Wang, F.; Cai, Y.; Liang, S.; Zhang, Z.; Jiang, X. Azeotropic distillation assisted fabrication of silver nanocages and their catalytic property for reduction of 4-nitrophenol. *Chem. Commun.* **2014**, *51*, 761–764. [CrossRef]
38. Bolouki, A.; Rashidi, L.; Vasheghani-Farahani, E.; Piravi-Vanak, Z. Study of mesoporous silica nanoparticles as nanocarriers for sustained release of curcumin. *Int. J. Nanosci. Nanotechnol.* **2015**, *11*, 139–146.

Article

Sensitization of Antibiotic-Resistant Gram-Negative Bacteria to Photodynamic Therapy via Perfluorocarbon Nanoemulsion

Peiyuan Niu †, Jialing Dai †, Zeyu Wang, Yueying Wang, Duxiang Feng, Yuanyuan Li * and Wenjun Miao *

School of Pharmaceutical Sciences, Nanjing Tech University, Nanjing 211816, China; niupy@njtech.edu.cn (P.N.); daijl@njtech.edu.cn (J.D.); wzyd@njtech.edu.cn (Z.W.); wangyueying@njtech.edu.cn (Y.W.); fengdx@njtech.edu.cn (D.F.)
* Correspondence: liyy@njtech.edu.cn (Y.L.); miaowj@njtech.edu.cn (W.M.); Tel.: +86-25-58139399 (W.M.)
† These authors contributed equally to this paper.

Abstract: With the merits of excellent efficacy, safety, and facile implementation, antibacterial photodynamic therapy (APDT) represents a promising means for treating bacterial infections. However, APDT shows an unsatisfactory efficacy in combating antibiotic-resistant Gram-negative bacteria due to their specific cell wall structure. In this work, we report a perfluorocarbon nanoemulsion (Ce6@FDC) used as a multifunctional nanocargo of photosensitizer and oxygen for sensitizing antibiotic-resistant Gram-negative bacteria to APDT. Ce6@FDC was fabricated via ultrasonic emulsification with good colloidal stability, efficient Ce6 and oxygen delivery, and excellent photodynamic activity. Meanwhile, Ce6@FDC could strongly bind with Gram-negative *Acinetobacter baumannii* (*A. baumannii*) and *Escherichia coli* (*E. coli*) via electrostatic interaction, thus leading to notable photodynamic bactericidal potency upon irradiation. In addition, oxygenated Ce6@FDC also exhibited a remarkable efficacy in eradicating Gram-negative bacteria biofilm, averaging five log units lower than the Ce6 group under identical conditions. Taken together, we demonstrate that photodynamic perfluorocarbon nanoemulsion with oxygen-delivery ability could effectively kill planktonic bacteria and remove biofilm, representing a novel strategy in fighting against antibiotic-resistant Gram-negative bacteria.

Keywords: photodynamic therapy; antibiotic-resistant; oxygen-delivery; Gram-negative bacteria; sensitization

1. Introduction

Antibiotics play an important role in the treatment of various pathogenic microbial infections. However, due to the abuse of antibiotics, the spread of antibiotic-resistant bacterial strains has arisen as one of the most worrying threats to public health in recent years, such as *E. coli*, *Staphylococcus aureus*, *Pseudomonas aeruginosa*, and *A. baumannii*, etc. [1–3]. These antibiotic-resistant bacteria could invalidate the primary treatment of traditional antibiotics through precise mutations, efflux pump expression, or up-regulation of defense-associated enzymes, and so on [4,5]. Although several dozen antibiotics are now in the R&D pipeline, most are modifications of existing antibiotic classes, for which progress is time-consuming and capital-demanding. Moreover, it is just a matter of time for microorganisms to develop resistance to these newly discovered antimicrobial agents, deriving from their single mode of action. Thus, the speed of new antibiotics development is far slower than that of multidrug-resistant bacterial evolution [6–8]. In addition, the formation of biofilm composed by Gram-negative bacteria and extracellular polymeric substances further results in high antibiotic resistance by restricting antibiotic penetration and inducing antibiotic inactivation [9,10]. Therefore, there is a pressing need to develop novel and effective therapies for combating bacterial infections and reducing the burden of antibacterial resistance via mechanisms different from those of traditional antibiotics in addition to the ongoing research to discover new antibiotics.

Antibacterial photodynamic therapy (APDT) is a promising alternative to antibiotic therapy because it induces few side effects, realizing precise treatment with outstanding biocompatibility [11,12]. APDT uses light at appropriate wavelengths to excite photosensitizers (PSs) to generate cytotoxic reactive oxygen species (ROS), predominantly singlet oxygen ($^1O_2$), to kill microbial pathogens. The generated $^1O_2$ can damage membrane lipids, proteins, and cellular DNA, which involves a variety of cell death mechanisms. Furthermore, bacteria can hardly develop resistance to APDT due to it causing irreparable damage to multiple targets in accordance with the positioning of PSs in bacteria, which means it has a highly effective bactericidal effect and eliminates bacteria completely before they can gain anti-oxidative damage mechanisms [11]. However, it has been reported that there is a basic distinction in the susceptibility to PDT for Gram-positive and Gram-negative bacteria because of their inherent cell wall structural differences. PSs can relatively easily penetrate through the cytoplasmic membrane of a Gram-positive bacterial cell by a relatively porous layer of peptidoglycan and lipoteichoic acid and thus effectively kill the bacteria upon light irradiation, whereas the outer membrane of Gram-negative bacteria forms a physically and functionally defensive barrier to numerous drug molecules, which greatly hampers the efficiency of APDT [13–15]. Worse still is the anoxic environment resulting from the insufficient O_2 supply, which is necessary in mediating the generation of cytotoxic ROS during APDT, further limiting the therapeutic effect of the oxygen-dependent PDT on Gram-negative bacteria.

Theoretically, the sensitivity of Gram-negative bacteria to PDT can be immensely improved via supplementing oxygen during the photodynamic process and relieving the anoxic or hypoxic environment. Recently, great efforts have been made to find a way out of this dilemma and to achieve a highly efficient APDT outcome for Gram-negative bacteria using organic or inorganic nanomaterials with oxygen- or $^1O_2$-generation functionalities [16,17]. For instance, transition metal oxides, especially nanostructured MnO_2, can react with endogenous hydrogen peroxide (H_2O_2) and H^+ to generate oxygen, which could relieve hypoxia circumstances. Nevertheless, the use of metal ions might raise potential concerns over toxicity [18–20]. Perfluorochemicals (PFCs), a kind of inert organic compound, have been extensively explored as blood substitutes for tissue oxygenation, owing to their high affinity to oxygen and excellent biocompatibility [21–24]. Thus, it is rational to hypothesize that photodynamic nanomedicine with oxygenation ability using PFCs might be preferable for the sensitizing of antibiotic-resistant Gram-negative bacteria to conventional APDT.

Inspired by the success of the oxygen-delivery strategy for the empowering of antibiotics and APDT [16,25], in this work, we propose an oxygen-affordable photodynamic nanomedicine (denoted Ce6@FDC) for the sensitizing of Gram-negative bacteria to APDT, especially clinically related drug-resistant strains (Figure 1). Among PFCs, perfluorodecalin (FDC), with its preeminent oxygen dissolving capacity and stability, was used here. Thus, Ce6@FDC, with the merits of low cost and ease of fabrication, is designed for oxygen delivery to increase the productivity of $^1O_2$ for mitigating or overcoming the inertia of Gram-negative bacteria to APDT. With a suitably positively charged surface and high oxygen-carrying capacity, Ce6@FDC is able to effectively associate with bacteria and release oxygen, thereby empowering the efficacy of APDT. We systemically evaluated the oxygen-delivery behavior and photodynamic effect of Ce6@FDC. Following that, the antibacterial potency of Ce6@FDC + O_2 on antibiotic-resistant *A. baumannii* and *E. coli* was investigated. By designing a late-model oxygen-delivery system, the sensitivity of Gram-negative bacteria to APDT could be enhanced comprehensively, which provides a strategic reference for the development of new antimicrobial agents.

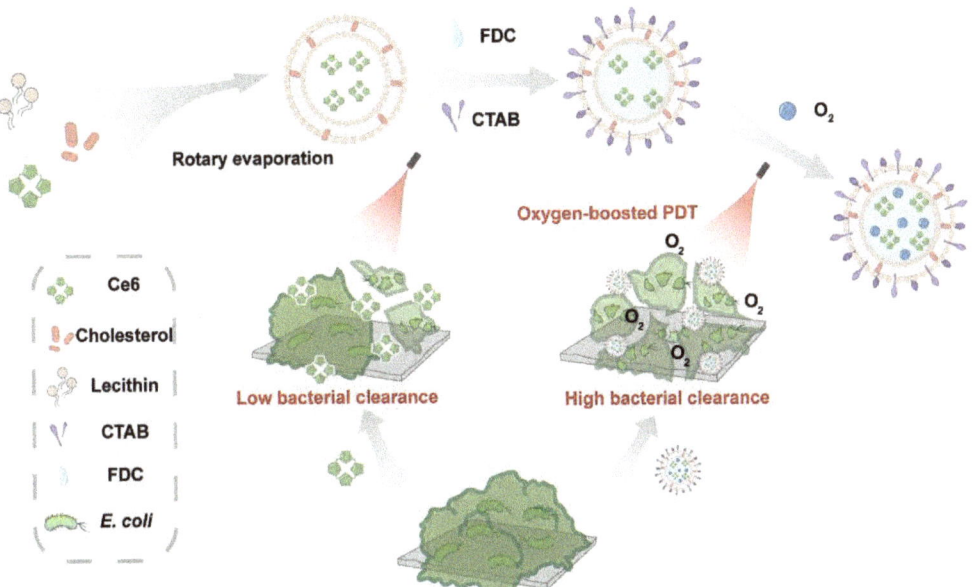

Figure 1. Schematic illustration for the fabrication process of Ce6@FDC nanoemulsion and its oxygen delivery for enhanced photodynamic antibacterial efficiency.

2. Results and Discussion

2.1. Preparation of Ce6@FDC

The Ce6@FDC nanoemulsion was fabricated through the conventional thin film hydration approach [26]. Various doses of Ce6 were added into the nanoemulsion and then the loading behavior of Ce6 in the nanoemulsion was investigated by a fluorescence spectrophotometer. As shown in Figure 2A,B, LC and EE was gradually elevated, which was proportional to the fed amounts of Ce6. The LC and EE reached 20.7% and 97.6% at 0.8 mg mL^{-1}, respectively. The zeta potential of the Ce6@FDC nanoemulsion was 35.4 ± 1.2 mV at 2 mg mL^{-1} of CTAB, displaying a good colloidal stability (Figure 2C). For maximizing oxygen loading, the impact of the FDC content on the hydrodynamic size of Ce6@FDC was also investigated. It was found that the size was gradually increased while increasing the FDC content, even up to 10%. The hydrodynamic diameter of the obtained Ce6@FDC was 113.5 ± 6.7 nm at 10% v/v of FDC with good colloidal stability (Figure 2D).

Compared with the blank FDC nanoemulsion, Ce6@FDC appeared to be dark greenish following Ce6 loading (Figure 3A). A TEM image clearly indicated the uniform spherical nanoemulsion formation. The average diameter of Ce6@FDC calculated from the TEM image was 87.5 ± 0.6 nm, which was slightly smaller than that from DLS measurement (Figure 3B). This might be attributed to shrinking of Ce6@FDC in a vacuum environment. As shown in Figure 3C, Ce6@FDC exhibited intense absorption in the visible and NIR regions with characteristic peaks at 400 nm and 660 nm originating from Ce6. Collectively, we have successfully fabricated Ce6@FDC for oxygen-affordable APDT.

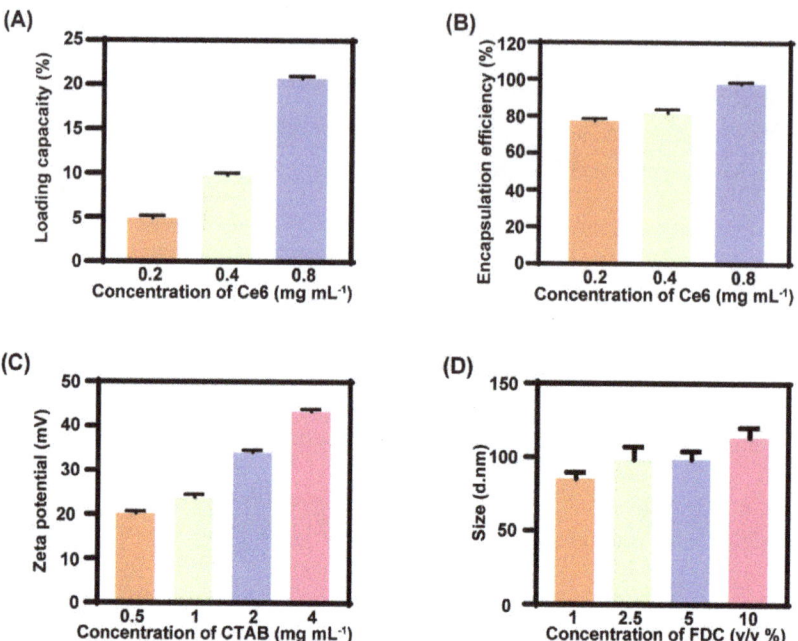

Figure 2. Formulation optimization of Ce6@FDC nanoemulsion. (**A**) Drug encapsulation efficiency and (**B**) loading capacity of Ce6@FDC with increasing feeding content of Ce6. (**C**) Zeta potentials of Ce6@FDC at different CTAB concentrations. (**D**) Hydrodynamic diameters of Ce6@FDC at different FDC concentrations. Data are shown as means ± S.D. (n = 3).

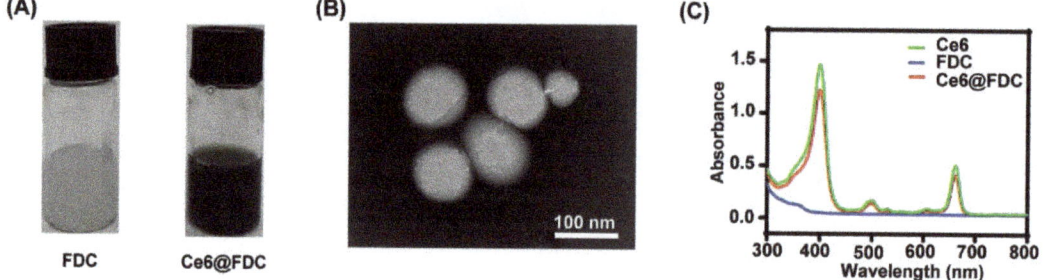

Figure 3. Characterization of Ce6@FDC nanoemulsion. (**A**) Photograph of FDC (left) and Ce6@FDC (right). (**B**) TEM image of Ce6@FDC. Scaler bar: 100 nm. (**C**) UV/Visible absorption spectra of Ce6, FDC, and Ce6@FDC.

2.2. Oxygen Loading and Releasing Behavior of Ce6@FDC

Owing to a strong oxygen-dissolving ability, FDC can be used as an excellent candidate for oxygen delivery to hypoxic biofilms. The oxygen content of Ce6@FDC + O_2 was 5.3 ± 0.7 mg mL^{-1} and 14.5 ± 0.5 mg mL^{-1} at 1% and 10% of FDC, respectively (Figure 4A), showing a positive correlation between its oxygen-loading capacity and the FDC content. Meanwhile, we found that the lipid content has no significant impact on the oxygen-loading capacity of the Ce6 nanoemulsion, which is approximately 3 mg mL^{-1} regardless of the lipid content, which was much lower than that of Ce6@FDC. This is likely due to the marginal oxygen-loading capacity of lipids. The oxygen-release behaviors of preoxygenated Ce6 nanoemulsion (Ce6 + O_2) and preoxygenated PBS (PBS + O_2) were similar when they were added into the deoxygenated DI water. In contrast, the preoxygenated Ce6@FDC (Ce6@FDC + O_2) exhibited notable oxygen release and the concentration of dissolved oxygen rapidly rose to 10.2 ± 0.5 mg mL^{-1} within 20 s in identical conditions (Figure 4B). Therefore, Ce6@FDC could serve as a preeminent nanocarrier to load and release oxygen in hypoxia regions, which can probably enhance the photodynamic effect for pathogen killing.

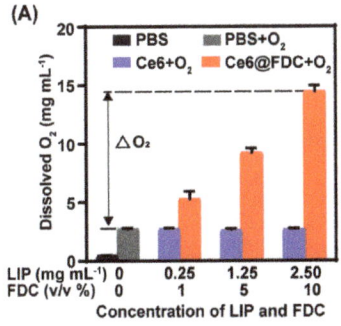 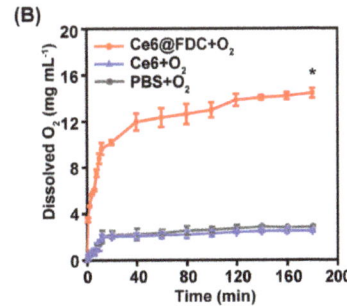

Figure 4. Oxygen-carrying and -releasing profile of Ce6@FDC. (**A**) The oxygen-loading capacity of Ce6@FDC. Dissolved oxygen concentration in PBS before and after adding Ce6@FDC at different concentrations. LIP: lipid. ΔO_2: enhanced O_2 concentration. (**B**) The oxygen-releasing behavior of Ce6 + O_2 or Ce6@FDC + O_2 in PBS. Data are shown as means ± S.D. (n = 3), * p < 0.05.

2.3. Enhanced Singlet Oxygen Generation

Next, the $^1O_2$-generation capacity of Ce6@FDC + O_2 was investigated using DPBF as a $^1O_2$ probe upon light irradiation (50 mW cm^{-2}). The bare $^1O_2$ production of PBS and Ce6 in deoxygenated conditions again verified the oxygen-dependence of PDT (Figure 5A,B). Meanwhile, we observed notable decay of DPBF absorption at 410 nm in the Ce6 + O_2 and Ce6@FDC + O_2 groups, implying their effective $^1O_2$-producing ability upon laser irradiation. However, Ce6@FDC + O_2 exhibited significantly higher DPBF consumption rates than those of Ce6 + O_2 and Ce6@FDC (Figure 5C–E). After irradiation for 60 s, the $^1O_2$-generation percentage of Ce6@FDC + O_2 achieved 80.0% vs. 21.1% in the case of Ce6 + O_2 (Figure 5F). It suggested that Ce6@FDC + O_2 would efficiently enhance the photosensitizing ability of Ce6 by oxygen affording, which is consistent with the results of previous oxygen-carrying PDTs [20,21,27].

2.4. Facilitated Bacteria Association of Ce6@FDC

To investigate the interaction between bacteria and Ce6@FDC, flow cytometry analysis was performed for *A. baumannii* after treatment with PBS, free Ce6, and Ce6@FDC. As shown in Figure 6A, free Ce6 was scarcely taken up or adsorbed by *A. baumannii*, exhibiting no difference with the PBS group. In contrast, Ce6@FDC displayed notably improved association with *A. baumannii*. Quantitative analysis revealed that the fluorescence intensity of Ce6@FDC-treated *A. baumannii* was 3.3 folds higher than those treated with free Ce6 (Figure 6B). Not surprisingly, *E. coli* showed a similar association profile with Ce6@FDC as *A. baumannii* (data not shown). Zeta potential analysis revealed that free-Ce6-treated bacteria remained negative (−12.4 mV for *E. coli*; −13.2 mV for *A. baumannii*), as did the PBS-treated groups. However, the values of *E. coli* and *A. baumannii* treated with Ce6@FDC were remarkably increased to 23.7 mV and 25.63 mV, respectively, displaying a positive charge surface of bacteria (Figure 6C). These results clearly demonstrated that Ce6@FDC could strongly bind with bacteria and alter their surface properties and potently deliver Ce6 and oxygen, which is greatly advantageous for efficient PDT against antibiotic-resistant Gram-negative bacteria.

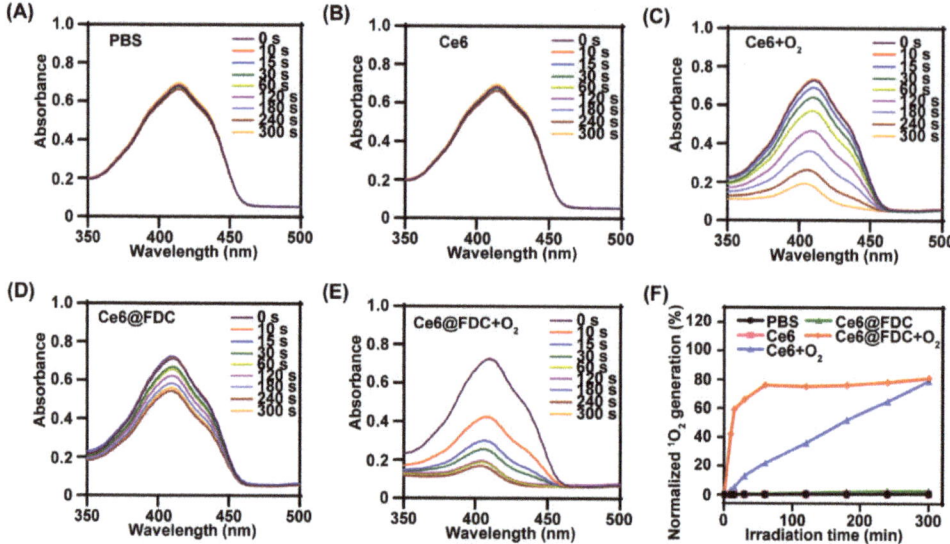

Figure 5. Light-triggered $^1O_2$-generation of Ce6@FDC + O_2 nanoemulsion. PBS (**A**), Ce6 (**B**), Ce6 + O_2 (**C**), Ce6@FDC (**D**), and Ce6@FDC + O_2 (**E**) were irradiated with 660 nm light at 50 mW cm^{-2} for 3 min in the presence of DPBF (100 µM) and their UV/vis spectra were monitored. (**F**) $^1O_2$-generation percentage according to the absorbance reduction at 410 nm.

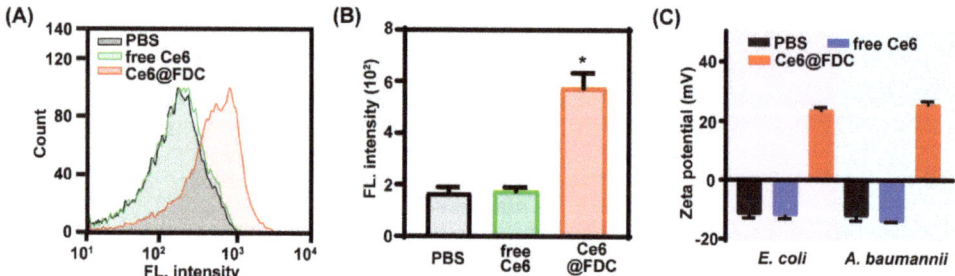

Figure 6. Enhanced bacterial association of Ce6@FDC nanoemulsion. (**A**) Flow cytometry analysis of *A. baumannii* treated with free Ce6 or Ce6@FDC (10 µg mL^{-1} of Ce6) for 30 min. (**B**) Corresponding fluorescence intensities of *A. baumannii* that received the indicated treatments. (**C**) Zeta potentials of *E. coli* and *A. baumannii* treated with PBS, free Ce6, or FDC@Ce6. Data are shown as means ± S.D. (n = 5), * $p < 0.05$, compared with free Ce6 group.

2.5. Photodynamic Bactericidal Performance against Antibiotic-Resistant Gram-Negative Bacteria

We next evaluated the photodynamic bactericidal effect of Ce6@FDC + O_2 using the conventional spread plate method. The number of viable bacteria gradually reduced with increasing the concentration of nanoemulsion in each group, displaying a dose-dependent manner both in *A. baumannii* and *E. coli*. Almost no colonies were found on the agar plate when *A. baumannii* and *E. coli* were treated with Ce6@FDC + O_2 (60 µg mL^{-1} of Ce6) plus light irradiation (100 mW cm^{-2} for 30 min), which was not observed in the Ce6 + O_2 or Ce6@FDC groups under identical conditions (Figure 7A,B). For quantitative evaluation, a log of survived bacteria was calculated according to the CFU values and shown in Figure 7C,D. When *A. baumannii* was treated with Ce6@FDC + O_2 (60 µg mL^{-1} of Ce6) plus light, the reduction in the log unit was 4.8, notably higher than 2.7 in the Ce6@FDC plus light group. Meanwhile, the analogical bactericidal effect for *E. coli* post treatments was also observed. Moreover, we observed that both the *A. baumannii* and *E. coli* treated with Ce6@FDC plus light appeared to have multiple lesions and holes, implying severe damage of the bacterial wall by oxygen-affording APDT (Figure S1, Supplementary Materials). The results demonstrated that Ce6@FDC + O_2 exhibited potent photodynamic antibacterial efficacy in combating Gram-negative bacteria via oxygen-affording and enhanced bacteria association.

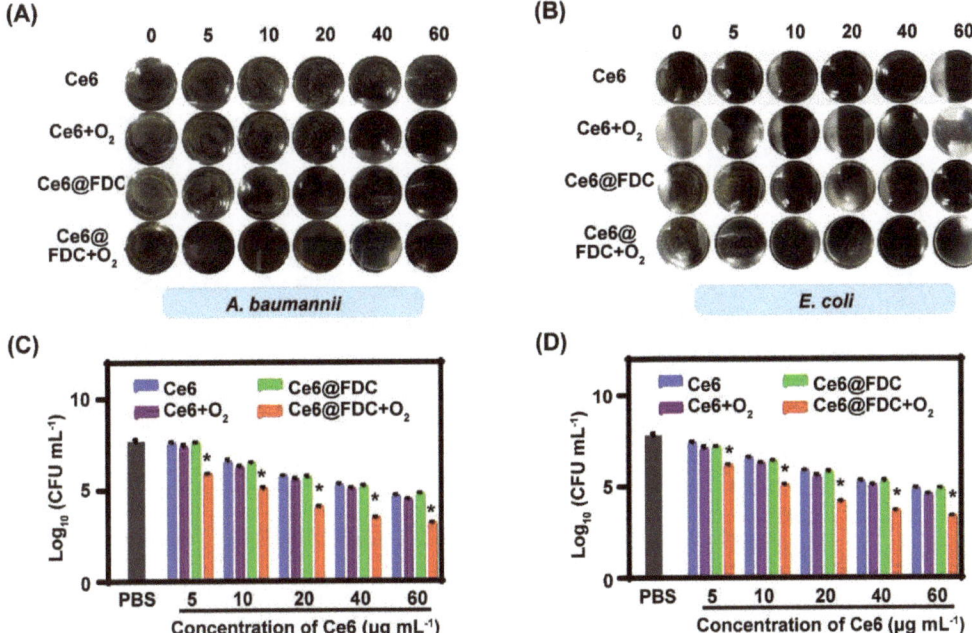

Figure 7. In vitro photodynamic antibacterial activity against Gram-negative strains. Photographs of plate samples of *A. baumannii* (**A**) and *E. coli* (**B**) incubated with Ce6, Ce6 + O_2, Ce6@FDC, and Ce6@FDC + O_2 at serial concentrations and then irradiated with a 660 nm laser (100 mW cm^{-2}, 20 min). Survived numbers of *A. baumannii* (**C**) and *E. coli* (**D**) that received various treatments as indicated. Data are shown as mean ± S.D. (n = 3), * $p < 0.05$ compared to Ce6 + O_2 and Ce6@FDC groups.

2.6. Potent Biofilm Ablation Ability In Vitro

Encouraged by the excellent oxygen-loading capacity and photodynamic antibacterial ability of Ce6@FDC + O_2, it is very interesting to know whether its antibiofilm performance can be enhanced. Under light irradiation, Ce6- or Ce6 + O_2-treated biofilms appeared to be purple both in the *A. baumannii* and *E. coli* biofilms without notable differences, whereas the color of the Ce6@FDC + O_2-treated ones was obviously lighter. Furthermore, we found that the structure of the biofilm in the Ce6@FDC + O_2 group was severely damaged, likely due to oxygen release from the nanoemulsion (Figure 8A,B). The treated biofilms were dispersed and the survived bacteria within the biofilm were quantitated by CFU counting. The log was calculated according to the CFU values and shown in Figure 8C,D. With light irradiation, a negligible antibiofilm performance of Ce6, Ce6 + O_2, and Ce6@FDC was found at concentrations of Ce6 from 5 to 60 μg mL^{-1}. Comparatively, Ce6@FDC + O_2 showed markedly dose-dependent antibiofilm profiles and these were much stronger than for the other groups under identical conditions. When treated with Ce6@FDC + O_2 (60 μg mL^{-1} of Ce6) plus light, the log value of bacteria dropped to 2.3 and 2.1 for

A. baumannii and *E. coli* biofilms, respectively, averaging five log units lower than the values for the Ce6@FDC plus light group. Taken together, Ce6@FDC + O_2 could release oxygen after efficient association with bacteria in the biofilm, leading to the sensitization of antibiotic-resistant Gram-negative bacteria biofilm to APDT.

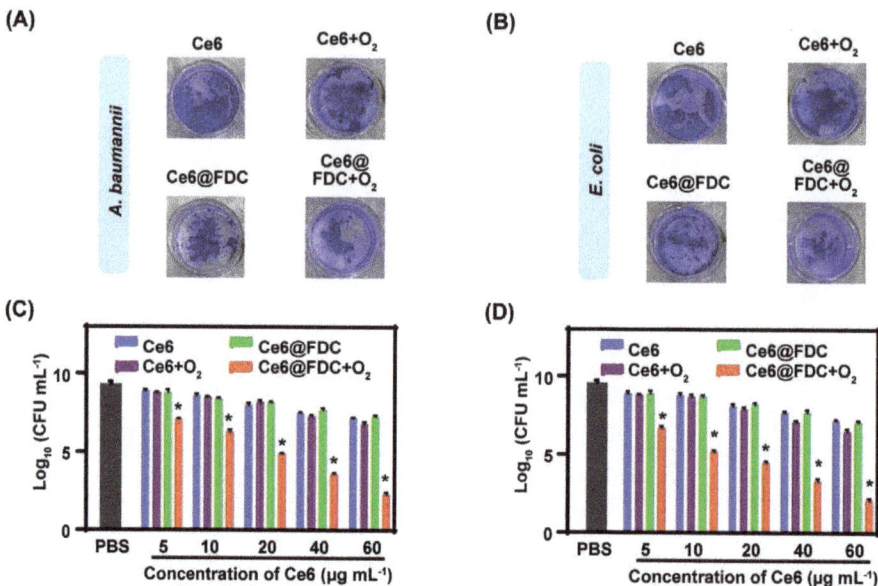

Figure 8. In vitro photodynamic biofilm eradiation by Ce6@FDC + O_2 nanoemulsion. Crystal violet-stained *A. baumannii* (**A**) and *E. coli* (**B**) biofilms having received the indicated treatment for 30 min followed by irradiation (660 nm, 100 mW cm^{-2} for 20 min). Corresponding log values of viable *A. baumannii* (**C**) and *E. coli* (**D**) cells in biofilm residues by colony forming unit counting. Data are shown as means ± S.D. (n = 3), * $p < 0.05$ compared to Ce6 + O_2 and Ce6@FDC groups.

3. Materials and Methods

3.1. Materials

Chlorin e6 (Ce6) was purchased from Frontier Scientific Inc. (West Logan, UT, USA). 1,3-diphenylisobenzofuran (DPBF) was obtained from Sigma-Aldrich (St. Louis, MO, USA). Perfluorodecalin (FDC), lecithin, and cholesterol were purchased from Aladdin (Shanghai, China). Yeast extract, tryptone, agar, and crystal violet (CV) were purchased from Sinopharm Chemical Reagent Co., Ltd. (Shanghai, China). Distilled water (DW) used in the experiments was made from a Milli-Q Direct 16 Water Purification System (Millipore Corporation, Bedford, MA, USA) with resistivity higher than 18.2 MΩ cm^{-1}. Other chemicals were supplied by Sinopharm Chemical Reagent Co., Ltd., China, and used as received. *E. coli* was purchased from Guangdong Microbial Culture Collection Center (Guangzhou, China) and *A. baumannii* was kindly supplied by Dr. Yishan Zheng at The Second Hospital of Nanjing.

3.2. Preparation of Ce6@FDC Nanoemulsion

Ce6@FDC nanoemulsion was fabricated by ultrasonic emulsification according to previous literature [16,21]. In brief, Ce6 was dissolved in methanol and lipid (lecithin: cholesterol = 4:1, w/w) was dissolved in chloroform at concentrations of 0.8 mg mL^{-1} and 2.5 mg mL^{-1}, respectively, and then mixed at different mass ratios. The resultant mixture was stirred for 10 min at room temperature and the remaining solvent was completely removed by rotary evaporation. Following that, 5 mL of hexadecyltrimethylammonium bromide (CTAB, 0.02% w/w in DW) and FDC (500 µL) were added to the mixture in sequence during sonication. Free Ce6 was removed by gel filtration using a dextran gel column (PD midi Trap G-25, GE Healthcare, UK). Purified Ce6@FDC nanoemulsion was obtained and stored at 4 °C for further use. Ce6 emulsion and FDC emulsion were prepared without FDC or Ce6, respectively. Finally, oxygenated Ce6@FDC, designated Ce6@FDC + O_2, was obtained by oxygen bubbling for 20 min.

3.3. Characterization

The morphology and size of Ce6@FDC nanoemulsion were observed using transmission electron microscopy (TEM) using a JSM-2100 electron microscope (JEOL, Japan) after staining with 1% phosphotungstic acid. The hydrodynamic diameters were analyzed by dynamic light scattering (DLS, Malvern Instruments, Malvern, UK) with a 10 mW He−Ne laser at 25 °C, and zeta potential values were determined by laser doppler microelectrophoresis at an angle of 22 °C using a Nano ZS90 Zetasizer (Malvern Instruments, UK). UV/vis absorption spectrum was acquired via microplate spectrophotometer (Multiskan GO, Thermo Fisher Scientific, Waltham, MA, USA). Fluorescence spectrophotometer (F7000, Hitachi, Japan) was used to acquire the fluorescence spectra. The content of Ce6 encapsulated in Ce6@FDC was calculated by absorbance at 660 nm according to the established calibration curve. The loading capacity (LC) and encapsulation efficiency (EE) of Ce6@FDC were calculated by the following equations: LC (%) = ((weight of loaded drug) /(total weight of emulsion)) × 100.; EE (%) = ((weight of loaded drug)/(weight of initially added drug)) × 100.

3.4. Oxygen Loading and Releasing of Ce6@FDC

For quantification of the oxygen-loading capacity, preoxygenated Ce6 and Ce6@FDC were added into deoxygenated phosphate buffered saline (PBS, 0.15 M, pH 7.4), and then the dissolved oxygen concentration of each sample was measured by portable dissolved oxygen meter at the time of oxygen equilibrium (~180 s, REX, JPF-605B, China). Meanwhile, to explore the oxygen-release behavior, Ce6 or Ce6@FDC saturated with oxygen was added into the deoxygenated PBS. The oxygen-release behavior was recorded by oxygen meter in real time. The oxygen releasing profile of oxygen-saturated PBS was measured as a negative control.

3.5. Light-Triggered $^1O_2$-Generation Ability of Ce6@FDC + O_2

Light-triggered $^1O_2$-generation potency of Ce6@FDC + O_2 nanoemulsion was quantitated using 1,3-diphenylisobenzofuran (DPBF) as probe [11]. For that, Ce6, Ce6 + O_2, Ce6@FDC, or Ce6@FDC + O_2 (5 µg mL^{-1} of Ce6) was mixed with deoxygenated dimethyl formamide (DMF) containing DPBF (100 µM) and irradiated with a 660 nm continuous wave diode laser beam (50 mW cm^{-2}, Rayan Tech., Changchun, China) for 3 min. The generated $^1O_2$ was detected with DPBF every 20 s by quantitating the reduction of its absorption at 410 nm via microplate spectrophotometer. DMF containing DPBF and deoxygenated PBS was used as a negative control.

3.6. Bacterial Association of Ce6@FDC

The interaction between Ce6@FDC nanoemulsion and bacteria was evaluated by flow cytometry. Briefly, E. coli or A. baumannii were cultured in Luria−Bertani (LB) broth medium at 37 °C with shaking overnight and harvested at the exponential growth phase. After being washed with PBS three times, bacteria were resuspended in PBS at 1×10^8 CFU mL^{-1}. Following that, bacteria suspension was treated with Ce6 or Ce6@FDC (10 µg mL^{-1}, equivalent to Ce6) at 37 °C for 30 min and washed with cold PBS for semiquantitative analysis by NovoCyte 2060R flow cytometry and ACEA NovoExpress software (ACEA Biosciences inc., San Diego, CA, USA). Meanwhile, the zeta potential of bacteria treated as above was measured by Zetasizer.

3.7. In Vitro Photodynamic Antibacterial Activity

The photodynamic bactericidal activity of Ce6@FDC + O_2 nanoemulsion against two Gram-negative strains, *E. coli* and *A. baumannii*, was evaluated by colony counting method. In short, bacteria (10^8 CFU mL^{-1}) were incubated with Ce6, Ce6 + O_2, Ce6@FDC, or Ce6@FDC + O_2 (concentration of Ce6 ranging from 0 to 60 μg mL^{-1}) at 37 °C for 30 min in dark and then exposed to 660 nm laser at an intensity of 100 mW cm^{-2} for 20 min. Finally, 100 μL of bacteria suspension was diluted several times and spread onto LB agar plate and grown at 37 °C overnight. PBS was set as the negative control. All experiments were independently performed in triplicate.

The morphology of Ce6@FDC + O_2-treated bacteria was observed by field emission scanning electron microscopy (FESEM). In brief, *A. baumannii* or *E. coli* suspension (10^8 CFU mL^{-1}) was incubated with Ce6@FDC + O_2 (20 μg mL^{-1} of Ce6) and irradiated as indicated above. Then bacteria were fixed with 2.5% glutaric dialdehyde for 2 h. After washing with PBS, samples were dehydrated by graded ethanol (30%, 50%, 70%, 80%, 90%, and 100% for 10 min each). The fixed bacteria were coated with gold and imaged with FESEM (JEOL, Osaka, Japan).

3.8. In Vitro Ablation Capacity of Bacterial Biofilm

The performance of biofilm ablation was assessed via crystal violet (CV) staining assay and plate counting assay. For biofilm construction, 10 μL of *E. coli* or *A. baumannii* (10^8 CFU mL^{-1}) were added to culture dishes supplemented with 1 mL of fresh LB + 1% sucrose solid medium in 24-well plates and incubated at 37 °C under static conditions for 24 h to form fresh biofilms [28]. Then, the formed biofilms were gently washed with PBS and treated with Ce6, Ce6 + O_2, Ce6@FDC, or Ce6@FDC + O_2 (concentration of Ce6 ranging from 0 to 60 μg mL^{-1}) at 37 °C for 30 min followed by irradiation (660 nm, 100 mW cm^{-2}) for 20 min. Biofilms subjected to PBS treatment as above were used as negative control. The viable bacteria in the biofilms were detached into sterile PBS via vortexing. Obtained bacteria suspension was diluted and spread onto LB agar plates and then incubated at 37 °C for 12 h. The CFU on different plates were imaged and counted.

Meanwhile, CV staining assay was carried out for quantitative analysis of the ablation effect of Ce6@FDC + O_2 nanoemulsion on *E. coli* or *A. baumannii* biofilm. Biofilms subjected to various treatments as above were washed with cold PBS and subsequently stained with CV at 37 °C for 30 min. The stained residual biofilms were photographed and resuspended with acetic acid (33%, w/v), and then the absorbance of supernatant was measured at 595 nm for quantitative analysis. All experiments were independently performed in triplicate.

3.9. Statistics Analysis

ANOVA was used to analyze all data with a Student−Newman−Keuls test for post hoc pairwise comparisons. All statistical analyses were performed using the GraphPad Prism software (version 8, Prism Software, Beijing, China). Differences with p-values < 0.05 were considered statistically significant.

4. Conclusions

In summary, we have successfully fabricated a photodynamic perfluorocarbon nanoemulsion, Ce6@FDC, which can simultaneously deliver photosensitizer and oxygen for the sensitization of antibiotic-resistant Gram-negative bacteria to APDT. Beyond the superior colloidal stability, Ce6@FDC also exhibits remarkable oxygen loading and releasing potency in an aqueous environment and excellent light-triggered singlet oxygen generation. Furthermore, we also demonstrated that Ce6@FDC could strongly bind with bacteria and alter their surface properties due to its positively charged surface, resulting in great bactericidal activity in combating antibiotic-resistant *A. baumannii* and *E. coli* compared with free Ce6. Ce6@FDC with oxygen loading also displayed a notable performance in the photodynamic eradication of Gram-negative bacteria biofilm. Therefore, Ce6@FDC, as a novel bactericidal agent enabling efficient delivery of Ce6 and oxygen, and enhanced APDT provide a new strategy for killing antibiotic-resistant Gram-negative bacteria. In future, the formulation of such a nanoemulsion needs to be further optimized for maximizing the safety and pathogen targetability, and its in vivo therapeutical efficacy is also yet to be investigated.

Supplementary Materials: The following are available online at https://www.mdpi.com/article/10.3390/ph15020156/s1, Figure S1. Field emission scanning electron microscopy images of bacteria after being treated with PBS or Ce6@FDC + O_2 plus light (660 nm, 100 mW, 20 min). Red arrows indicate lesions and holes on bacterial wall. Scale bar: 1 µm.

Author Contributions: Conceptualization, P.N., J.D. and W.M.; formal analysis, P.N., Z.W. and D.F.; investigation, P.N. and Y.W.; writing—original draft, P.N. and W.M.; writing—review and editing, Y.L. and W.M.; funding acquisition, W.M.; methodology, P.N. and Y.L.; supervision, W.M. All authors have read and agreed to the published version of this manuscript.

Funding: This research was supported by research grants from the National Natural Science Foundation of China (project number: 51603101) and the Postgraduate Research and Practice Innovation Program of Jiangsu Province for project SJCX21_0545.

Institutional Review Board Statement: Not applicable.

Informed Consent Statement: Not applicable.

Data Availability Statement: Data is contained within the article and supplementary material.

Conflicts of Interest: The authors declare no competing financial interests.

References

1. Timsit, J.F.; Bassetti, M.; Cremer, O.; Daikos, G.; Waele, J.; Kallil, A.; Kipnis, E.; Kollef, M.; Laupland, K.; Paiva, J.; et al. Rationalizing Antimicrobial Therapy in the ICU: A Narrative Review. *Intens. Care Med.* **2019**, *45*, 172–189. [CrossRef] [PubMed]
2. Hu, D.; Zou, L.; Gao, Y.; Jin, Q.; Ji, J. Emerging Nanobiomaterials Against Bacterial Infections in Postantibiotic Era. *View* **2020**, *1*, 20200014. [CrossRef]
3. Bassetti, M.; Poulakou, G.; Ruppe, E.; Bouza, E.; Van Hal, S.J.; Brink, A. Antimicrobial Resistance in the Next 30 Years, Humankind, Bugs and Drugs: A Visionary Approach. *Intens. Care Med.* **2017**, *43*, 1464–1475. [CrossRef] [PubMed]
4. Wang, Y.; Yang, Y.; Shi, Y.; Song, H.; Yu, C. Antibiotic-Free Antibacterial Strategies Enabled by Nanomaterials: Progress and Perspectives. *Adv. Mater.* **2019**, *32*, 1904106. [CrossRef]
5. Zhao, Y.; Chen, L.; Wang, Y.; Song, X.; Li, K.; Yan, X.; Yu, L.; He, Z. Nanomaterial-based Strategies in Antimicrobial Applications: Progress and Perspectives. *Nano Res.* **2021**, *14*, 4417–4441. [CrossRef]
6. Tacconelli, E.; Carrara, E.; Savoldi, A.; Harbarth, S.; Mendelson, M.; Monnet, D.L.; Celine, P.; Gunnar, K.; Jan, K.; Yehuda, C.; et al. Discovery, Research, and Development of New Antibiotics: The WHO Priority List of Antibiotic-resistant Bacteria and Tuberculosis. *Lancet Infect. Dis.* **2018**, *18*, 318–327. [CrossRef]
7. Fjell, C.D.; Hiss, J.A.; Hancock, R.E.W.; Schneider, G. Designing Antimicrobial Peptides: Form Follows Function. *Nat. Rev. Drug Discov.* **2012**, *11*, 37–51. [CrossRef]
8. Chen, H.; Jin, Y.; Wang, J.; Wang, Y.; Jiang, W.; Dai, H.; Pang, S.; Lei, L.; Ji, J.; Wang, B. Design of Smart Targeted and Responsive Drug Delivery Systems with Enhanced Antibacterial Properties. *Nanoscale* **2018**, *10*, 20946–20962. [CrossRef]
9. Ermini, M.L.; Voliani, V. Antimicrobial Nano-agents: The Copper Age. *ACS Nano* **2021**, *15*, 6008–6029. [CrossRef]
10. Hu, D.; Deng, Y.; Jia, F.; Jin, Q.; Ji, J. Surface Charge Switchable Supramolecular Nanocarriers for Nitric Oxide Synergistic Photodynamic Eradication of Biofilms. *ACS Nano* **2020**, *14*, 347–359. [CrossRef]
11. Zhang, R.; Li, Y.; Zhou, M.; Wang, C.; Feng, P.; Miao, W.; Huang, H. Photodynamic Chitosan Nano-assembly as a Potent Alternative Candidate for Combating Antibiotic-resistant Bacteria. *ACS Appl. Mater. Inter.* **2019**, *11*, 26711–26721. [CrossRef]
12. Zhao, J.; Xu, L.; Zhang, H.; Zhuo, Y.; Weng, Y.; Li, S.; Yu, D.H. Surfactin-methylene Blue Complex Under LED Illumination for Antibacterial Photodynamic Therapy: Enhanced Methylene Blue Transcellular Accumulation Assisted by Surfactin. *Colloids Surf. B Biointerfaces* **2021**, *207*, 111974. [CrossRef]
13. Demidova, T.N.; Hamblin, M.R. Effect of Cell-photosensitizer Binding and Cell Density on Microbial Photoinactivation. *Antimicrob. Agents Chemother.* **2005**, *49*, 2329–2335. [CrossRef]
14. Bourre, L.; Giuntini, F.; Eggleston, I.M.; Mosse, C.A.; Macrobert, A.J.; Wilson, M. Effective Photoinactivation of Gram-positive and Gram-negative bacterial Strains Using An HIV-1 Tat Peptide-porphyrin Conjugate. *Photochem. Photobiol. Sci.* **2010**, *9*, 1613–1620. [CrossRef]
15. Sun, Y.D.; Zhu, Y.X.; Zhang, X.; Jia, H.R.; Xia, Y.; Wu, F.G. Role of Cholesterol Conjugation in the Antibacterial Photodynamic Therapy of Branched Polyethylenimine-containing Nanoagents. *Langmuir* **2019**, *35*, 14324–14331. [CrossRef]

16. Hu, D.; Zou, L.; Yu, W.; Jia, F.; Han, H.; Yao, K.; Jin, Q.; Ji, J. Relief of Biofilm Hypoxia Using An Oxygen Nanocarrier: A New Paradigm for Enhanced Antibiotic Therapy. *Adv. Sci.* **2020**, *7*, 2000398. [CrossRef]
17. Zhang, L.; Wang, D.; Yang, K.; Sheng, D.; Tan, B.; Wang, Z.; Ran, H.; Yi, H.; Zhong, Y.; Lin, H. Mitochondria-targeted Artificial "Nano-RBCs" for Amplified Synergistic Cancer Phototherapy by A Single NIR Irradiation. *Adv. Sci.* **2018**, *5*, 1800049. [CrossRef]
18. Chen, Z.; Wang, Z.; Ren, J.; Qu, X. Enzyme Mimicry for Combating Bacteria and Biofilms. *Accounts Chem. Res.* **2018**, *51*, 789–799. [CrossRef]
19. Yin, Z.; Chen, D.; Zou, J.; Shao, J.; Tang, H.; Xu, H.; Si, W.; Dong, X. Tumor Microenvironment Responsive Oxygen-self-generating Nanoplatform for Dual-imaging Guided Photodynamic and Photothermal Therapy. *ChemistrySelect* **2018**, *3*, 4366–4373. [CrossRef]
20. Chen, Q.; Feng, L.; Liu, J.; Zhu, W.; Dong, Z.; Wu, Y.; Liu, Z. Intelligent Albumin-MnO$_2$ Nanoparticles as pH-/H$_2$O$_2$-responsive Dissociable Nnanocarriers to Modulate Tumor Hypoxia for Effective Combination Therapy. *Adv. Mater.* **2016**, *28*, 7129–7136. [CrossRef]
21. Sheng, D.; Liu, T.; Deng, L.; Zhang, L.; Li, X.; Xu, J.; Hao, L.; Li, P.; Ran, H.; Chen, H.; et al. Perfluorooctyl Bromide & Indocyanine Ggreen co-loaded Nanoliposomes for Enhanced Multimodal Imaging-guided Phototherapy. *Biomaterials* **2018**, *165*, 1–13. [PubMed]
22. Gao, M.; Liang, C.; Song, X.; Chen, Q.; Jin, Q.; Wang, C.; Liu, Z. Erythrocyte-membrane-enveloped Perfluorocarbon as Nanoscale Artificial Red Blood Cells to Relieve Tumor Hypoxia and Enhance Cancer Radiotherapy. *Adv. Mater.* **2017**, *29*, 1701429. [CrossRef] [PubMed]
23. Cheng, Y.; Cheng, H.; Jiang, C.; Qiu, X.; Wang, K.; Huan, W.; Yuan, A.; Wu, J.; Hu, Y. Perfluorocarbon Nanoparticles Enhance Eeactive Oxygen Levels and Tumor Growth Inhibition in Photodynamic Therapy. *Nat. Commun.* **2015**, *6*, 8785. [CrossRef] [PubMed]
24. Zhu, B.; Wang, L.; Huang, J.; Xiang, X.; Tang, Y.; Cheng, C.; Feng, Y.; Lang, M.; Li, Q. Ultrasound-triggered Perfluorocarbon-derived Nanobombs for Targeted Therapies of Rheumatoid Arthritis. *J. Mater. Chem. B* **2019**, *7*, 4581–4591. [CrossRef]
25. Zou, L.Y.; Hu, D.F.; Wang, F.J.; Jin, Q.; Ji, J. The Relief of Hypoxic Microenvironment Using An O$_2$ Self-sufficient Fluorinated Nanoplatform for Enhanced Photodynamic Eradication of Bacterial Biofilms. *Nano Res.* **2022**, *15*, 1636–1644. [CrossRef]
26. Zhao, C.; Tong, Y.; Li, X.; Shao, L.; Chen, L.; Lü, J.; Deng, X.; Wang, X.; Wu, Y. Photosensitive Nanoparticles Combining Vascular-independent Intratumor Distribution and On-demand Oxygen-depot Delivery for Enhanced Cancer Photodynamic Therapy. *Small* **2018**, *14*, e1703045. [CrossRef]
27. Yu, M.; Xu, X.; Cai, Y.; Zou, L.; Shuai, X. Perfluorohexane-cored Nanodroplets for Stimulations-responsive Ultrasonography and O$_2$-potentiated Photodynamic Therapy. *Biomaterials* **2018**, *175*, 61–71. [CrossRef]
28. Ma, W.; Chen, X.; Fu, L.; Zhu, J.; Fan, M.; Chen, J.; Yang, C.; Yang, G.; Wu, L.; Mao, G.; et al. Ultra-efficient Antibacterial System Based on Photodynamic Therapy and CO Gas Therapy for Synergistic Antibacterial and Ablation Biofilms. *ACS Appl. Mater. Inter.* **2020**, *12*, 22479–22491. [CrossRef]

Article

In Vitro Antibacterial Susceptibility of Different Pathogens to Thirty Nano-Polyoxometalates

Ștefana Bâlici [1], Dan Rusu [2], Emőke Páll [3], Miuța Filip [4], Flore Chirilă [5], Gheorghe Zsolt Nicula [1], Mihaela Laura Vică [1,*], Rodica Ungur [6], Horea Vladi Matei [1] and Nicodim Iosif Fiț [5]

[1] Department of Cell and Molecular Biology, Faculty of Medicine, "Iuliu Hațieganu" University of Medicine and Pharmacy, 400349 Cluj-Napoca, Romania; sbalici@umfcluj.ro (Ș.B.); gnicula@umfcluj.ro (G.Z.N.); hmatei@umfcluj.ro (H.V.M.)

[2] Department of Physical-Chemistry, Faculty of Pharmacy, "Iuliu Hațieganu" University of Medicine and Pharmacy, 400349 Cluj-Napoca, Romania; drusu@umfcluj.ro

[3] Department of Reproduction, Obstetrics and Veterinary Gynecology, Faculty of Veterinary Medicine, University of Agricultural Science and Veterinary Medicine, 400372 Cluj-Napoca, Romania; emoke.pall@usamvcluj.ro

[4] Analytical and Environmental Chemistry Laboratory, "Raluca Ripan" Institute for Research in Chemistry, "Babeș-Bolyai" University, 400294 Cluj-Napoca, Romania; filip_miuta@yahoo.com

[5] Department of Microbiology and Immunology, Faculty of Veterinary Medicine, University of Agricultural Science and Veterinary Medicine, 400372 Cluj-Napoca, Romania; flore.chirila@usamvcluj.ro (F.C.); nfit@usamvcluj.ro (N.I.F.)

[6] Department of Medical Rehabilitation, Faculty of Medicine, "Iuliu Hațieganu" University of Medicine and Pharmacy, 400347 Cluj-Napoca, Romania; ungurmed@yahoo.com

* Correspondence: mvica@umfcluj.ro

Abstract: Due to their unique properties, nano-polyoxometalates (POMs) can be alternative chemotherapeutic agents instrumental in designing new antibiotics. In this research, we synthesized and characterized "smart" nanocompounds and validated their antibacterial effects in order to formulate and implement potential new drugs. We characterized thirty POMs in terms of antibacterial activity–structure relationship. The antibacterial effects of these compounds are directly dependent upon their structure and the type of bacterial strain tested. We identified three POMs that presented sound antibacterial activity against *S. aureus*, *B. cereus*, *E. coli*, *S. enteritidis* and *P. aeruginosa* strains. A newly synthesized compound $K_6[(VO)SiMo_2W_9O_{39}]\cdot 11H_2O$ (POM 7) presented antibacterial activity only against *S. aureus* (ATCC 6538P). Twelve POMs exerted antibacterial effects against both Gram-positive and Gram-negative strains. Only one POM (a cluster derivatized with organometallic fragments) exhibited a stronger effect compared to amoxicillin. New studies in terms of selectivity and specificity are required to clarify these extremely important aspects needed to be considered in drug design.

Keywords: nano-polyoxometalates; UV; FTIR and NMR spectroscopy; drug designs; antibacterial activity; Gram-positive bacteria; Gram-negative bacteria

1. Introduction

Polyoxometalates (POMs) are a class of anionic polynuclear metal-oxo compounds of early transition metals synthesized by the "all in one pot" method [1,2], based on self-assembling mechanisms [3,4], or by two-stage methods involving ligand synthesis followed by cluster formation. They have multiple applications in fields such as catalysis [5,6], magnetism [7–10], electrochemistry [3,11–13], materials science [1,5,7,12–14], biology [15,16] and medicine [16–18] (presenting antidiabetic [19–22], antitumor [23,24], antiviral [25–27], or antibacterial [28–31] activities), due to their particular properties: high negative charge, redox behavior, shape, size, high solubility in water, etc. [12,14,16,18].

Metals (M) most often used in the synthesis of this class of compounds are vanadium (V), molybdenum (Mo) and tungsten (W), less frequent compounds involving tantalum (Ta) and niobium (Nb). POMs can be divided into two major classes: isopolyanions with the general formula $[M_mO_y]^{p-}$ and heteropolyanions with the general formula $[X_xM_mO_y]^{q-}$, the latter category including heteroatoms (X) such as Si, P, Ge, Sb, As, Bi [1,2,7,14,16]. These compounds self-assemble into complete (saturated) archetypal structures such as the Anderson-Evans, Keggin or Wells-Dawson ones, or incomplete structures presenting 1–3 lacunes that could be occupied by identical or different metal ions [1,7,12,16,18]. Several parameters need to be strictly controlled during the syntheses. The process is influenced by temperature, pH, reducing agents, reactant concentrations, heteroatom nature and concentration, type of metal oxide anion involved, presence/absence of mixed addenda atoms and of additional ligands [1,3,12]. In order to design new biocompatible compounds with medical applications, POMs need to be derivatized/functionalized as these inorganic nanocompounds are not highly compatible with living organisms. This must be the goal of future research [14,16,18]. For over two decades POMs have been intensively studied for their antibacterial activity against both Gram-positive and Gram-negative reference strains, bacterial strains resistant to various antibiotics-including those in the β-lactam class (such as penicillins and cephalosporins) [16,18,28–31]. The finding of new natural [32–34] or chemically synthesized compounds [16,18,22,30] with antibacterial activity is a continuous challenge. As alternative chemotherapeutic agents, POMs have shown high antibacterial activity, being instrumental in creating new drugs to combat the antibiotic resistance of the bacteria.

Here we characterize the antibacterial activity of thirty nanoPOMs we synthesized in relation to their chemical structure. They include several Keggin-type nanocompounds presenting saturated (with/without mixed addenda atoms), mono-/tri-lacunary (with/without mixed addenda atoms, either or not sandwich type) structures, as well as one mono-lacunary Wells-Dawson polyoxometalate with mixed addenda atoms and also six clusters. The synthesized POMs were characterized using elemental and thermal analysis, UV and FTIR spectroscopy, while their potential antibacterial activity was evaluated against five bacterial species, two Gram-positive bacteria: *Staphylococcus aureus* (ATCC 6538P), *Bacillus cereus* (ATCC 14579) and three Gram-negative bacteria: *Escherichia coli* (ATCC 10536), *Salmonella enteritidis* (ATCC 13076), *Pseudomonas aeruginosa* (ATCC 27853), compared to amoxicillin (25 µg; Oxoid Ltd., Basingstoke, UK), a broad-spectrum antibiotic.

2. Results

2.1. Chemistry of Polyoxometalates

All thirty POMs we synthesized and characterized are presented in Table 1. One compound, a mono-lacunary Keggin with mixed addenda atoms, with a new formula proposed here, $K_6[(VO)SiMo_2W_9O_{39}] \cdot 11H_2O$ (POM 7), was first synthesized by us according to the methodology described in the Supplementary Material 1. Its characterization can be seen there.

The results of the chemical elemental analysis, thermal analysis, along with UV and FTIR spectroscopy data (in the Supplementary Material 2), are illustrated in Table 2.

Table 1. The structure of all the synthesized polyoxometalates.

POM No.	Chemical Formula of POMs	Structure Types of POMs
1.	$Na_4[Fe^{III}(H_2O)PMo_{11}O_{39}] \cdot 18H_2O$	mono-lacunary Keggin
2.	$Na_9[Fe_3(H_2O)_3(PMo_9O_{34})_2]$	tri-lacunary Keggin/sandwich type
3.	$Na_8[SiW_{11}O_{39}] \cdot 12H_2O$	mono-lacunary Keggin
4.	$Na_{11}[Fe_3(H_2O)_3(SiW_9O_{34})_2] \cdot 25H_2O$	tri-lacunary Keggin/sandwich type
5.	$K_3[(VO)_3PMo_9O_{34}] \cdot 14H_2O$	tri-lacunary Keggin
6.	$Na_6[PMo_9^{VI}V_3^{V}O_{40}] \cdot 16H_2O$	Keggin with mixed addenda atoms
7.	$K_6[(VO)SiMo_2W_9O_{39}] \cdot 11H_2O$	mono-lacunary Keggin with mixed addenda atoms
8.	$K_{10}[(VO)_4(PW_9O_{34})_2] \cdot 26H_2O$	tri-lacunary Keggin/sandwich type
9.	$K_{10}[(VO)_4(AsW_9O_{34})_2] \cdot 21H_2O$	tri-lacunary *pseudo*-Keggin/sandwich type
10.	$K_{11}H[(VO)_3(Sb^{III}W_9O_{33})_2] \cdot 27H_2O$	tri-lacunary *pseudo*-Keggin/sandwich type

Table 1. Cont.

POM No.	Chemical Formula of POMs	Structure Types of POMs
11.	$Na_{12}[Sb_2W_{22}O_{74}(OH)_2] \cdot 38H_2O$	cluster
12.	$H_4[SiW_{12}O_{40}] \cdot 14H_2O$	saturated Keggin
13.	$H_3[PW_{12}O_{40}] \cdot 12H_2O$	saturated Keggin
14.	$H_3[PMo_{12}O_{40}] \cdot 13H_2O$	saturated Keggin
15.	$Na_9[SbW_9O_{33}] \cdot 19,5H_2O$	tri-lacunary *pseudo*-Keggin
16a.	$Na_{10}[SiW_9O_{34}] \cdot 24H_2O$	tri-lacunary Keggin
16b.	$Na_{10}[SiW_9O_{34}] \cdot 24H_2O$–recryst.	tri-lacunary Keggin
17.	$Na_{27}[NaAs_4W_{40}O_{140}] \cdot 42H_2O$	cluster
18.	$Na_8H[PW_9O_{34}] \cdot 20H_2O$	tri-lacunary Keggin
19.	$(NBu_4)_{27}[NaAs_4Mo_{40}O_{140}] \cdot 12H_2O$	cluster
20.	$(Bu_3Sn)_{18}[NaSb_9W_{21}O_{86}]$	cluster
21.	$K_6[Co(H_2O)SiMo_2W_9O_{39}] \cdot 14H_2O$	mono-lacunary Keggin with mixed addenda atoms
22.	$K_{10}[Co(H_2O)Si_2MoW_{16}O_{61}] \cdot 18H_2O$	mono-lacunary Wells-Dawson with mixed addenda atoms
23a.	$Na_5[Fe^{III}(H_2O)SiW_{11}O_{39}] \cdot 24H_2O$	mono-lacunary Keggin
23b.	$Na_5[Fe^{III}(H_2O)SiW_{11}O_{39}] \cdot 24H_2O$–recryst.	mono-lacunary Keggin
24a.	$Na_5[Fe^{III}(H_2O)GeW_{11}O_{39}] \cdot 26H_2O$	mono-lacunary Keggin
24b.	$Na_5[Fe^{III}(H_2O)GeW_{11}O_{39}] \cdot 26H_2O$–recryst.	mono-lacunary Keggin
25.	$Na_{10}[Mn_4(H_2O)_2(AsW_9O_{34})_2] \cdot 27H_2O$	tri-lacunary *pseudo*-Keggin/sandwich type
26.	$Na_{12}[Co_3(H_2O)_3(BiW_9O_{33})_2] \cdot 37H_2O$	tri-lacunary *pseudo*-Keggin/sandwich type
27.	$Na_{14}[Mn_3(H_2O)_3(SiW_9O_{34})_2] \cdot 28H_2O$	tri-lacunary Keggin/sandwich type
28.	$(NH_4)_4(NBu_4)_5[Na(BuSn)_3Sb_9W_{21}O_{86}] \cdot 17H_2O$	cluster
29.	$K_{27}[NaAs_4W_{40}O_{140}] \cdot 52H_2O$	cluster
30.	$K_6[SiV^{IV}W_{11}O_{40}] \cdot 12H_2O$	mono-lacunary Keggin

Table 2. Physico-chemical data of polyoxometalates.

POM No.	Elemental Analysis and TG Data (Found (Calcd.))	UV (H_2O) Data (nm/cm^{-1}): $\nu_2(M=O_t)$ and $\nu_1(M-O_{c,e}-M)$	FTIR Spectral Data (ν_{max} (cm^{-1}) and Their Contribution in the POMs' Structure)
1.	M = 2200.38; Na (4.20 (4.18)); Fe (2.58 (2.54)); P (1.39 (1.41)); Mo (47.98 (47.96)); H_2O (15.62 (15.55)).	$\nu_2 = 210/47,619$ and $\nu_1 = 228/43,859$.	1128 (w, ν_{as}(P-O_i)); 1049 (sh, ν_{as}(P-O_i)); 924 (vs, sh, ν_{as}(Mo=O_t)); 887 (vs, ν_{as}(Mo-O_c-Mo)); 847 (s, sp ν_{as}(Mo-O_e-Mo)); 658 (s, br, ν_{as}(Mo-O_e-Mo)); 621 (s, br, δ(P-O_i)); 577 (s, br, δ(P-O_i)); 546 (m, sh, δ(Mo-O-Mo)); 486 (m, ν(Fe-O)).
2.	M = 3737.68; Na (5.57 (5.54)); Fe (4.50 (4.48)); P (1.63 (1.66)); Mo (46.24 (46.20)); H_2O (13.10 (13.01)).	$\nu_2 = 219/45,662$ and $\nu_1 = 271/36,900$.	1180–1044 (s, sp, ν_{as}(P-O_i)); 997 (vs, sp, ν_{as}(Mo=O_t)); 978 (vs, sp, ν_{as}(Mo-O_c-Mo)); 775 (m, b ν_{as}(Mo-O_e-Mo)); 667 (w, ν_{as}(Mo-O_b-Mo)/sandwich); 514 (m, sp, δ(Mo-O-Mo)).
3.	M = 3074.40; Na (6.04 (5.98)); Si (0.88 (0.91)); W (63.58 (65.78)); H_2O (7.10 (7.03)).	$\nu_2 = 206/48,544$ and $\nu_1 = 258/38,759$.	3446 (vs, br, ν_{as}(O-H)); 1635 (w, δ(O-H)); 1005 (vw, sh, ν_{as}(Si-O_i)); 962 (s, sp, ν_{as}(W=O_t)); 910 (vs, sp, ν_{as}(W-O_c-W)); 798 (vs, br, ν_{as}(W-O_e-W)); 517 (vs, br, δ(W-$O_{c,e}$-W)).
4.	M = 5378.10; Na (4.72 (4.70)); Fe (3.15 (3.12)); Si (1.02 (1.04)); W (61.58 (61.53)); H_2O (9.41 (9.38)).	$\nu_2 = 200/50,000$ and $\nu_1 = 257/38,911$.	1190–1063 (w, sp, ν_{as}(Si-O_i)); 964 (vs, sp, ν_{as}(W=O_t)); 910 (vs, sp, ν_{as}(W-O_c-W)); 879 (m, sh, ν_{as}(W-O_c-W)); 787 (vs, vbr, ν_{as}(W-O_e-W)); 708 (m, sh, ν(W-O_b-W)/sandwich); 542 (vw, br, δ(W-$O_{c,e}$-W)); 499 (m, sp, ν(Fe-O)); 403 (m, sp, ν(Fe-O)).

Table 2. Cont.

POM No.	Elemental Analysis and TG Data (Found (Calcd.))	UV (H$_2$O) Data (nm/cm^{-1}): ν_2(M=O$_t$) and ν_1(M-O$_{c,e}$-M)	FTIR Spectral Data (ν_{max} (cm^{-1}) and Their Contribution in the POMs' Structure)
5.	M = 2008.74; K (5.87 (5.84)); V (7.64 (7.61)); P (1.51 (1.54)); Mo (43.05 (42.99)); H$_2$O (12.62 (12.56)).	ν_2 = 218/45,871 and ν_1 = 305/32,787.	1180–1088 (s, sp, ν_{as}(P-O$_i$)); 989 (vs, sp, ν_{as}(V=O$_t$)); 941 (vs, sp, ν_{as}(Mo=O$_t$)); 879 (s, br, ν_{as}(Mo-O$_c$-Mo)); 796 (m, br, ν_{as}(Mo-O$_e$-Mo)); 726 (m, ν_{as}(Mo-O$_e$-Mo)); 625 (s, sp, δ(M-O-M)); 513 (w, sp, δ(Mo-O$_{c,e}$-Mo)).
6.	M = 2113.42; Na (6.56 (6.53)); P (1.45 (1.47)); Mo (40.92 (40.86)); V (7.26 (7.23)); H$_2$O (13.70 (13.64)).	ν_2 = 221/45,249 and ν_1 = 305/32,786.	1190–1063 (vs, sp, ν_{as}(P-O$_i$)); 989 (m, sh, ν_{as}(V=O$_t$)); 962 (vs, sp, ν_{as}(Mo=O$_t$)); 866 (vs, br, ν_{as}(V-O$_c$-V)+ ν_{as}(Mo-O$_c$-Mo)); 785 (vs, vbr, ν_{as}(V-O$_e$-V)+ ν_{as}(Mo-O$_e$-Mo)); 619 (vs, sp, δ(P-O$_i$); 519 (vw, br, δ(V-O$_{c,e}$-V) + δ(Mo-O$_{c,e}$-Mo).
7.	M = 2998.20; K (7.84 (7.82)); V (1.73 (1.70)); Si (0.91 (0.94)); Mo (6.44 (6.40)); W (55.26 (55.19)); H$_2$O (6.62 (6.61)).	ν_2 = 199/50,251 and ν_1 = 258/38,759.	1109 (w, ν_{as}(Si-O$_i$)); 968 (s, ν_{as}(W=O$_t$) + ν_{as}(Mo=O$_t$)); 906 (vs, ν_{as}(W-O$_c$-W) + ν_{as}(Mo-O$_c$-Mo)); 783 (vs, ν_{as}(W-O$_e$-W) + (Mo-O$_e$-Mo)); 669 (m, ν_{as}(W-O$_e$-W)+ ν_{as}(Mo-O$_e$-Mo)).
8.	M = 5586.17; K (7.03 (6.99)); V (3.68 (3.65)); P (1.08 (1.11)); W (59.28 (59.24)); H$_2$O (8.45 (8.38)).	ν_2 = 201/49,751 and ν_1 = 248/40,323.	3437 (vs, br, ν_{as}(O-H)); 1624 (m, sp, δ(O-H)); 1186 (vs, sp, ν_{as}(P-O$_i$)); 1103 (vs, sp, ν_{as}(P-O$_i$)); 987 (sh, ν_{as}(W=O$_t$)); 968 (vs, br, ν_{as}(W=O$_t$)); 891 (s, br, ν_{as}(W-O$_c$-W)); 850 (sh, ν_{as}(W-O$_c$-W)); 791 (vs, vbr, ν_{as}(W-O$_e$-W)); 719 (s, sp, ν(V-O$_b$-W)); 619 (vs, sp, ν_s(P-O$_i$)); 514 (m, br, δ(W-O-W)); 463 (m, br, δ(W-O-W)).
9.	M = 5583.99; K (7.05 (7.00)); V (3.66 (3.65)); As (2.65 (2.68)); W (59.29 (59.26)); H$_2$O (6.81 (6.78)).	ν_2 = 201/49,751 and ν_1 = 256/39,062.	3419 (vs, br, ν_{as}(O-H)); 1626 (m, sp, δ(O-H)); 1045 (sh, ν_{as}(As-O$_i$)); 931 (vs, br, ν_{as}(W=O$_t$)); 874 (s, sp, ν_{as}(W-O$_c$-W)); 831 (m, ν_{as}(W-O$_c$-W)); 796 (m, sp, ν_{as}(W-O$_e$-W)); 712 (s, sh, ν(V-O$_b$-W)/sandwich); 621 (m, br, δ(W-O-W)); 553 (m, br, δ(W-O-W)).
10.	M = 5726.92; K (7.55 (7.51)); V (2.70 (2.67)); Sb (4.22 (4.25)); W (57.83 (57.78)); H$_2$O (8.55 (8.49)).	ν_2 = 202/49,505 and ν_1 = 251/39,841.	3423 (vs, br, ν_{as}(O-H)); 1697 (m, δ(H-O-H)); 1667 (m, br, δ(H-O-H)); 995 (m, sp, ν_{as}(V=O$_t$)); 930 (m, sp, ν_{as}(W=O$_t$)); 857 (s, sp, ν_{as}(W-O$_c$-W)); 833 (vs, ν_{as}(W-O$_c$-W)); 743 (m, sp, ν_{as}(Sb-O$_i$)); 697 (s, ν_{as}(W-O$_e$-W)); 553 (m, br, δ(W-O$_{c,e}$-W)).
11.	M = 6466.43; Na (4.31 (4.27)); Sb (3.75 (3.77)); W (62.59 (62.55)); H$_2$O (10.65 (10.59)).	ν_2 = 200/50,000 and ν_1 = 255/39,216.	3332 (vs, br, ν_{as}(O-H)); 1619 (m, sp, δ(H-O-H)); 1385 (s, sp ν_{as}(NO$_3^-$)); 943 (vs, sp, ν_{as}(W=O$_t$)); 887 (vs, ν_{as}(W-O$_c$-W)); 864 (s, sh ν_{as}(Sb-O$_i$)); 837 (vs, ν_{as}(W-O$_c$-W)); 800 (s, sh ν_{as}(W-O$_e$-W)); 771 (vs, br, ν_{as}(W-O$_e$-W)); 673 (s, br, ν_{as}(W-O$_b$-W) +δ(O-Sb-O)); 507 (w, br, δ(W-O$_{c,e}$-W)).
12.	M = 3130.39; Si (0.88 (0.90)); W (70.52 (70.47)); H$_2$O (8.11 (8.06)).	ν_2 = 207/48,309 and ν_1 = 263/38,023.	1020 (m, sh, ν_{as}(Si-O$_i$)); 982 (s, ν_{as}(W=O$_t$)); 926 (vs, sp, ν_s(Si-O$_i$)); 883 (m, sp, ν_{as}(W-O$_c$-W)); 787 (vs, br, ν_{as}(W-O$_e$-W)); 538 (m, δ(W-O-W)).
13.	M = 3096.24; P (0.98 (1.00)); W (71.28 (71.25)); H$_2$O (7.00 (6.98)).	ν_2 = 201/49,751 and ν_1 = 248/40,323.	1080 (vs, sp, ν_{as}(P-O$_i$); 984 (vs, ν_{as}(W=O$_t$)); 889 (vs, sp, ν_{as}(W-O$_c$-W)); 808 (vs, sp, ν_{as}(W-O$_e$-W)); 596 (w, sp, δ(W-O$_c$-W)); 525 (m, δ(W-O$_e$-W)).

Table 2. Cont.

POM No.	Elemental Analysis and TG Data (Found (Calcd.))	UV (H_2O) Data (nm/cm^{-1}): ν_2(M=O_t) and ν_1(M-$O_{c,e}$-M)	FTIR Spectral Data (ν_{max} (cm^{-1})) and Their Contribution in the POMs' Structure)
14.	M = 2059.45; P (1.48 (1.50)); Mo (55.93 (55.90)); H_2O (11.40 (11.37)).	ν_2 = 193/51,550 and ν_1 = 270/37,000.	1065 (vs, sp, ν_{as}(P-O_i)); 962 (vs, sp, ν_{as}(Mo=O_t)); 870 (s, vbr, ν_{as}(Mo-O_c-Mo)); 787 (vs, br, ν_{as}(Mo-O_e-Mo)); 595 (w, δ(Mo-O-Mo)); 509 (vw, δ(Mo-O-Mo)).
15.	M = 2862.51; Na (7.26 (7.23)); Sb (4.22 (4.25)); W (57.87 (57.80)); H_2O (12.33 (12.27)).	ν_2 = 207/48,309 and ν_1 = 238/42,017.	920 (vs, sp, ν_{as}(W=O_t)); 890 (vs, sp, ν_{as}(W-O_c-W)); 767 (s, ν_{as}(W-O_e-W)); 743 (s, sp, ν_{as}(Sb-O_i)); 715 (s, ν_{as}(W-O_e-W)); 505 (w, br, δ(W-O-W)).
16a.	M = 2888.89; Na (7.98 (7.96)); Si (0.96 (0.97)); W (57.29 (57.27)); H_2O (15.01 (14.97)).	ν_2 = 208/48,077 and ν_1 = 265/37,736.	1635 (m, δ(O-H)); 987 (m, sp, ν_{as}(W=O_t)); 937 (s, ν_{as}(Si-O_i)); 878 (vs, ν_{as}(W-O_c-W)); 844 (vs, ν_{as}(W-O_c-W)); 810 (vs, ν_{as}(W-O_e-W)); 723 (s, ν_{as}(W-O_e-W)); 618 (s, ν_s(Si-O_i)); 528 (m, δ(W-O-W)).
16b.	M = 2888.89; Na (7.98 (7.96)); Si (0.96 (0.97)); W (57.29 (57.27)); H_2O (14.91 (14.97)).	ν_2 = 208/48,077 and ν_1 = 265/37,736.	1635 (m, δ(O-H)); 987 (m, sp, ν_{as}(W=O_t)); 937 (s, ν_{as}(Si-O_i)); 878 (vs, ν_{as}(W-O_c-W)); 844 (vs, ν_{as}(W-O_c-W)); 810 (vs, ν_{as}(W-O_e-W)); 723 (s, ν_{as}(W-O_e-W)); 618 (s, ν_s(Si-O_i)); 528 (m, δ(W-O-W)).
17.	M = 11293.56; Na (5.73 (5.70)); As (2.63 (2.65)); W (65.15 (65.11)); H_2O (6.75 (6.70)).	ν_2 = 200/50,000 and ν_1 = 243/41,152.	951 (vs, sp, ν_{as}(W=O_t)); 876 (vs, b ν_{as}(As-O_i)+ν_{as}(W-O_c-W)); 793 (vs, sp ν_{as}(W-O_c-W)); 710 (vs, b ν_{as}(W-O_e-W)); 634 (s, b ν_s(As-O_i)); 577 (m, b, δ(W-O-W)).
18.	M = 2774.75; Na (6.65 (6.63)); P (1.10 (1.12)); W (59.68 (59.63)); H_2O (13.05 (12.99)).	ν_2 = 208/48,077 and ν_1 = 245/40,816.	1054 (s, sp, ν_{as}(P-O_i)); 1014 (w, ν_{as}(P-O_i)); 937 (vs, sp, ν_{as}(W=O_t)); 881 (vs, sp, ν_{as}(W-O_c-W)); 741 (vs, b, ν_{as}(W-O_e-W)); 503 (vw, b, δ(W-O-W)).
19.	M = 13162.90; Na (0.20 (0.17)); C (39.46 (39.42)); H (7.66 (7.63)); N (2.88 (2.87)); As (2.26 (2.28)); Mo (29.21 (29.15)); H_2O (1.67 (1.64)).	ν_2 = 209/47,847 and ν_1 = 228/43,859.	3446 (vs, br, ν_{as}(O-H)); > 2800 (vs, br, ν_{as}(C-H)); 1483 (vs, br, ν_{as}(C-N)); 1617 (w, b, δ(H-O)); 943 (vs, sh, ν_{as}(Mo=O_t)); 924 (vs, sp, ν_{as}(Mo=O_t)); 904 (vs, sp, ν_{as}(As-O_i)+ν_{as}(Mo-O_c-Mo)); 879 (s, sh, ν_{as}(Mo-O_c-Mo)); 854 (vs, ν_{as}(Mo-O_c-Mo)); 806 (vs, b, ν_{as}(Mo-O_e-Mo)); 764 (vs, sh, ν_{as}(Mo-O_e-Mo)); 735 (vs, sh, ν_{as}(Mo-O_e-Mo)); 706 (vs, b, ν_{as}(Mo-O_e-Mo)); 663 (s, ν_s(As-O_i)); 584 (m, δ(Mo-O-Mo)); 557 (w, b, δ(Mo-O-Mo)); 517 (w, b, δ(Mo-O-Mo)).
20.	M = 11576.37; Na (0.22 (0.20)); C (22.44 (22.41)); H (4.25 (4.23)); Sn (18.48 (18.46)); Sb (9.45 (9.47)); W (33.39 (33.35)).	ν_2 = 200/50,000 and ν_1 = 254/39,370.	949 (vs, sp, ν_{as}(W=O_t)); 862 (s, b, ν_{as}(Sb-O_i) + ν_{as}(W-O_c-W)); 796 (s, ν_{as}(W-O_e-W)); 739 (vs, ν_{as}(W-O_e-W)); 749 (vs, ν_{as}(W-O_e-W)); 657 (s, δ(Sb-O_i)); 577 (w, ν_{as}(Sb-O_i)); 505 (w, ν(C-Sn-O)); 493 (w, δ(Sb-O)); the presence of bands due to the stretching and deformation vibrations of the C-H and C-C bonds of the butyl groups in the ranges 1000–1300, 1700–1950 and >2800 cm^{-1} is also observed in the spectrum.
21.	M = 3062.25; K (7.70 (7.66)); Co (1.94 (1.92)); Si (0.90 (0.92)); Mo (6.30 (6.27)); W (54.08 (54.03)); H_2O (8.87 (8.82)).	ν_2 = 203/49,261 and ν_1 = 253/39,526.	995 (s, sp, ν_{as}(Si-O_i)); 953 (vs, sp, ν_{as}(Mo=O_t)); 901 (vs, sp, ν_{as}(W=O_t)); 798 (vs, b, ν_{as}(Mo-O_c-Mo)+ν_{as}(W-O_c-W)); 739 (vs, b, ν_{as}(Mo-O_e-Mo)+ν_{as}(W-O_e-W)); 704 (s, vb, ν_{as}(Mo-O_e-Mo)+ν_{as}(W-O_e-W)); 538 (m, sh, δ(W-O-W)); 524 (m, b, δ(W-O-W)) + δ(Mo-O-Mo)); 482 (m, sh δ(W-O)).

Table 2. Cont.

POM No.	Elemental Analysis and TG Data (Found (Calcd.))	UV (H_2O) Data (nm/cm^{-1}): ν_2(M=O_t) and ν_1(M-$O_{c,e}$-M)	FTIR Spectral Data (ν_{max} (cm^{-1})) and Their Contribution in the POMs' Structure
22.	M = 4861.72; K (8.08 (8.04)); Co (1.24 (1.21)); Si (1.14 (1.16)); Mo (1.98 (1.97)); W (60.53 (60.50)); H_2O (7.10 (7.04)).	ν_2 = 203/49,261 and ν_1 = 253/39,526.	995 (sh, sp, ν_{as}(Si-O_i)); 952 (vs, sp, ν_{as}(Mo=O_t)); 901 (vs, b ν_{as}(W=O_t)); 798 (s, b ν_{as}(W-O_c-W) + ν_{as}(Mo-O_c-Mo)); 739 (vs, b ν_{as}(W-O_c-W)); 704 (s, ν_{as}(W-O_e-W)); 525 (s, b, δ(W-O-W) + δ(Mo-O-Mo));); 482 (sh, b ν_s(W-O_c-Co) + ν_s(Mo-O_c-Co)).
23a.	M = 3295.48; Na (3.50 (3.49)); Fe (1.70 (1.69)); Si (0.82 (0.85)); W (61.38 (61.36)); H_2O (13.68 (13.67)).	ν_2 = 200/50,000 and ν_1 = 259/38,610.	1088 (m, ν_{as}(Si-O_i); 1005 (sh, ν_{as}(Si-O_i); 964 (s, ν_{as}(W=O_t)); 910 (vs, b, ν_s(Si-O_i)+ν_{as}(W-O_c-W)); 876 (sh, ν_{as}(W-O_c-W)); 787 (vs, b, ν_{as}(W-O_e-W)); 704 (sh, ν_{as}(W-O_e-W)); 538 (m, δ(W-O_c-W)); 519 (m, b, δ(W-O_e-W)); 418 (m, sh, ν(Fe-O)).
23b.	M = 3295.48; Na (3.50 (3.49)); Fe (1.70 (1.69)); Si (0.82 (0.85)); W (61.38 (61.36)); H_2O (13.58 (13.67)).	ν_2 = 200/50,000 and ν_1 = 259/38,610.	1088 (m, ν_{as}(Si-O_i); 1005 (sh, ν_{as}(Si-O_i); 964 (s, ν_{as}(W=O_t)); 910 (vs, b, ν_s(Si-O_i)+ν_{as}(W-O_c-W)); 876 (sh, ν_{as}(W-O_c-W)); 787 (vs, b, ν_{as}(W-O_e-W)); 704 (sh, ν_{as}(W-O_e-W)); 538 (m, δ(W-O_c-W)); 519 (m, b, δ(W-O_e-W)); 418 (m, sh, ν(Fe-O)).
24a.	M = 3376.06; Na (3.42 (3.40)); Fe (1.67 (1.65)); Ge (2.12 (2.15)); W (59.92 (59.90)); H_2O (14.42 (14.41)).	ν_2 = 202/49,505 and ν_1 = 255/39,216.	982 (vs, sp ν_{as}(W=O_t)); 903 (vs, sh, ν_{as}(W-O_c-W)); 876 (vs, b, ν_{as}(W-O_c-W)); 814 (s, sh, ν_{as}(Ge-O) + ν_{as}(W-O_e-W)); 771 (vs, b, ν_{as}(Ge-O_i) + ν_{as}(W-O_e-W)); 525 (w, b, δ(W-$O_{c,e}$-W)).
24b.	M = 3376.06; Na (3.42 (3.40)); Fe (1.67 (1.65)); Ge (2.12 (2.15)); W (59.92 (59.90)); H_2O (14.38 (14.41)).	ν_2 = 202/49,505 and ν_1 = 255/39,216	982 (vs, sp ν_{as}(W=O_t)); 903 (vs, sh, ν_{as}(W-O_c-W)); 876 (vs, b, ν_{as}(W-O_c-W)); 814 (s, sh, ν_{as}(Ge-O) + ν_{as}(W-O_e-W)); 771 (vs, b, ν_{as}(Ge-O_i) + ν_{as}(W-O_e-W)); 525 (w, b, δ(W-$O_{c,e}$-W))
25.	M = 5519.02; Na (4.18 (4.17)); Mn (3.99 (3.98)); As (2.68 (2.72)); W (59.98 (59.96)); H_2O (9.51 (9.47)).	ν_2 = 201/49,751 and ν_1 = 248/40,323.	3421 (vs, b, ν_{as}(O-H)); 1624 (vs, sp, δ(H-O-H)); 957 (vs, sp, ν_{as}(W=O_t)); 877 (vs, b ν_{as}(As-O_i)+ν_{as}(W-O_c-W)); 839 (s, sp, ν_{as}(W-O_c-W)); 768 (vs, ν_{as}(W-O_e-W)); 712 (s, ν_{as}(W-O_e-W)+ν_{as}(W-O_b-W)/sandwich); <514 (m, b, δ(W-O-W)).
26.	M = 5956.33; Na (4.66 (4.63)); Co (2.98 (2.97)); Bi (7.00 (7.02)); W (55.60 (55.56)); H_2O (12.15 (12.10)).	ν_2 = 194/51,500 and ν_1 = 256/38,991.	946 (s, ν_{as}(W=O_t)); 867 (vs, vb, ν_{as}(W-O_c-W)); 839 (s, sp, ν_{as}(Bi-O_i)); 795 (vs, ν_{as}(W-O_e-W)); 740 (s, b, ν_{as}(W-O_e-W)); 740 (s, b, ν_{as}(W-O_b-W)); 508 (w, δ(W-O-W)).
27.	M = 5498.39; Na (5.88 (5.85)); Mn (3.05 (3.00)); Si (1.00 (1.02)); W (60.25 (60.18)); H_2O (10.12 (10.16)).	ν_2 = 213/46,948 and ν_1 = 256/39,066.	1631 (m, δ(H_2O)); 1568 (m, δ(H_2O)); 987 (m, ν_{as}(W=O_t)); 940 (s, ν_{as}(Si-O_i)); 893 (vs, ν_{as}(W-O_c-W)); 807 (vs, ν_{as}(W-O_e-W)); 722 (s, ν_{as}(W-O_e-W)); 682 (s, ν_s(Si-O_i-W)); 519 (vw, δ(W-$O_{c,e}$-W)); 350 (s, ν(Mn-$O_{c,e}$-W)).

Table 2. Cont.

POM No.	Elemental Analysis and TG Data (Found (Calcd.))	UV (H$_2$O) Data (nm/cm^{-1}): ν_2(M=O$_t$) and ν_1(M-O$_{c,e}$-M)	FTIR Spectral Data (ν_{max} (cm^{-1}) and Their Contribution in the POMs' Structure)
28.	M = 8473.62; C (13.06 (13.04)); H (3.10 (3.06)); N (1.54 (1.49)); Na (0.28 (0.27)); Sb (12.94 (12.93)); Sn (4.35 (4.20)); W (45.61 (45.56)); H$_2$O (3.64 (3.61)).	ν_2 = 191/52,356 and ν_1 = 275/36,363.	3485 (s, ν_{as}(hydrogen bond from lattice water)); 3373 (vs, ν_{as}(hydrogen bond from lattice water)); 3171 (m, b, ν(N-H) from NH$_4^+$); 1648 (m, δ(O-H)); 1621 (sh, δ(O-H)); 1404 (s, δ(N-H) from NH$_4^+$); 1293 (m, ν_{as}(C-N) from NBu$_4$); 958 (s, ν_{as}(W=O$_t$)); 927 (m, ν_{as}(W=O$_t$)); 881 (s, ν_{as}(W-O$_c$-W)); 871 (s, ν_{as}(W-O$_c$-W)); 851 (s, ν_{as}(W-O$_c$-W)); 800 (vs, ν_{as}(W-O$_e$-W)); 766 (vs, ν_{as}(W-O$_e$-W)); 701(sh, ν_{as}(C-N) from NBu$_4$); 681 (s, ν_{as}(Sb-O$_i$) + ν_{as}(Sn-O) + ν(C-Sn-O)); 613 (m, ν_{as}(Sb-O$_i$) + ν_{as}(Sn-O) + ν(C-Sn-O)); 549 (s, ν_{as}(Sb-O$_i$) + ν_{as}(Sn-O) + ν(C-Sn-O)); 489 (m, ν_{as}(Sn-C) + δ(W-O-W)); 431 (w, δ(Sb-O)); 418 (m, ν_{as}(Sn-C)).
29.	M = 11908.64; K (8.90 (8.86)); Na (0.21 (0.19)); As (2.50 (2.52)); W (61.81 (61.75)); H$_2$O (7.92 (7.87)).	ν_2 = 201/49,751 and ν_1 = 254/39,370.	966 (vs, sp, ν_{as}(W=O$_t$)); 883 (vs, b, ν_{as}(As-O$_i$)+(W-O$_c$-W)); 783 (vs, b, ν_{as}(W-O$_e$-W)); 733 (s, sh, ν_{as}(W-O$_e$-W)); 671 (vs, b, ν_{as}(As-O$_i$)); 553 (m, b, δ(W-O-W)).
30.	M = 3192.02; K (7.38 (7.35)); Si (0.86 (0.88)); V (1.62 (1.60)); W (63.39 (63.35)); H$_2$O (6.62 (6.77)).	ν_2 = 198/50,505 and ν_1 = 257/38,910.	1054 (w, sp, ν_{as}(Si-O$_i$)); 1000 (w, sp, ν_s(Si-O$_i$)); 965 (s, sp, ν_{as}(W=O$_t$)); 989 (m, sp, ν_{as}(V=O)); 884 (vs, ν_{as}(W-O$_c$-W)); 805 (vs, ν_{as}(W-O$_e$-W)); 741 (vs, vb, ν_{as}(W-O$_e$-W)); 661 (m, δ(O$_i$-Si-O$_i$)); 518 (w, δ(W-O$_{c,e}$-W)).

2.2. Antimicrobial Activity of Polyoxometalates

To compare the antimicrobial activity of all synthesized POMs versus amoxicillin, we measured the diameters of the inhibition zones (Halo Zone in mm) employing the disk diffusion method. Results are presented in Table 3.

Of the thirty POMs whose antibacterial activity was tested with the disk diffusion method, nine nanocompounds exhibited no effects against the five bacterial strains: POMs 3, 5, 6, 9, 10, 12, 16a, 16b, 21 and 22.

Table 3. POM antibacterial activity as measured by the disk diffusion method.

POM No.	Effect of POMs on Microorganisms (Halo Zone Test/mm)				
	S. aureus	B. cereus	S. enteritidis	E. coli	P. aeruginosa
1.	12 ± 0.50 R [1]	7 ± 0.30 7 ± 0.22 [2]	6 ± 0.24 R	R	9 ± 0.22 R
2.	R	12 ± 0.30 12 ± 0.44	R	R	R
3.	R	R	R	R	R
4.	8 ± 0.23 R	7 ± 0.45 R	R	R	R
5.	R	R	R	R	R
6.	R	R	R	R	R
7.	15 ± 0.50 13 ± 0.50	R	R	R	R
8.	10 ± 0.50 R	10 ± 0.20 R	R	R	R

Table 3. *Cont.*

9.	R	R	R	R	R	
10.	R	R	R	R	R	
11.	11 ± 0.55 R	R	10 ± 0.20 R	R	R	
12.	R	R	R	R	R	
13.	8 ± 0.12 12 ± 0.5	8 ± 0.22 7 ± 0.25	10 ± 0.50 10 ± 0.22	10 ± 0.50 12 ± 0.25	R	
14.	8 ± 0.22 7 ± 0.25	R	12 ± 0.25 6 ± 0.32	12 ± 0.35 8 ± 0.25	12 ± 0.50 8 ± 0.42	
15.	32 ± 0.22 18 ± 0.50	23 ± 0.25 12 ± 0.50	26 ± 0.25 12 ± 0.50	R	R	
16a. [3]	R	R	R	R	R	
16b. [4]	R	R	R	R	R	
17.	R	10 ± 0.25 R	18 ± 0.25 10 ± 0.50	R	R	
18.	R	R	8 ± 0.22 R	R	R	
19.	20 ± 0.50 12 ± 0.30	14 ± 0.50 8 ± 0.65	25 ± 0.23 19 ± 0.18	R R	R R	
20.	30 ± 0.10 13 ± 0.25	24 ± 0.15 14 ± 0.22	22 ± 0.10 10 ± 0.22	12 ± 0.25 8 ± 0.25	12 ± 0.22 18 ± 0.25	
21.	R	R	R	R	R	
22.	R	R	R	R	R	
23a. [3]	14 ± 0.25 13 ± 0.25	R	R	R	R	
23b. [4]	14 ± 0.22 13 ± 0.12	R	R	R	R	
24a. [3]	12 ± 0.15 10 ± 0.25	R	R	R	R	
24b. [4]	10 ± 0.25 R	R	R	R	R	
25.	13 ± 0.25 16 ± 0.55	R	R	R	R	
26.	18 ± 0.55 16 ± 0.10	20 ± 0.55 12 ± 0.15	18 ± 0.55 15 ± 0.15	16 ± 0.25 14 ± 0.22	15 ± 0.25 22 ± 0.50	
27.	14 ± 0.50 10 ± 0.35	14 ± 0.37 10 ± 0.22	R	R	R	
28.	40 ± 0.50 20 ± 0.22	30 ± 0.50 12 ± 0.55	30 ± 0.52 23 ± 0.23	20 ± 0.23 16 ± 0.27	16 ± 0.45 8 ± 0.56	
29	12 ± 0.50 12 ± 0.22	R	18 ± 0.50 R	R	R	
30	18 ± 0.55 11 ± 0.25	6 ± 0.51 6 ± 0.45	12 ± 0.56 7 ± 0.52	14 ± 0.57 8 ± 0.45	R	
+ive C [5]	19 ± 0.52	12 ± 0.37	18 ± 0.33	18 ± 0.26	R	
−ive C [6]	R	R	R	R	R	

[1] R = resistant; [2] the retest values of the halo zone test (from the same solution, 6 months after the initial preparation) are written in the second row for each POM; [3] only POMs 16, 23 and 24 required recrystallizations: POM (number) a = original POM, [4] POM (number) b = recrystallized POM; [5] +ive C = positive control (Amoxicillin, 25 µg); [6] −ive C = negative control (0.15 M NaCl solution).

POMs which initially presented no activity against one strain, or another were not retested after 6 months against the respective bacterial strains. POMs 20, 26 and 28 were the only ones to present sound antibacterial activity against all five bacterial strains tested by the disk diffusion method, but only the latter constantly exhibited a stronger effect compared to amoxicillin. The antibacterial activity of POM 28 against all bacterial strains is illustrated in Figure 1.

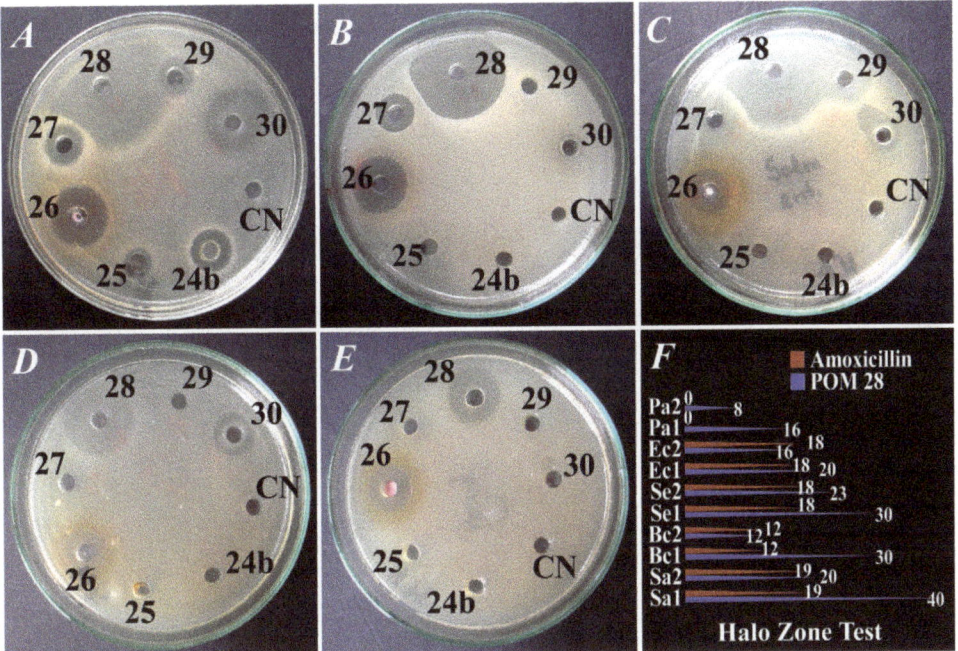

Figure 1. Antibacterial effects (assessed by the disk diffusion method) of various POMs (in black Arabic numerals) against (**A**). *Staphylococcus aureus* ATCC 6538P (abbreviated as Sa in panel F); (**B**). *Bacillus cereus* ATCC 14,579 (abbreviated as Bc in panel F); (**C**). *Salmonella enteritidis* ATCC 13,076 (abbreviated as Se in panel F); (**D**). *Escherichia coli* ATCC 10,536 (abbreviated as Ec in panel F); (**E**). *Pseudomonas aeruginosa* ATCC 27,853 (abbreviated as Pa in panel F); (**F**). Diameters of the inhibition zone (mm, values marked in white) of POM 28 (blue line), $(NH_4)_4(NBu_4)_5[Na(BuSn)_3Sb_9W_{21}O_{86}] \cdot 17H_2O$ compared to Amoxicillin (red line); Sa1, Bc1, Se1, Ec1, Pa1–initial testing; Sa2, Bc2, Se2, Ec2, Pa2–retesting after 6 months. The values of the Halo Zone Test (mm) are highlighted in white on the right side of panel F.

The sensitivity of several Gram-positive and Gram-negative bacterial strains to the twenty-one POMs that exhibited antibacterial effects (via the disk diffusion method) was established by determining their minimum inhibitory concentration (MIC), our results being presented in Table 4 and in the Supplementary Material 2.

We also determined the sensitivity of several Gram-positive and Gram-negative bacterial strains to the twenty-one POMs that exerted antibacterial effects, their minimum bactericidal concentration (MBC) being presented in Table 5. In terms of the MBC, POMs that demonstrated antibacterial effects (via the disk diffusion method) were selectively tested against those bacterial strains on which they exerted bactericidal action.

Table 4. Minimum inhibitory concentration of active POMs.

POM No.	Minimum Inhibitory Concentration (mg/L)				
	S. aureus	B. cereus	S. enteritidis	E. coli	P. aeruginosa
1.	0.625	1.25	1.25	-	0.625
2.	-	1.25	-	-	-
4.	0.625	1.25	-	-	-
7.	1.25	-	-	-	-
8.	1.25	1.25	-	-	-
11.	0.039	-	0.156	-	-
13.	1.25	2.5	0.312	0.078	-
14.	0.156	-	0.312	0.156	0.312
15.	1.25	0.312	0.625	-	-
17.	-	0.625	1.25	-	-
18.	-	-	1.25	-	-
19.	0.156	0.625	0.312	-	-
20.	0.039	0.039	0.156	0.156	0.625
23a.	0.078	-	-	-	-
23b.	0.625	-	-	-	-
24a.	0.078	-	-	-	-
24b.	0.625	-	-	-	-
25.	0.156	-	-	-	-
26.	0.312	0.078	0.156	0.312	0.625
27.	0.156	0.156	-	-	-
28.	0.625	0.0048	0.019	0.078	0.039
29.	0.625	-	0.625	-	-
30.	0.078	0.312	0.312	0.156	-

Table 5. Minimum bactericidal concentration of active POMs.

POM no.	Minimum Bactericidal Concentration (mg/L)				
	S. aureus	B. cereus	S. enteritidis	E. coli	P. aeruginosa
1.	1.25	2.5	-	-	-
2.	-	-	-	-	-
4.	2.5	2.5	-	-	-
7.	1.25	-	-	-	-
8.	2.5	2.5	-	-	-
11.	1.25	-	1.25	-	-
13.	0.625	-	1.25	1.25	-
14.	0.625	-	1.25	1.25	0.625
15.	2.5	0.625	0.625	-	-
17.	-	1.25	2.5	-	-
18.	-	-	2.5	-	-
19.	1.25	1.25	0.625	-	-
20.	1.25	0.625	2.5	1.25	2.5
23a.	1.25	-	-	-	-
23b.	1.25	-	-	-	-
24a.	1.25	-	-	-	-
24b.	1.25	-	-	-	-
25.	2.5	-	-	-	-
26.	2.5	0.625	1.25	1.25	1.25
27.	1.25	0.625	-	-	-
28.	0.625	0.312	0.625	1.25	0.625
29.	2.5	-	0.625	-	-
30.	0.625	0.625	0.312	1.25	-

3. Discussion

3.1. Chemistry of Polyoxometalates

The synthesis of polyoxometalates is a complex process involving molecular reorganization, the reaction mechanisms leading to the formation of new compounds being rather difficult to establish [14,26]. However, several synthesis possibilities are already well documented, such as the indirect synthesis replacing one polyoxoanion with another, or the "all in one pot" synthesis. All parameters need to be strictly controlled, including pH and temperature (essential in triggering the self-assembling mechanisms), the POMs composition and organic/inorganic nature of the solvent(s) or the presence of reducing agents, ionic strength, reflux, hydrothermal or ambient conditions, the type and concentration of oxoanions, the presence/absence and concentration of certain heteroatoms or addenda atoms [26]. For example, the pH of an aqueous solution of POMs needs to be increased (for V) or decreased (for W or Mo) during the more efficient "all in one pot" synthesis in order to increase the nuclearity of the oxoanion fragments [1,3,14,16,26,35]. Incorporation of heteroatoms, lacunary fragments, organic/organo-metallic fragments, transition-metal cations, or ligands significantly contributes to controllable structural changes in POMs' size, shape and architecture, explaining their chemical variability [12,13,22,30].

The results of the **chemical elemental analysis** data are in agreement with the proposed chemical formula (see Table 1), and with the theoretical compositions (see Table 2).

POMs' thermogravimetric analysis revealed the presence of two types of water molecules: coordinated water molecules (in our nanocompounds 1, 2, 4, 11, 21–27) and crystalization water molecules (in a specific number for each POM). The latter are eliminated first, in the second stage the removal of the coordinated water molecules paralleling the POMs' decomposition [12,16]. POM 20 (a butyltin salt cluster) presented neither type, while in POM 2 only coordinated water molecules were observed.

UV electronic spectra of all POMs (in 5×10^{-5} M aqueous solutions) exhibited two characteristic charge-transfer bands of high intensity [22,30] in the region of interest for polyoxoanions. Their contributions (as shown in Table 2) were attributed to specific POM bonds. The very small spectral displacements of the more intense bands, centered at $\nu_2 \sim 200 \pm 10$ nm, corresponding to the $p_\pi(O_t) \rightarrow d_\pi(M)$ transition, were attributed to the M=O_t double bonds, explaining the non-involvement of terminal oxygen atoms in the coordination structure for saturated/lacunary Keggin or *pseudo*-Keggin compounds, as well as their involvement in the coordination structure of all clusters (e.g., in MO_6 octahedra). The broader spectral band displacements, generally centered at $\nu_1 \sim 250 \pm 20$ nm (except for POM 1 and POM 19–at 228 nm, POM 5 and POM 6–at 305 nm), corresponding to the charge transfer transition $p_\pi(O_{c,e}) \rightarrow d_{\pi^*}(M)$, were attributed to the tri-centric bonds M-$O_{c,e}$-M (bridge oxygen atoms connecting MO_6 octahedra via their corners–O_c, and via their edges–O_e, respectively). They also explain the involvement of different oxygen atoms in coordination, as well as the spatial arrangement in each POM structure (lacunary/non-lacunary Keggin/*pseudo*-Keggin compounds, sandwich type or clusters). The second band was generally shifted towards lower energy levels in all cluster structures compared to their ligands because distortion of the MO_6 octahedra due to an intensified inequivalence in these bonds decreases their original spatial symmetry. Our results are similar to other literature data [22,26,30].

The recorded **FTIR vibrational spectra** of POM salts fixed in KBr pellets as potassium salts for the $(VO)^{2+}$ ions, and as sodium, butyltin, ammonium or tetrabutyl-ammonium salts for other POMs, exhibited characteristic bands for their structures [3,14,16,21,22,30]. For each POM, FTIR data were recorded for the 4000–400 cm^{-1} domain, and the polyoxoanion fingerprint region was found to be 1200–400 cm^{-1} [12,36–39]. Their contributions were assigned to specific POM bonds in correlation with their structures, as shown in Table 2. FTIR spectra similarities were observed between POMs of the same class, the shifting and splitting variations in our studied nanocompounds being explained by the influence exerted by addenda atoms, various heteroatoms or transitional metal cations

3.2. POM Pharmacology and Antimicrobial Activity

We found that the initial antibacterial effect of compounds 28, 15, 20, 19 against *S. aureus* was higher than that of amoxicillin. However, another test from the same solutions six months later revealed a drastic decrease of the antibacterial effect, only compound 28 maintaining a stronger activity than that of amoxicillin, although its effect was halved.

This proves that the antibacterial effect of such compounds dissolved in saline buffer (0.15 M NaCl) is severely decreasing in time. Of the other POMs whose activity was lower than that of amoxicillin, compounds 13, 25, and 29 maintained or increased their antibacterial effect over the 6 months interval.

The initial antimicrobial activity upon *B. cereus* was stronger (in compounds 28, 20, 15, 26, 27 and 19, in descending order) or equal (POM 2) to that of amoxicillin. After six months, only POM 20 exerted a greater effect than the reference, POMs 28, 26, 15 and 2 presenting a similar influence. Compounds 1, 13 and 30 maintained their initially lower than the reference antibacterial activity after the follow-up interval, while POMs 4, 8, 17 completely lost their action.

Concerning Gram-negative bacteria, we found that POMs 28, 15, 19, 20 initially exhibited stronger antimicrobial activity against *S. enteritidis* than amoxicillin. Retesting 6 months later evidenced compounds 28 and 19 as stronger than amoxicillin in respect of their antimicrobial activity. In contrast, compound 29 completely lost its action.

The only POM that initially exerted higher antibacterial activity than amoxicillin against *E. coli* was compound 28, but its effect fell below the reference level after six months. The compounds that maintained their antibacterial activity against *E. coli* after the six months interval were POMs 26, 30, 14, 20, 13, however to a much lesser extent than amoxicillin.

We found that five compounds, 28, 26, 14, 20 and 1 (listed in descending order of their antibacterial effects), were the most effective upon *P. aeruginosa*, a bacterium resistant to amoxicillin. Six months after preparing the POMs solutions, compound 28 maintained its activity (even if its effect was halved), compounds 26 and 20 exhibited greater antibacterial effect than in the initial stage, while compound 14 diminished its activity and compound 1 completely lost its action.

POMs 26 and 20 presented higher activity than amoxicillin upon all tested strains except for *E. coli*. Compounds 19 and 15 exerted a better antibacterial action than amoxicillin against strains of *S. aureus*, *B. cereus* and *S. enteritidis*, but *E. coli* and *P. aeruginosa* were found to be resistant to their action. In one-on-one comparisons, compound 2 presented similar activity to amoxicillin against *B. cereus*, as did compound 30 against *S. aureus* and compound 17 against *S. enteritidis*. Presumably higher concentrations of these compounds, closer to amoxicillin levels, would enhance their antibacterial action. Only nine POMs of the thirty tested presented no antibacterial activity in concentrations of 5 µg/well, i.e., compounds 3, 12, 16a, 16b, 21 and 22 (with Si as heteroatom), compounds 5, 6 (with P as heteroatom in their structures), and 9 and 10 (with As and Sb), respectively.

We found out that the strongest inhibitory concentration against *S. aureus* strains was exerted by compounds 20, 11, 23a, 24a and 30. Concerning cultures of *B. cereus*, the most powerful effect was observed in compounds 28, 20, 26 and 27, respectively. Good inhibitory concentrations against *B. cereus* strains were also found for compounds 15, 30, 17 and 19. The most powerful effect on *S. enteritidis* strains was once again exhibited by POM 28, followed by POMs 11, 20, 26. On cultures of *E. coli*, the best effect was validated for compounds 13, 28, and for 14, 20, 30, respectively. Of the five bacterial strains, POMs fared worst against *P. aeruginosa*. Still, a good inhibitory concentration was exhibited by compounds 28, 14, 1, 20, 26.

We found out that the minimum bactericidal concentration against *S. aureus* strains was rather high. The most pronounced bactericidal effect was induced by compounds

13, 14, 28 and 30, seconded by the group comprising compounds 1, 7, 11, 19, 20, 23a, 23b, 24a, 24b and 27. Against *B. cereus*, the minimum bactericidal effect was observed in POMs 28, 15, 20, 26, 27 and 30. Against *S. enteritidis,* the most effective compound proved to be no. 30, 15, 19 and 29. The most pronounced bactericidal effect against *E. coli* was noted in six compounds, 13, 14, 20, 26, 28 and 30. Only five POMs exhibited antibacterial activity against *P. aeruginosa* strain, four of which exerted bactericidal action, and one presented only bacteriostatic effect. On this Gram-negative strain, POMs 14, 28, 26 and 20 were the most efficient.

To conclude, in the initial testing three of the thirty analyzed "smart" nanocompounds (28, 20 and 26) presented antibacterial activity against all five bacterial strains tested, while POMs 1, 13, 14 and 30 missed one target (*E. coli* and *B. cereus*, respectively). POMs 2, 4, 7, 8, 23a, 23b, 24a, 24b, 25 and 27 exerted strong antibacterial effects against one or two Gram-positive strains, while POMs 18 presented antibacterial effects against one Gram-negative strain. All POMs exerting antibacterial effects on Gram-negative strains (1, 11, 13, 14, 15, 17, 18, 19, 20, 26, 28, 29, 30) were active against Gram-positive strains as well. When retesting after six months, we found out that some soluted POMs completely lost their antibacterial activity.

The new compound synthesized and characterized by us, $K_6[(VO)SiMo_2W_9O_{39}]\cdot 11H_2O$ (POM 7) presented antibacterial action only against *S. aureus*, with relatively high MIC and MBC (both 1.25 mg/L).

All results obtained using the disk diffusion method were concordant to the results obtained in terms of MIC and MBC and were in close interdependence with the POMs' structures and the bacterial strain on which they were tested. In these terms, the selectivity and specificity of new antibacterial agents (POMs) is extremely important in drugs design.

3.3. POM Structure-Antibacterial Activity Relationship

We managed to characterize the thirty tested "smart" nanocompounds (POMs) in terms of their antibacterial activity–structure relationship. The antibacterial effects of these compounds are directly dependent on their structure and the type of bacterial strain tested. In several compounds presenting monolacunary, trilacunary or trilacunary/sandwich Keggin structures (POMs 3, 5, 6, 9, 10, 16a, 16b) bacterial growth was not inhibited, as bacteria proved to be resistant to their action. Moreover, the monolacunary Keggin (POM 21) and the monolacunary Wells-Dawson (POM 22), both with mixed addenda atoms, did not exhibit antibacterial activity.

In small amounts (5 µg) cluster structures (POMs 20, 28) and trilacunary/sandwich *pseudo-*Keggin structures (POM 26) exhibited the strongest antibacterial effect on all bacterial strains tested, proving to be more efficient than the tested antibiotic (Amoxicillin, 25 µg). Other polyoxometalates such as nos. 1, 2, 4, 7, 8, 11, 13, 14, 17, 18, 23a, 23b, 24a, 24b, 25, 27, 29 or 30 presented selective and specific antibacterial effect against the bacterial strains tested. They may have had some bacteriostatic effect but are not bactericidal.

Compared to uncomplexed salts, POM 20 (as tributyltin salt), POM 26 (as natrium salt), and POM 28 (cluster structure incorporating organo-metallic fragments, crystalized as tetrabutyl-ammonium and ammonium salt) presented an enhanced antibacterial effect against all five bacterial strains tested, including the Gram-negative *P. aeruginosa* strain against which amoxicillin had no effect.

Our results indicate that suppression of bacterial cell proliferation was initially inoculation-dependent and decreased in parallel with the progressive decrease of the POMs' concentration, according to literature data [18,30,31,40].

Although mono-lacunary Keggin species were presumed to be more efficient than their saturated structures, our mono-lacunary Keggin (POM 21) and mono-lacunary Wells-Dawson (POM 22), both with mixed addenda atoms, did not present antibacterial effects. However, the antibacterial activity is not conditioned by the existence of the lacuna [31,40], because it was found that substituted lacunary Keggin structures present a higher antibacterial action than the original mono-lacunary structures, the effect being due to the transitional metal MT^{n+} occupying the lacuna [30].

The antibacterial activity of POMs was demonstrated on several Gram-positive and Gram-negative bacterial strains resistant to β-lactam antibiotics (penicillin, cephalosporins) against some reference bacterial strains. Yamase and co-workers noted that, in combination with β-lactamase inhibitors, POMs restore antibiotic effectiveness. They concluded that POMs with Keggin complete or lacunary structures, Wells-Dawson structures, or double sandwich Keggin structures enhance antibacterial activity against methicillin- (MRSA) and vancomycin-resistant *S. aureus* (VRSA) [31].

Similarly, polyoxoanions such as $[KAs_4W_{40}O_{140}]^{27-}$ or $[KSb_9W_{21}O_{86}]^{18-}$ demonstrated antibacterial activity against the *Helicobacter pylori* strain, which is resistant to metronidazole and clarithromycin. Ultrastructural changes into coccoid forms under the action of $K_{27}[KAs_4W_{40}O_{140}] \cdot H_2O$ were evidenced by scanning electron microscopy [28]. *Helicobacter pylori* (ID3023) is a major pathogen associated with the development of duodenal and gastric cancer, duodenal ulcers, gastritis, and gastric ulcers, affecting about half of the world's population [28]. World Health Organization 2018 data ranked this pathogen on top of the list in terms of incidence, mortality and prevalence of cancers attributable to infections [41]. Several possible mechanisms of POMs antibacterial actions have been described. For instance, these compounds may cross the peptidoglycans layer [30,40], or may penetrate the bacterial membrane via WtpABC and TupABC transporters [42], thus leading to the disintegration of the peptidoglycan layer and the dissolution of their membranes [28]. Inhibition of the DNA to RNA transcription by directly interacting with DNA molecules was also suggested for low-molecular-weight compounds able to electrostatically disrupt the cell envelope and penetrate into the bacterial cell [43]. We will address this problem in further studies.

As Yamase and co-workers postulated, some properties and characteristics of POMs, such as their redox behaviors, strong negative charges and chemical stability, are responsible for their strong antibacterial activity [18,30,44].

These nanocompounds destined to be the active ingredient of the potential pharmaceutical product are considered "smart" because of the way the huge clusters are formed (self-assembling of "block" units which then combine in "wheel" or "ball" structures) [12,45–49] and how they are modeled to specifically recognize targeted biological substrates. As a result of such properties, about 100 research studies relating POMs to cancer were published in the last decade [50]. These syntheses generate low amounts of chemical residues, being environmentally friendly, as literature data point out that the nanoPOMs' maximal efficiency concentrations are often in the nanomolar range [22,23,51].

We studied the toxicity of two of our nanoPOMs (10, 28) in previous works. Based on our in vivo studies [21], we concluded that POM 28 and particularly POM 10 presented significant hypoglycemiant activity following oral treatment of rats with streptozotocin-induced diabetes. The main cause seems to be the prevention of pancreatic β-cells apoptosis, as observed by transmission electron microscopy (TEM), but our data also revealed stimulation of insulin synthesis by pancreatic β-cells in diabetic rats. Our TEM ultrastructural studies demonstrated the ability of POM 10 and POM 28 to prevent the hepatotoxicity of streptozotocin. Our in vitro studies revealed the significant biological activity of POM 10 and POM 28 as active stimuli for the differentiation of stem cells into insulin-producing cells [22]. MTT assay on human umbilical vein endothelial cells and human bone marrow adult mesenchymal stem cells proved the low, dose-dependent toxicity of POM 10 and POM 28 [22]. In view of the design of future drugs, identifying new compounds with strong antibacterial effects is an ongoing challenge addressing today's technological advances. POMs have a real potential towards such a goal and our study opens new directions in future research. The limitations of our study are due to the scarcity of data on POMs' stability in the culture media of different bacterial strains and at physiological pH (a prerequisite for drugs to be administered to humans), and of toxicity studies (a shortcoming that needs to be addressed taking into account that they are inorganic compounds).

4. Materials and Methods

4.1. Synthesis and Physico-Chemical Characterizations of Polyoxometalates

4.1.1. Reagents and Chemical Materials

All solvents of analytical purity and all chemical substances used in the synthesis of polyoxometalates were purchased from Sigma-Aldrich (N.V./S.A., Bornem, Belgium or Co. LLC., St. Louis, MO, USA) and Merck (KGaA, Darmstadt, Germany), respectively.

The list of analytically pure reagents used in the syntheses included acetone, acetonitrile, glacial acetic acid, HCl (37%), H_3PO_4 (85%), $HClO_4$ (70%), sodium acetate trihydrate, trisodium citrate dihydrate, n-butyltin trichloride, tributyltin chloride, tetrabutylammonium bromide, KCl, NaCl, Na_3VO_4, $VOSO_4 \cdot 5H_2O$, $H_3[PMo_{12}O_{40}] \cdot 13H_2O$, $H_3[PW_{12}O_{40}] \cdot 12H_2O$, $H_4[SiW_{12}O_{40}] \cdot 14H_2O$, $Na_2HAsO_4 \cdot 7H_2O$, $Na_2MoO_4 \cdot 2H_2O$, $Na_2HPO_4 \cdot 12H_2O$, $Na_2GeO_3 \cdot H_2O$, $Na_2WO_4 \cdot 2H_2O$, $Bi(NO_3)_3 \cdot 5H_2O$, Sb_2O_3, $CoCl_2 \cdot 6H_2O$, $MnCl_2 \cdot 4H_2O$, $FeCl_3 \cdot 6H_2O$, $FeCl_2 \cdot 4H_2O$ etc.

Bi-distilled and deionized water produced with a FI-Streem III Cyclon Glass Still Bi-Distiller (Sanyo/Gallenkamp PLC, Cambridge, UK) was used for all solutions prepared during the POMs syntheses. All POMs were recrystallized from a minimum volume of bi-distilled water, crystals of different shapes and various colors were obtained. Only three POMs (16, 23 and 24) required double recrystallization.

4.1.2. Synthesis of Polyoxometalates

POMs are generally difficult to be synthesized, involving either solvothermal or hydrothermal methods requiring high pressure and temperature (implying high energy consumption) or ones using toxic organic solvents (incompatible with the development of green chemistry) [35,39,52–54]. We carefully selected the method based on each POM's specifics, sometimes modifying protocols mentioned in literature [3,12,14,16,18,22,30,53].

Initially we employed the **two-step method**, involving the ligand synthesis followed by the cluster formation, for synthesizing ten POMs, i.e., the Keggin (saturated/monolacunary) structures with mixed addenda atoms, the mono-lacunary Wells-Dawson polyoxometalate with mixed addenda atoms, and the six clusters-type structures. Certain molar ratios of organic and organometallic fragments or transition-metal cations were carefully calculated according to the reactions' stoichiometry [12,22,30,36,54]. An important aspect in the first stage was to obtain a ligand precipitate of high purity. It took up to five days for some POM crystals to precipitate, the process yields being below 70%. The reaction products were then recrystallized to achieve a desirable purity. The new compound, $K_6[(VO)SiMo_2W_9O_{39}] \cdot 11H_2O$ (POM 7), was also synthesized via this method.

For twenty POMs we employed the **"all in one pot" method** based on self-assembling mechanisms. Crystalline powders of different metal ions, e.g., germanium(IV) oxide, vanadyl sulfate, molybdates, vanadates, tungstates, sodium germanate, antimony(III) oxide, were stoichiometricaly mixed with HCl (6 M), the salt of a transition metal cation (TM) if needed, and NaCl (for most POMs, excepting those with vanadyl ions where KCl was added) to produce POM in aqueous solutions [1,14–16,22,30] which precipitated in up to two days. The precipitates were then filtered and desiccated, beautifully colored POMs crystals being obtained with reaction yields of 71–84%. Some POMs required recrystallization (in order) to reach the desired purity.

All syntheses of nanoPOMs were detailed in Supplementary Material 3.

4.1.3. Physico-Chemical Characterizations of Polyoxometalates

The instrumental methods employed in the physico-chemical characterization of the synthesized POMs were elemental chemical analysis, thermogravimetric analysis, electronic spectroscopy in the UV range, FTIR and NMR spectroscopy.

Elemental chemical analysis served to determine the composition of the various elements. We determined the presence of P, As, Sb and Si by atomic absorption spectrometry using a Perkin-Elmer 3030 AA spectrophotometer (Perkin-Elmer, Norwalk, CT, USA). For the C, N and H atoms from organic and organometallic fragments, a Vario EL analyzer

(Elementar Analysensysysteme GmbH, Hanau, Germany) was employed. The contents of Sn, Mo, V, Sb and W were determined by inductively coupled plasma atomic emission spectroscopy using a RigaKu Spectro CIROS^{CCD} spectrometer (RigaKu Co., Tokyo, Japan). Finally, Na and K were determined by flame photometry with an Eppendorf FEP flame photometer (Eppendorf GmbH, Hamburg, Germany).

The thermogravimetric analysis was conducted to determine the water content (crystallization/lattice water and coordination water molecules) for each POM, using a Mettler-Toledo TG/S DTA 851 thermogravimeter (Mettler-Toledo GmbH, Greifensee, Switzerland) with a platinum crucible, 20 mL/min N_2 flow, and 5 °C/min heating rate.

Electronic spectra in the UV range were recorded on a Shimadzu Specord UV-VIS-75 (Shimadzu Europe GmbH, Duisburg, Germany), using quartz cells, with a path length of 1 cm. All POM samples were used in 5×10^{-5} M aqueous solutions.

Vibrational spectroscopy served to establish the presence of certain bond types in the POMs' structures. A Jasco 610 FTIR spectrophotometer (Jasco Int. Co Ltd., Gross-Umstadt, Germany) set at a resolution of 0.5 cm^{-1}, in the wavenumber range between 7800–350 cm^{-1}, was used in the process. The FTIR absorption spectra were recorded in KBr pellets, and all FTIR spectra were analyzed using a Jasco Spectra Manager Version 2.05.03 software (Jasco Int. Co. Ltd., Gross-Umstadt, Germany).

NMR spectra for all POMs with organic or organometallic fragments have been published in our previous work [22]. They were recorded at room temperature, using a Varian Gemini-300 spectrophotometer (Varian Inc. NMR Systems, Palo Alto USA) operating at 300 MHz for ^{1}H spectra and at 75.47 MHz for ^{13}C spectra, respectively. To record the spectra, each POM was dissolved in CDCl$_3$ (solvent), while TMS (Si(CH$_3$)$_4$) and n-butyltin trichloride (n-C$_4$H$_9$SnCl$_3$) served as standard reference.

4.2. Antimicrobial Activity of Polyoxometalates

4.2.1. Reagents and Materials

Five reference microbial strains, two Gram-positive (*S. aureus* ATCC 6538P, *B. cereus* ATCC 14579) and three Gram-negative (*E. coli* ATCC 10536, *S. enteritidis* ATCC 13076, *P. aeruginosa* ATCC 27853), were used for in vitro susceptibility testing. These reference microbial strains were obtained from the American Type Culture Collection (ATCC, Manassas, VA, USA). In this experiment we used ultrapure water produced by a Millipore Milli-Q50 Ultra-Pure Water System, with 18.00 MΩ·cm (Millipore S.A., Molsheim, France). Sterile media and various consumables were also needed for the antimicrobial characterization of the POMs.

4.2.2. Disk Diffusion Method

The antimicrobial activity of all the synthesized POMs was qualitatively determined using the **disk diffusion susceptibility method**, according to the standards developed by the Clinical and Laboratory Standards Institute [55,56], as previously described in literature [57–59], that have been adapted for the purposes of this screening.

For each of the five species an initial suspension of bacterial cultures was inoculated on nutrient agar plates (Merck KGaA, Darmstadt, Germany), incubated for 24 h at 37 ± 2 °C and resuspended in a physiological saline buffer to a 10^6 CFU/mL concentration (on a 0.5 McFarland scale) and was further inoculated on Muëller Hinton agar plates (Merck KGaA, Darmstadt, Germany). The initial inoculum was similar to that prepared for the classical antibiotic susceptibility test, so POMs' effects (in sensitivity terms) were comparable to those of the antibiotic tested, i.e., Amoxicillin (25 µg; Oxoid Ltd., Basingstoke, UK).

After inoculation, the medium surface was dried and a number of eight wells were radially drilled 1.5 cm from the outer edge, 3 cm apart. From aliquot samples of 1 mg/mL of each nanocompound dissolved in physiological solution (NaCl 0.15 M) 5 µL were placed in each of the eight wells and let 30 min to diffuse into the agar plates. The plates were then incubated for 24 h at 37 ± 2 °C. Amoxicillin served as a positive control, while the physiological solution (NaCl 0.15 M) was used as a negative control. Readings were conducted by measuring the diameter of the inhibition zone (Halo Zone Test, in mm). All tests were triplicated, and the measured diameters of the inhibition zone were expressed (in mm) as mean ± standard deviation.

In order to observe their stability in physiological solutions and to check the evolution of their antimicrobial activity, the POM solutions were retested six months after their preparation. The diameters of the inhibition zone determined during retesting are highlighted below the initial values in Table 3.

4.2.3. Minimum Inhibitory Concentration (MIC)

In order to quantify their effectiveness, the **Minimum Inhibitory Concentration** of these POMs on bacterial activity was determined using the broth microdilution method described by Quinn et al. [60], Markey et al. [57], the Clinical and Laboratory Standards Institute [61,62], adapted for this experiment.

Testing was conducted on the same Gram-positive and Gram-negative bacteria. Microorganisms' suspensions in saline buffer (NaCl solution 0.15 M) obtained according to Section 4.2.2 were inoculated. Ten successive dilutions of the POM solutions (1/2 to 1/1024) in nutrient broth (Merck KGaA, Darmstadt, Germany) were performed. As a result, amounts of 2.5 to 0.0048 mg/well of each active POM were placed in sterile microplates (one each for every nanocompound). The microplates thus prepared were incubated at 37 ± 2 °C for 24 h.

Comparisons of the amount of bacterial growth in each well containing POM solutions with the amount of growth in the growth-control wells were performed and the maximal dilution for which the tested POMs inhibited bacterial growth was established.

4.2.4. Minimum Bactericidal Concentration (MBC)

The minimum bactericidal activity (**Minimum Bactericidal Concentration**) of POMs for the five bacterial species described above was established using the microdilution method. 5 µL of POMs solutions from each of the wells where the inhibitory effect was observed were introduced in the same nutrient broth as mentioned above and inoculations on nutrient agar sterile plates (Merck KGaA, Darmstadt, Germany) were performed for similar dilutions. The plates thus prepared were incubated for 24 h at 37 ± 2 °C and bacterial growth was observed. The reading of the results was performed at 24 h by observing the evolution of the bacterial growth on the solid medium.

Polyoxometalates were classified as bactericidal if they prevented growth on this medium. The minimum bactericidal concentration was determined for the lowest dilution at which bacterial growth was blocked.

5. Conclusions

This research achieved its goals to synthesize "smart" nanocompounds based on different structures, to characterize them in terms of structure-property relations, to investigate their molecular mechanisms and to test in vitro their antibacterial effects, in order to formulate and implement potential drugs meant to replace similar products obtained via organic syntheses. Herein, we characterized in terms of chemical structure-antimicrobial activity relationship the following types: a. the Keggin series–polyoxometalates with structures such as saturated/mono- and trilacunary Keggin, mono- and trilacunary *pseudo*-Keggin, saturated/monolacunary Keggin with mixed addenda atoms, trilacunary Keggin/*pseudo*-Keggin by sandwich type; b. one monolacunary Wells-Dawson structure with mixed addenda atoms; c. six clusters. For all active compounds the stability of their antimicrobial effects was also investigated.

We identified three POMs that presented sound antibacterial activity against all five bacterial strains tested: POMs 26 (a tri-lacunary *pseudo*-Keggin sandwich type), 20 (a butyltin salt cluster) and 28 (a cluster derivatized with organometallic fragments). Of these, only POM 28 has constantly exhibited a stronger effect compared to amoxicillin (including after 6 months retesting, even if its effects were diminished by half). POMs

1, 13, 14 and 30 were effective against four bacterial strains. POMs 2, 4, 7, 8, 23a, 23b, 24a, 24b, 25 and 27 exhibited antibacterial effects against Gram-positive strains, while POMs 18 was effective against one Gram-negative strain, *S. enteritidis*. All other POMs exerting antibacterial effects on Gram-negative strains (1, 11, 13, 14, 15, 17, 19, 20, 26, 28, 29, 30) were active against Gram-positive strains as well. In contrast, nine compounds (3, 5, 6, 9, 10, 12, 16a, 16b, 21 and 22) had no antibacterial actions at all. Presumably higher concentrations of these compounds, closer to the amoxicillin level, would enhance their antibacterial action. New studies in terms of selectivity and specificity of these potential antibacterial agents are required to clarify these extremely important aspects needed to be considered in drug designs. The new compound synthesized and characterized by us, $K_6[(VO)SiMo_2W_9O_{39}] \cdot 11H_2O$ (POM 7), a mono-lacunary Keggin with mixed addenda atoms, presented antibacterial activity only against *S. aureus*. These nanocompounds present the disadvantage of being essentially inorganic substances and their toxicity is a matter of concern. Nevertheless, the nano-revolution will inevitably transfer spectacular new technological advances into life sciences. All these novel steps leading to functionalized biocompatible, non-toxic, inorganic compounds, stable in physiological conditions, modelled for exerting maximal antimicrobial effects, can open an alternative approach to the classical treatment of infectious diseases.

A true success in POMs chemistry and pharmacology would lead to the synthesis of pharmaceutical nanocompounds mimicking the behavior of biomacromolecules, with remarkable antibacterial activity effective against certain pathogens with acquired antibiotic resistance, able to regenerate animal and human tissue while annihilating the infection.

Supplementary Materials: The following are available online at https://www.mdpi.com/article/10.3390/ph15010033/s1, In Supplementary Material 1, the synthesis method and physico-chemical characterization of $K_6[(VO)SiMo_2W_9O_{39}] \cdot 11H_2O$ (POM 7) and its ligand $K_8[SiMo_2W_9O_{39}] \cdot 14H_2O$ (L 7); Figure S1: TG/DTG/DTA curves of $K_6[(VO)SiMo_2W_9O_{39}] \cdot 11H_2O$; Figure S2: The proposed structure for POM 7: $K_6[(VO)SiMo_2W_9O_{39}] \cdot 11H_2O$; Figure S3: FTIR spectra of POM 7, $K_6[(VO)SiMo_2W_9O_{39}] \cdot 11H_2O$, and his ligand (L7), $K_8[SiMo_2W_9O_{39}] \cdot 14H_2O$; Figure S4: UV (upper right window) and VIS electronic spectra of $K_6[(VO)SiMo_2W_9O_{39}] \cdot 11H_2O$; Figure S5: ESR spectrum of $K_6[(VO)SiMo_2W_9O_{39}] \cdot 11H_2O$ complex powder at room temperature (solid line) and its simulated spectrum (dashed line); Table S1. FTIR spectral data (M = metal atoms); Table S2. Spectral data from UV electronic spectrum of $K_6[(VO)SiMo_2W_9O_{39}] \cdot 11H_2O$. In Supplementary Material 2, FTIR vibrational spectra of 27 nanoPOMs (Figures S6–S32); UV electronic spectra of 27 nanoPOMs (Figures S33–S59); Bacterial inoculi microplates prepared from various reference strains inhibited by the 21 analyzed nanoPOMs (Figures S60–S64). In Supplementary Material 3, all syntheses of nanoPOMs (Pages 1–12).

Author Contributions: Conceptualization, Ş.B., M.L.V., H.V.M. and N.I.F.; synthesis and analysis of POMs, Ş.B. and D.R.; FTIR analysis of POMs, M.F. and Ş.B.; biological investigation, E.P. and F.C.; biological methodology, R.U., G.Z.N. and M.L.V.; biological experiments validation, R.U. and M.L.V.; biological supervisors, H.V.M. and N.I.F.; writing—original draft preparation, Ş.B. and M.L.V.; writing—review and editing, H.V.M. and N.I.F. All authors have read and agreed to the published version of the manuscript.

Funding: This research received no external funding.

Institutional Review Board Statement: Not applicable.

Informed Consent Statement: Not applicable.

Data Availability Statement: Data is contained within the article and all three Supplementary Materials.

Conflicts of Interest: The authors declare no conflict of interest.

References

1. Long, D.L.; Tsunashima, R.; Cronin, L. Polyoxometalates: Building Blocks for Functional Nanoscale Systems. *Angew. Chem. Int. Ed.* **2010**, *49*, 1736–1758. [CrossRef] [PubMed]

2. Pope, M.T. *Heteropoly and Isopoly Oxometalates*, 1st ed.; Springer: Berlin/Heidelberg, Germany, 1983; pp. 32–65.
3. Fan, D.; Hao, J. Polyoxometalate-Based Assembly. In *Self-Assembled Structures. Properties and Applications in Solution and on Surfaces*, 1st ed.; Hao, J., Ed.; CRC Press: Boca Raton, FL, USA; Taylor&Francis Group: Boca Raton, FL, USA, 2011; pp. 141–174.
4. Hervé, G.; Tézé, A.; Contant, R. General Principles of the Synthesis of Polyoxometalates in Aqueous Solution. In *Polyoxometalate Molecular Science NATO Science Series II*, 1st ed.; Borrás-Almenar, J.J., Coronado, E., Müller, A., Pope, M.T., Eds.; Kluwer Academic Publishers: Dordrecht, The Netherlands, 2003; Volume 98, pp. 33–54. [CrossRef]
5. Ivanova, S. Hybrid Organic-Inorganic Materials Based on Polyoxometalates and Ionic Liquids and Their Application in Catalysis. *ISRN Chem. Eng.* **2014**, *2014*, 963792. [CrossRef]
6. Hill, C.L. Progress and Challenges in Polyoxometalate-Based Catalysis and Catalytic Materials Chemistry. *J. Mol. Catalysis A Chem.* **2007**, *262*, 2–6. [CrossRef]
7. Pope, M.T.; Müller, A. Polyoxometalate Chemistry: An Old Field with New Dimensions in Several Disciplines. *Angew. Chem. Int. Ed.* **1991**, *30*, 34–48. [CrossRef]
8. Katsoulis, D.E. A Survey of Applications of Polyoxometalates. *Chem. Rev.* **1998**, *98*, 359–387. [CrossRef] [PubMed]
9. Clemente-Juan, J.; Coronado, E. Magnetic Clusters from Polyoxometalate Complexes. *Coord. Chem. Rev.* **1999**, *193*, 361–394. [CrossRef]
10. Yamase, T.; Abe, H.; Ishikawa, E.; Nojiri, H.; Ohshima, Y. Structure and Magnetism of [n-BuNH$_3$]$_{12}$[Cu$_4$(GeW$_9$O$_{34}$)$_2$]·14H$_2$O Sandwiching a Rhomblike Cu$_4^{8+}$ Tetragon through α-Keggin Linkage. *Inorg. Chem.* **2009**, *48*, 138–148. [CrossRef] [PubMed]
11. Keita, B.; Nadjo, L. Electrochemistry of Isopoly and Heteropoly Oxometalates. In *Encyclopedia of Electrochemistry*, 1st ed.; Bard, A.J., Stratmann, M., Eds.; Wiley-VCH: Weinheim, Germany, 2006; Volume 7, pp. 607–700.
12. Rusu, D.; Bâlici, Ş. *Polioxometalații. Aplicații Biomedicale*, 1st ed.; Casa Cărții de Știință: Cluj-Napoca, România, 2013.
13. Marcu, G.; Rusu, M. *Chimia Polioxometalaților*, 1st ed.; Tehnică: București, România, 1997; pp. 22–55.
14. Long, D.L.; Burkholder, E.; Cronin, L. Polyoxometalate Clusters, Nanostructures and Materials: From Self-Assembly to Designer Materials and Devices. *Chem. Soc. Rev.* **2007**, *36*, 105–121. [CrossRef]
15. Cronin, L. From the Molecular to the Nanoscale: Synthesis, Structure, and Properties. In *Comprehensive Coordination Chemistry. II: From Biology to Nanotechnology*, 2nd ed.; McCleverty, J.A., Meyer, T.J., Eds.; Elsevier Science: Amsterdam, The Netherlands, 2004; Volume 7, pp. 7–52.
16. Bernold, H. Polyoxometalates: Introduction to a Class of Inorganic Compounds and Their Biomedical Applications. *Front. Biosci.* **2005**, *10*, 275–287. [CrossRef]
17. Rhule, J.T.; Hill, C.L.; Judd, D.A.; Schinazi, R.F. Polyoxometalate in Medicine. *Chem. Rev.* **1998**, *98*, 327–357. [CrossRef]
18. Yamase, T. Anti-Tumor, -Viral, and -Bacterial Activities of Polyoxometalates for Realizing an Inorganic Drug. *J. Mater. Chem.* **2005**, *15*, 4773–4782. [CrossRef]
19. Nomiya, K.; Torii, H.; Hasegawa, T.; Nemoto, Y.; Nomura, K.; Hashino, K.; Uchida, M.; Kato, Y.; Shimizu, K.; Oda, M. Insulin Mimetic Effect of a Tungstate Cluster. Effect of Oral Administration of Homo-Polyoxotungstates and Vanadium-Substituted Polyoxotungstates on Blood Glucose Level of STZ Mice. *J. Inorg. Biochem.* **2001**, *86*, 657–667. [CrossRef]
20. Ilyas, F.; Shah, H.S.; Al-Oweini, R.; Kortz, U.; Iqbal, J. Antidiabetic Potential of Polyoxotungstates: *In Vitro* and *In Vivo* Studies. *Metallomics* **2014**, *6*, 1521–1526. [CrossRef] [PubMed]
21. Bâlici, Ş.; Wankeu-Nya, M.; Rusu, D.; Nicula, G.Z.; Rusu, M.; Florea, A.; Matei, H. Ultrastructural Analysis of *In Vivo* Hypoglycemiant Effect of Two Polyoxometalates in Rats with Streptozotocin-Induced Diabetes. *Microsc. Microanal.* **2015**, *21*, 1236–1248. [CrossRef] [PubMed]
22. Bâlici, Ş.; Şușman, S.; Rusu, D.; Nicula, G.Z.; Sorițău, O.; Rusu, M.; Biris, A.S.; Matei, H. Differentiation of Stem Cells into Insulin-Producing Cells under the Influence of Nanostructural Polyoxometalates. *J. Appl. Toxicol.* **2016**, *36*, 373–384. [CrossRef]
23. Prudent, R.; Moucadel, V.; Laudet, B.; Barette, C.; Lafanechere, L.; Hasenknopf, B.; Li, J.; Bareyt, S.; Lacote, E.; Thorimbert, S.; et al. Identification of Polyoxometalates as Nanomolar Noncompetitive Inhibitors of Protein Kinase CK2. *Chem. Biol.* **2008**, *15*, 683–692. [CrossRef]
24. Ogata, A.; Yanagie, H.; Ishikawa, E.; Mitsui, S.; Yamashita, A.; Hasumi, K.; Takamoto, S.; Yamase, T.; Eriguchi, M. Antitumour Effect of Polyoxomolybdates: Induction of Apoptotic Cell Death and Autophagy in *In Vitro* and *In Vivo* Models. *Br. J. Cancer* **2008**, *98*, 399–409. [CrossRef]
25. Wang, J.; Qu, X.; Qi, Y.; Li, J.; Song, X.; Li, L.; Yin, D.; Xu, K.; Li, J. Pharmacokinetics of Anti-HBV Polyoxometalate in Rats. *PLoS ONE* **2014**, *9*, e98292. [CrossRef]
26. Yamase, T. Polyoxometalates for Molecular Devices: Antitumor Activity and Luminescence. In *Polyoxometalates: From Platonic Solids to Anti-Retroviral Activity*, 1st ed.; Pope, M.T., Müller, A., Eds.; Kluwer Academic: Dordrecht, The Netherlands, 1994; pp. 337–358.
27. Sarafianos, S.G.; Kortz, U.; Pope, M.T.; Modak, M.J. Mechanism of Polyoxometalate-Mediated Inactivation of DNA Polymerases: An Analysis with HIV-1 Reverse Transcriptase Indicates Specificity for the DNA-Binding Cleft. *Biochem. J.* **1996**, *319*, 619–626. [CrossRef]
28. Inoue, M.; Segawa, K.; Matsunaga, S.; Matsumoto, N.; Oda, M.; Yamase, T. Antibacterial Activity of Highly Negative Charged Polyoxotungstates, K$_{27}$[KAs$_4$W$_{40}$O$_{140}$] and K$_{18}$[KSb$_9$W$_{21}$O$_{86}$], and Keggin-Structural Polyoxotungstates against *Helicobacter pylori*. *J. Inorg. Biochem.* **2005**, *99*, 1023–1031. [CrossRef]

29. Tajima, Y. The Effects of Tungstophosphate and Tungstosilicate on Various Stress Promoters Transformed in *Escherichia Coli*. *J. Inorg. Biochem.* **2003**, *94*, 155–160. [CrossRef]
30. Bâlici, Ș.; Niculae, M.; Pall, E.; Rusu, M.; Rusu, D.; Matei, H. Antibiotic-Like Behavior of Polyoxometalates. In Vitro Comparative Study: Seven Polyoxotungstates—Nine Antibiotics against Gram-Positive and Gram-Negative Bacteria. *Rev. Chim.* **2016**, *67*, 485–490.
31. Inoue, M.; Suzuki, T.; Fujita, Y.; Oda, M.; Matsumoto, N.; Yamase, T. Enhancement of Antibacterial Activity of β-Lactam Antibiotics by $[P_2W_{18}O_{62}]^{6-}$, $[SiMo_{12}O_{40}]^{4-}$, and $[PTi_2W_{10}O_{40}]^{7-}$ against Methicillin-Resistant and Vancomycin-Resistant Staphylococcus aureus. *J. Inorg. Biochem.* **2006**, *100*, 1225–1233. [CrossRef] [PubMed]
32. Vică, M.L.; Glevitzky, M.; Tit, D.M.; Behl, T.; Heghedus-Mindru, R.C.; Zaha, D.C.; Ursu, F.; Popa, M.; Glevitzky, I.; Bungau, S. The Antimicrobial Activity of Honey and Propolis Extracts from the Central Region of Romania. *Food Biosci.* **2021**, *41*, 101014. [CrossRef]
33. Vică, M.L.; Glevitzky, I.; Glevitzky, M.; Siserman, C.V.; Matei, H.V.; Teodoru, C.A. Antibacterial Activity of Propolis Extracts from the Central Region of Romania against Neisseria gonorrhoeae. *Antibiotics* **2021**, *10*, 689. [CrossRef] [PubMed]
34. Junie, L.M.; Vică, M.L.; Glevitzky, M.; Matei, H.V. Physico-Chemical Characterisation and Antibacterial Activity of Different Types of Honey Tested on Stains Isolated from Hospitalised Patients. *J. Apic. Sci.* **2016**, *60*, 5–17. [CrossRef]
35. Yan, T.-T.; Xuan, Z.-X.; Wang, S.; Zhang, X.; Luo, F. Facile One-Pot Construction of Polyoxometalate-Based Lanthanide-Amino Acid Coordination Polymers for Proton Conduction. *Inorg. Chem. Commun.* **2019**, *105*, 147–150. [CrossRef]
36. Sablinskas, V.; Steiner, G.; Hof, M. Section II: Methods 1: Optical Spectroscopy, Applications. In *Handbook of Spectroscopy*, 1st ed.; Gauglitz, G., Vo-Dinh, T., Eds.; WILEY-VCH, Verlag: Weinheim, Germany, 2003; Volume 1, pp. 98–103.
37. Nakamoto, K. Applications in Inorganic Chemistry. In *Infrared and Raman Spectra of Inorganic and Coordination Compounds Part A: Theory and Applications in Inorganic Chemistry*, 5th ed.; Nakamoto, K., Ed.; John Wiley & Sons Inc: New York, NY, USA, 1997; pp. 38–52.
38. Čolović, M.B.; Lacković, M.; Lalatović, J.; Mougharbel, A.S.; Kortz, U.; Krstić, D.Z. Polyoxometalates in Biomedicine: Update and Overview. *Curr. Med. Chem.* **2020**, *27*, 362–379. [CrossRef]
39. Gumerova, N.I.; Rompel, A. Polyoxometalates in Solution: Speciation under Spotlight. *Chem. Soc. Rev.* **2020**, *49*, 7568–7601. [CrossRef]
40. Inoue, M.; Suzuki, T.; Fujita, Y.; Oda, M.; Matsumoto, N.; Iijima, J.; Yamase, T. Synergistic Effect of Polyoxometalates in Combination with Oxacillin against Methicillin-Resistant and Vancomycin-Resistant *Staphylococcus Aureus*: A High Initial Inoculum of 1×10^8 cfu/ml for In Vivo Test. *Biomed. Pharmacother.* **2006**, *60*, 220–226. [CrossRef]
41. World Health Organization (WHO); International Agency for Research on Cancer (IARC). [Internet]. Cancers Attributable to Infections. Available online: https://gco.iarc.fr/causes/infections/tools-pie?mode=2&sex=0&population=who&continent=0&country=0&population_group=0&cancer=0&key=attr_cases&lock_scale=0&pie_mode=1&nb_results=5 (accessed on 30 September 2021).
42. Aguilar-Barajas, E.; Díaz-Pérez, C.; Ramírez-Díaz, M.I.; Riveros-Rosas, H.; Cervantes, C. Bacterial Transport of Sulfate, Molybdate, and Related Oxyanions. *Biometals* **2011**, *24*, 687–707. [CrossRef]
43. Bijelic, A.; Aureliano, M.; Rompel, A. The Antibacterial Activity of Polyoxometalates: Structures, Antibiotic Effects and Future Perspectives. *Chem. Commun.* **2018**, *54*, 1153–1169. [CrossRef] [PubMed]
44. Fukuda, N.; Yamase, T.; Tajima, Y. Inhibitory Effect of Polyoxotungstates on the Production of Penicillin-binding Proteins and β-Lactamase against Methicillin-resistant Staphylococcus aureus. *Biol. Pharm. Bull.* **1999**, *22*, 463–470. [CrossRef]
45. Müller, A.; Kögerler, P.; Bögge, H. Pythagorean Harmony in the World of Metal Oxygen Clusters of the Mo11 Type: Giant Wheels and Spheres both Based on a Pentagonal Type Unit. In *Molecular Self-Assembly: Organic versus Inorganic Approaches (Structure and Bonding, Book Series: Structure)*, 1st ed.; Fujita, M., Ed.; Springer: Berlin/Heidelberg, Germany, 2000; Volume 96, pp. 203–236.
46. Müller, A.; Kögerler, P.; Dress, A.W.M. Giant Metal-Oxide-Based Spheres and Their Topology: From Pentagonal Building Blocks to Keplerates and Unusual Spin Systems. *Coord. Chem. Rev.* **2001**, *222*, 193–218. [CrossRef]
47. Müller, A.; Kögerler, P.; Kuhlmann, C. A Variety of Combinatorially Linkable Units as Disposition:† From a Giant Icosahedral Keplerate to Multi-Functional Metal–Oxide Based Network Structures. *Chem. Commun.* **1999**, *15*, 1347–1358. [CrossRef]
48. Müller, A.; Serain, C. Soluble Molybdenum Blues-des Pudels Kern. *Acc. Chem. Res.* **2000**, *33*, 2–10. [CrossRef]
49. Prescott, D.M. *Cells: Principles of Molecular Structure and Function*; Jones and Bartlett Publishers: Boston, MA, USA, 1988; pp. 23–58.
50. Aureliano, M.; Gumerova, N.; Sciortino, G.; Garribba, E.; Rompel, A.; Crans, D. Polyoxovanadates with Emerging Biomedical Activities. *Coord. Chem. Rev.* **2021**, *447*, 214143. [CrossRef]
51. Yamamoto, N.; Schols, D.; de Clercq, E.; Debyser, Z.; Pauwels, R.; Balzarini, J.; Nakashima, H.; Baba, M.; Hosoya, M.; Snoeck, R.; et al. Mechanism of Anti-Human Immunodeficiency Virus Action of Polyoxometalates, A Class of Broad-Spectrum Antiviral Agents. *Mol. Pharmacol.* **1992**, *42*, 1109–1117. [PubMed]
52. Liu, R.; Cao, K.; Clark, A.H.; Lu, P.; Anjass, M.; Biskupek, J.; Kaiser, U.; Zhang, G.; Streb, C. Top-Down Synthesis of Polyoxometalate-Like Subnanometer Molybdenum-Oxo Clusters As High-Performance Electrocatalysts. *Chem. Sci.* **2020**, *11*, 1043–1051. [CrossRef]
53. Itul, D.; Marcu, G. Co (II) Complexes with Lacunary Heteropolywolfram Ligands and Mixed Addendum Atoms—A Study on the Optimum Formation pH Range. *Rev. Chim.* **1995**, *46*, 143–146.

54. Sadowska, J.M.; Genoud, K.J.; Kelly, D.J.; O'Brien, F.J. Bone Biomaterials for Overcoming Antimicrobial Resistance: Advances in Non-Antibiotic Antimicrobial Approaches for Regeneration of Infected Osseous Tissue. *Mater. Today* **2021**, *46*, 136–154. [CrossRef]
55. Clinical and Laboratory Standards Institute (CLSI). *Performance Standards for Antimicrobial Disk Susceptibility Tests*, 12th ed.; Approved Standard-Twelfth Edition; CLSI Document M02-A12; Clinical and Laboratory Standards Institute: Wayne, NJ, USA, 2015; Volume 35, pp. 10–15.
56. Clinical and Laboratory Standards Institute (CLSI). *Performance Standards for Antimicrobial Disk Susceptibility Tests*, 13th ed.; Approved Standard-Thirteenth Edition; CLSI Document M02; Clinical and Laboratory Standards Institute: Wayne, NJ, USA, 2018; Volume 36, pp. 21–65.
57. Markey, B.K.; Leonard, F.C.; Archambault, M.; Cullinane, A.; Maguire, D. *Clinical Veterinary Microbiology.*, 2nd ed.; Elsevier: Edinburgh, UK, 2013; pp. 3–48, 105–288.
58. Bauer, A.W.; Kirby, W.M.M.; Sherris, J.C.; Turck, M. Antibiotic Susceptibility Testing by A Standardized Single Disk Method. *Am. J. Clin. Pathol.* **1966**, *45*, 493–496. [CrossRef] [PubMed]
59. Kronvall, G.; Giske, C.G.; Kahlmeter, G. Setting Interpretive Breakpoints for Antimicrobial Susceptibility Testing Using Disk Diffusion. *Int. J. Antimicrob. Agents.* **2011**, *38*, 281–290. [CrossRef] [PubMed]
60. Quinn, P.J.; Markey, B.K.; Leonard, F.C.; FitzPatrick, E.S.; Fanning, S.; Hartigan, P.J. *Veterinary Microbiology and Microbial Disease*; Wiley-BlackWell: Hoboken, NJ, USA, 2011; pp. 149–292.
61. Clinical and Laboratory Standards Institute (CLSI). *Methods for Dilution Antimicrobial Susceptibility Tests for Bacteria That Grow Aerobically*, 11th ed.; CLSI Standard M07; Clinical and Laboratory Standards Institute: Wayne, NJ, USA, 2018; pp. 15–53.
62. Clinical and Laboratory Standards Institute (CLSI). *Methods for Dilution Antimicrobial Susceptibility Tests for Bacteria That Grow Aerobically*, 9th ed.; Approved Standard-Ninth Edition; CLSI Standard M07-A9; Clinical and Laboratory Standards Institute: Wayne, NJ, USA, 2012; Volume 32, pp. 12–58.

MDPI
St. Alban-Anlage 66
4052 Basel
Switzerland
Tel. +41 61 683 77 34
Fax +41 61 302 89 18
www.mdpi.com

Pharmaceuticals Editorial Office
E-mail: pharmaceuticals@mdpi.com
www.mdpi.com/journal/pharmaceuticals

www.ingramcontent.com/pod-product-compliance
Lightning Source LLC
LaVergne TN
LVHW070731100526
838202LV00013B/1209